Edward P. Weeks

A Commentary on the Mining Legislation of Congress with a Preliminary Review of the Repealed Sections of the Mining Act of 1866

Edward P. Weeks

A Commentary on the Mining Legislation of Congress with a Preliminary Review of the Repealed Sections of the Mining Act of 1866

ISBN/EAN: 9783337253042

Printed in Europe, USA, Canada, Australia, Japan

Cover: Foto ©Suzi / pixelio.de

More available books at **www.hansebooks.com**

A

COMMENTARY

ON THE

Mining Legislation of Congress

WITH

A PRELIMINARY REVIEW OF THE REPEALED SECTIONS
OF THE MINING ACT OF 1866,

THE WHOLE CONSTITUTING A COMPLETE STATEMENT OF THE

LAW AND PRACTICE RELATIVE TO MINES, MINING, AND MINERAL LANDS, UNDER UNITED STATES LAWS, AS CONSTRUED BY THE GENERAL LAND OFFICE, THE SECRETARY OF THE INTERIOR, AND THE COURTS FROM 1866 TO THE PRESENT TIME.

BY
EDWARD P. WEEKS,
COUNSELLOR AT LAW.

SAN FRANCISCO:
SUMNER WHITNEY & CO.
1877.

PREFACE.

The following pages are the result of an attempt to systematize and condense the materials which constitute the law relative to the Mineral Lands of the United States, from the date of the first Mining Act, in 1866, to the present time. Those materials consist, in the main, of the various acts of Congress and the Revised Mining Statutes of the United States, the decisions and instructions of the Commissioner of the General Land Office and the Secretary of the Interior, the opinions of the United States Attorneys-General, and the decisions of the Courts, so far as the Mining Acts have been the subjects of judicial construction. The cases in the regular reports relative to these acts have not hitherto been numerous, but they have been of great importance, and have affected interests of large pecuniary value.

The subject of the binding effect and the value as authority of the decisions of the Land Department has been discussed in its legal aspect in a chapter in the body of the work. It is sufficient to say, here, that they are the decisions of that Department, and of that branch of the executive portion of the Government, to which Congress has left the practical administration of the laws relative to the mineral lands of the nation. In that Department, at least, they are the law. There, they are of acknowledged force and authority. Under them the miner must proceed when he is in quest of the Government title to his mine; and a departure from them, or a want of attention to them, may completely invalidate his entire proceedings, or imperil his title. This is deemed sufficient to demonstrate their practical importance. It is enough to show that attorneys and clients interested in any way in mineral lands should be familiar with these decisions, and, being thus acquainted with them, should follow them. They have, therefore, been admitted among the materials which form the basis of this work.

The plan pursued has been to take a section, or, when upon the same topic, a group of sections, of the United States Revised Statutes relative to the subject-matter, and, using it as a text for a chapter, follow with a commentary upon that particular branch of the statutory law. The order pursued is the one adopted in the Revised Statutes themselves. If not the most logical order that could have been devised, it has been thought that it is the order in which the practitioner, already familiar with it in the statutes, will naturally look for, and expect to find pursued, in the commentary. This consideration—one addressed to the convenience of the practitioner—has outweighed other considerations in the author's mind. Nor is the arrangement sufficiently illogical to call loudly for a disturbance of the statutory sequence of the sections.

The book commences with an introductory chapter relative to the first Act, that of 1866, in which Congress undertook to regulate mines and mining, and provide for the sale of mineral lands. The commentary in this chapter is especially directed to a consideration of the repealed sections of that Act, both with a view to illustrate the changes which have taken place, and also as of value in questions of rights vested under them before their repeal. This is followed by a chapter on Reservations and Exceptions of Mineral Lands in Grants by the Government, with a review of the policy of Congress in reserving mineral lands from pre-emption, sale, and entry. Chapter three treats of the Rights of Exploration and Purchase of Valuable Mineral Deposits, and the Occupation and Purchase of Mineral Lands, together with the Citizenship required, and the Proof thereof. Chapters four, five, six, and seven treat of the Dimensions of Claims and Locations upon Veins or Lodes; the Locator's Right of Possession and Enjoyment of the Surface-ground, and of the Lode, Tunnel Rights, Regulations and Customs, Expenditures and Improvements, Surveys and Boundaries of Claims. Chapter eight is devoted to the important subject of Patents to Mineral Lands, and the mode of procuring Government Title to Mines; and chapter nine to Adverse Claims to the Lands applied for by the Claimants to these Patents, and the subsequent proceedings in the Courts after those adverse claims have been filed. Chapter ten covers the subject of Placer Claims, their Survey, Entry, and Patenting, their Dimensions and the Subdivision of the Land into Ten-acre Tracts, together with the requisite Evidence of Possession. Chapter eleven is upon the subject of Surveys of Mineral Lands and the Duties of the Surveyor-General. Chapter twelve treats of the Intersection of Veins; chapter thirteen, of Mill-sites and Patents to nonmineral Lands; and chapter fourteen, of Water and other Vested Rights, the Right of Way for Canals and Ditches, Easements, Drainage, State and Territorial Legislation, and the Sutro Tunnel Act. Chapter fifteen embraces the subject of Homesteads and Town-sites. Chapter sixteen treats of the Segregation of Mineral and Agricultural Lands, and Withdrawal of Mineral Lands from Agricultural Entry. Chapter seventeen is an Abstract of the Law as to Coal Lands; and chapter eighteen includes the consideration of various miscellaneous topics, such as the General Power and Authority of the Land Department, Hearings in Contested Cases, Appeals, Evidence, etc., etc.

The subject is one of recent growth and development. New, and perhaps perplexing questions, must inevitably arise. The author has confined himself to a statement, as accurate and clear as possible, of what the law is at the date of writing. The domain of speculation as to what it will be in future has been carefully avoided. If there were any temptations to trespass upon such a territory, the inclination would certainly be restrained, if not extinguished, in reflecting on the diametrical difference of opinion existing between one of the most learned of the State Supreme Courts, and the Supreme Court of the United States, upon questions of pre-eminent gravity, in regard to the Land Laws of the United States Government. When a State Court of acknowledged ability, and of highest resort in the State, has been frequently and recently reversed by the highest tribunal in the country upon such topics, a writer may well be deterred from anything more than a bare statement of the law as already construed, and will certainly not be expected to express any opinion as to what future adjudications may be. "*Ita lex scripta est*," is sufficient. Whether the law will in future be written in any other way, is not within the scope of this book.

SACRAMENTO, June, 1877.

Table of Parallel Reference.

ACTS OF 1866, 1870 AND 1872.	REVISED STATUTES.
Act of July 26th, 1866, 14 U. S. Stats. 252:	Sec. 2319..Sec. 1, Act of 1866, and Sec. 1, Act of 1872.
Revised Statutes.	Sec. 2320..Sec. 4, Act of 1866, and Sec. 2, Act of 1872.
Sec. 1.....................Sec. 2319	Sec. 2322..Sec. 3, Act of 1872.
Sec. 2.....................Sec. 2325	Sec. 2323..Sec. 4, Act of 1872.
Sec. 3.....................Sec. 2325	Sec. 2324..Sec. 5, Act of 1872.
Sec. 4.....................Sec. 2320	Sec. 2325..Secs. 2, 3, Act of 1866, and Sec. 6, Act of 1872.
Sec. 5.....................Sec. 2338	
Sec. 6.....................Sec. 2326	Sec. 2326..Sec. 6, Act of 1866, and Sec. 7, Act of 1872.
Sec. 7.....................Sec. 2343	
Sec. 8.....................Sec. 2344	Sec. 2327..Sec. 8, Act of 1872.
Sec. 9.....................Sec. 2339	Sec. 2328..Sec. 9, Act of 1872.
Sec. 10....................Sec. 2341	Sec. 2330..Sec. 12, Act of 1870.
Sec. 11....................Sec. 2342	Sec. 2331..Sec. 16, Act of 1870, and Sec. 10, Act of 1872.
Act of July 9th, 1870, 16 U. S. Stats. 218:	Sec. 2332..Sec. 13, Act of 1870.
Sec. 12....................Sec. 2330	Sec. 2333..Sec. 11, Act of 1872.
Sec. 13....................Sec. 2332	Sec. 2334..Sec. 12, Act of 1872.
Sec. 14....................Sec. 2335	Sec. 2335..Sec. 13, Act of 1872, and Sec. 14, Act of 1870.
Sec. 15....................Sec. 2338	
Sec. 16....................Sec. 2331	Sec. 2336..Sec. 14, Act of 1872.
Sec. 17....................Sec. 2340	Sec. 2337..Sec. 15, Act of 1872.
Act of May 10th, 1872, 17 U. S. Stats. 91:	Sec. 2338..Sec. 5, Act of 1866, and Sec. 15, Act of 1870.
Sec. 1.....................Sec. 2319	Sec. 2339..Sec. 9, Act of 1866.
Sec. 2.....................Sec. 2320	Sec. 2340..Sec. 17, Act of 1870.
Sec. 3.....................Sec. 2322	Sec. 2341..Sec. 10, Act of 1866.
Sec. 4.....................Sec. 2323	Sec. 2342..Sec. 11, Act of 1866.
Sec. 5.....................Sec. 2324	Sec. 2343..Sec. 7, Act of 1866.
Sec. 6.....................Sec. 2325	Sec. 2344. Sec. 8, Act of 1866, and Sec. 16, Act of 1872.
Sec. 7.....................Sec. 2326	
Sec. 8.....................Sec. 2327	
Sec. 9.....................Sec. 2328	
Sec. 10....................Sec. 2331	
Sec. 11....................Sec. 2333	
Sec. 12....................Sec. 2334	
Sec. 13....................Sec. 2335	
Sec. 14....................Sec. 2336	
Sec. 15....................Sec. 2337	
Sec. 16....................Sec. 2344	

TABLE OF CONTENTS.

CHAPTER I.

INTRODUCTORY—THE FIRST MINING ACT.

§ 1. The Act of 1866—The repealed sections.
§ 2. Section one—License without title.
§ 3. Duties of registers and receivers.
§ 4. Title and patent—The second section.
§ 5. Limitation of the right to obtain patents.
§ 6. Applications for patents.
§ 7. Citizenship required.
§ 8. Entry and diagram.
§ 9. Defects in the Instructions.
§ 10. The application.
§ 11. Publication of the notice.
§ 12. Duties of claimants, registers, and receivers.
§ 13. What a patent conveyed.
§ 14. Diagram, notice, survey, and patent.
§ 15. Survey.
§ 16. Posting the notice of application.
§ 17. Effect of irregularities—Notice of application—Requisites.
§ 18. Fees of surveyors.
§ 19. Size of locations—Adjustment of surveys.
§ 20. Duties of deputy-surveyors.
§ 21. Following the vein to any depth.
§ 22. Mode of survey—Quantity and restriction to one claim.
§ 23. Deviation from the rectangular form of survey.
§ 24. Number of feet located.
§ 25. Adverse claims and contests.
§ 26. Proceedings on adverse claims.
§ 27. Miscellaneous.

CHAPTER II.

RESERVATIONS AND EXCEPTIONS OF MINERAL LANDS IN GRANTS BY THE GOVERNMENT.

§ 28. Mineral lands reserved.
§ 29. Mineral lands in certain States not excepted.
§ 30. Exception from certain grants.

§ 31. The policy of the Government in reserving or excepting mineral lands.
§ 32. Excepting clause in placer and agricultural patents.
§ 33. Saline lands.
§ 34. School lands containing mineral.
§ 35. School lands in Nevada.
§ 36. Mineral lands in railroad grants.

CHAPTER III.

RIGHT OF EXPLORATION AND PURCHASE OF VALUABLE MINERAL DEPOSITS, AND THE OCCUPATION AND PURCHASE OF MINERAL LANDS—CITIZENSHIP AND PROOF THEREOF.

§ 37. Right to purchase.
§ 38. Valuable deposits.
§ 39. The general rule stated.
§ 40. Borax deposits.
§ 41. Mineral deposits.
§ 42. What is a mineral vein?
§ 43. Mineral veins, classifications.
§ 44. Definitions of terms in common use.
§ 45. Who may acquire patents.
§ 46. Application by aliens.
§ 47. Citizenship.
§ 48. Proof of citizenship.
§ 49. Affidavit of citizenship.
§ 50. Foreign corporation.
§ 51. Restriction as to proof.

CHAPTER IV.

DIMENSIONS OF CLAIMS AND LOCATIONS UPON VEINS OR LODES.

§ 52. Length and width of lode-claims.
§ 53. Veins or lodes of quartz or other rock in place.
§ 54. Location previous to the Mining Acts.
§ 55. Width of lode-claims—Rights granted by the patent.
§ 56. Survey must conform to the patent.
§ 57. Manner of locating prior to 1872.
§ 58. Several locations may be made.
§ 59. Local regulations.

CHAPTER V.

LOCATOR'S RIGHTS OF POSSESSION AND ENJOYMENT OF THE SURFACE GROUND AND OF THE LODE.

§ 60. Locator's rights of possession and enjoyment.
§ 61. Status of lode-claims previously located.
§ 62. Patents for veins or lodes previously issued.
§ 63. Priority of location—Importance of.

CHAPTER VI.

TUNNEL RIGHTS.

§ 64. Owners of tunnel rights.
§ 65. Patenting tunnel rights.
§ 66. Expenditures upon tunnel.

CHAPTER VII.

REGULATIONS AND CUSTOMS — EXPENDITURES AND IMPROVEMENTS — SURVEYS AND BOUNDARIES.

§ 67. Regulations and customs.
§ 68. Definition of "claim."
§ 69. Annual expenditures on placer-claims.
§ 70. Annual expenditures on lode-claims.
§ 71. Neglect of co-claimants to contribute.
§ 72. Re-located mines—expenditure.
§ 73. Amount of expenditure shown upon plat and field-notes.
§ 74. Location and survey—Boundaries.
§ 75. Certificate as to improvements.
§ 76. Fixed monuments—Courses—Distances.

CHAPTER VIII.

PATENTS TO MINERAL LANDS—MODE OF PROCURING GOVERNMENT TITLE.

§ 77. Patents for vein or lode-claims, how obtained.
§ 78. Details of procedure.
§ 79. Duties of registers and receivers.
§ 80. Nature of the patent.
§ 81. Impeachment of patent.
§ 82. Adverse possession as against a patent.
§ 83. What is granted.
§ 84. Who may apply.
§ 85. Evidence of ownership—Deraigning title—Identity of applicant—Transfers.
§ 86. Claim through an executor—Where an alien is grantee of a claim.
§ 87. United applications—Unincorporated associations.
§ 88. Several claims cannot be embraced in one application.
§ 89. Grantee of several locators may obtain patent for the whole tract.
§ 90. Conflicting patents.
§ 91. Errors in description in patent—Relinquishment—Calls for the relinquishment of land inadvertently patented.
§ 92. Second patent—Entries of mineral lands by settlers and corporations.
§ 93. Minerals discovered after agricultural patent.
§ 94. Setting aside patent.

- § 95. Number of patents.
- § 96. Protests against issuance of patents—Status of protestants.
- § 97. An illegal location invalidates subsequent proceedings.
- § 98. Location by a minor.
- § 99. Application for several lodes and a mill-site—Claim partly in one district and partly in another.
- § 100. Delaying action at request of Congressional Committees.
- § 101. The affidavit—Proper party to make it.
- § 102. Verification of affidavits.
- § 103. The location notice.
- § 104. Parol evidence to aid the notice.
- § 105. Plat must show the boundaries of the claim.
- § 106. Surveys to show exterior boundaries.
- § 107. Specific surface ground.
- § 108. Posting on claim, and proof thereof.
- § 109. Publication of the notice.
- § 110. Time of publication.
- § 111. Counting the sixty days.
- § 112. Proof of publication.
- § 113. The newspaper in which the notice is to be published.
- § 114. Defects in the published notice.
- § 115. Discrepancies between final survey and patent and the application and published notice.
- § 116. Discrepancies between the published notice and the notice and diagram filed.
- § 117. Discrepancies between the published notice and the diagram and posted notice.
- § 118. Discrepancies between the final survey and patent and the application.
- § 119. New survey, pending another application.
- § 120. Discrepancies between survey and diagram.
- § 121. Discrepancies between survey and notice, matter of description.
- § 122. Errors in survey.
- § 123. When application will be rejected.
- § 124. Sworn statement.
- § 125. Approval of survey—Jurisdiction of Surveyor-General.
- § 126. Proof of citizenship.
- § 127. Miscellaneous.

CHAPTER IX.

ADVERSE CLAIMS—PROCEEDINGS IN COURT.

- § 128. Adverse claims.
- § 129. Adverse claims under Act of 1866.
- § 130. Adverse claims under statutes now in force—details of procedure.
- § 131. Who may file.
- § 132. Verification of adverse claim.
- § 133. Verification of adverse claims by agents of companies.
- § 134. Time of filing.
- § 135. Commencing second suit—Dismissal of former suit.
- § 136. What constitutes an adverse claim.
- § 137. Necessary allegations.
- § 138. What adverse claimant must show

§ 139. Form of adverse claim.
§ 140. Prima facie adverse claim.
§ 141. Sufficient filing.
§ 142. Adverse claim must be accompanied by certified survey.
§ 143. The object of giving notice by publication.
§ 144. Jurisdiction of the Land Office over adverse claims.
§ 145. Notice of suit.
§ 146. Authority of register to dismiss.
§ 147. Proceedings in Court—proper party to commence suit.
§ 148. Possession as equivalent to adverse claim—parties to institute suit.
§ 149. What are Courts of competent jurisdiction.
§ 150. Contests in Court—Jurisdiction.
§ 151. Jurisdiction of State Courts.
§ 152. Transfer of causes to United States Courts—Jurisdiction of mining causes.
§ 153. Cancelation of entry pending suit.
§ 154. Stay of proceedings.
§ 155. Filing consent to judgment.
§ 156. Laches in bringing suit.
§ 157. Prosecution of suits—reasonable diligence.
§ 158. Abandonment of portion of adverse claim.
§ 159. Abandonment of surface ground.
§ 160. Cross-applications—delay.
§ 161. Fees on filing adverse claim.
§ 162. Miscellaneous.

CHAPTER X.

PLACER CLAIMS—SURVEY, ENTRY, AND PATENT—DIMENSIONS OF CLAIMS—SUBDIVISIONS OF TEN-ACRE TRACTS—EVIDENCE OF POSSESSION—MODE OF OBTAINING PATENT.

§ 163. Conformity of placer claims to surveys—Limits and boundaries.
§ 164. Subdivision of ten-acre tracts—Extent of placer locations.
§ 165. Survey of placer claims—Limitations.
§ 166. Evidence of possession—Sufficient to establish right to patent.
§ 167. Proceedings for patent for placer claims.
§ 168. Details of procedure.
§ 169. Description in the notice.
§ 170. Entry and survey of placer claims under the Act of 1866.
§ 171. Survey of placer claims under the Acts of 1866, 1870.
§ 172. Survey and entry under the Act of 1870.
§ 173. Quantity of placer ground subject to location.
§ 174. Proofs necessary to establish possessory rights.
§ 175. Placer ground located after May 10th, 1872.
§ 176. Conflicting claims—Placer and lode claims.
§ 177. Miscellaneous provisions.

CHAPTER XI.

PUBLIC SURVEYS OVER MINERAL LANDS—SURVEYS OF MINING CLAIMS—DUTIES OF SURVEYOR-GENERAL—APPOINTMENT OF DEPUTIES.

§ 178. Appointment of surveyors of mining claims by Surveyor-General.
§ 179. Public surveys extended over mineral lands.
§ 180. Description of vein claims on surveyed and unsurveyed lands.
§ 181. Appointment of deputies.
§ 182. Charges for surveys and publications.
§ 183. Special instructions to deputies.
§ 184. Authority of deputies outside the district.

CHAPTER XII.

INTERSECTION OF VEINS.

§ 185. Intersection of veins.
§ 186. Conflicts as to surface ground.
§ 187. Identity of lodes.
§ 188. Interference of claims.
§ 189. Abandonment of surface ground.

CHAPTER XIII.

MILL SITES—PATENTS FOR NON-MINERAL LANDS.

§ 190. Patents for non-mineral lands.
§ 191. Location of mill sites.
§ 192. Procuring patent.
§ 193. A mill site must be non-mineral in character.
§ 194. Improvements.
§ 195. Mill sites in railroad grants.

CHAPTER XIV.

WATER AND OTHER VESTED RIGHTS—RIGHT OF WAY FOR CANALS AND DITCHES—EASEMENTS—DRAINAGE—STATE AND TERRITORIAL LEGISLATION—PATENTS SUBJECT TO VESTED RIGHTS—SUTRO TUNNEL ACT.

§ 196. State and Territorial legislation—Easements—Drainage, etc.
§ 197. Conditions inserted in the patent.
§ 198. Vested rights to use of water—Right of way for canals.
§ 199. Patents subject to vested water rights.
§ 200. Possessory water rights confirmed.
§ 201. Local water rights protected.

§ 202. Conditions as to vested water rights inserted in patent.
§ 203. Mining ditch in railroad grant.
§ 204. Conflicting rights of ditch-owners and miners.
§ 205. Exercise of eminent domain for a private ditch company's use.
§ 206. Water rights in California under the Codes.
§ 207. Existing water rights obtained by patent, how affected.
§ 208. Effect of the acts upon previous diversion of water upon patented lands.
§ 209. Recognition of the doctrine of prior appropriation.
§ 210. Effect of the statute upon prior appropriation without Government title.
§ 211. Construction of flumes over public lands.
§ 212. Rights of ditch-owners on public lands.
§ 213. Sutro Tunnel Act.
§ 214. Conditions inserted in patents for mines on Comstock Lode, Nevada.
§ 215. Claim rejected.

CHAPTER XV.

HOMESTEADS AND TOWN SITES — HOMESTEAD RIGHTS ON NON-MINERAL LANDS — TOWN-SITE ENTRIES.

§ 216. Non-mineral lands open to homesteads.
§ 217. Pre-emption of homesteads on agricultural lands formerly designated as mineral.
§ 218. Homestead entries including mineral deposits.
§ 219. Rights of pre-emptioners and homestead claimants.
§ 220. Conflicts between homestead and mill-site claimants.
§ 221. Title to town lots subject to mineral rights.
§ 222. Conflicts between mineral and town-site claimants.

CHAPTER XVI.

SEGREGATION OF MINERAL AND AGRICULTURAL LANDS — WITHDRAWAL FROM AGRICULTURAL ENTRY.

§ 223. Manner of setting apart mineral lands as agricultural.
§ 224. Segregation of agricultural from mineral lands.
§ 225. Mineral affidavits.
§ 226. Mineral affidavits on timber land.
§ 227. Segregation under Acts of 1866 and 1870.
§ 228. Withdrawal of certain lands from agricultural entry.
§ 229. Surveyors' returns.
§ 230. Their prima facie accuracy.
§ 231. Hearings to determine the character of land — Publication.
§ 232. What is mineral land.
§ 233. Burden of proof.
§ 234. Evidence as to agricultural character of land.
§ 235. The testimony.
§ 236. Proof as to mineral character of land.
§ 237. Discovery of mines on agricultural lands.
§ 238. Agricultural patent covering mines already worked.
§ 239. Fraud in pre-emption entry.
§ 240. Compromises between miners and settlers.

§ 241. Attempt by railroad to disprove mineral character of lands.
§ 242. Non-mineral proof by settlers on lands within railroad limits.

CHAPTER XVII.

COAL LANDS—RIGHT OF ENTRY AND OF PRE-EMPTION—PRESENTATION OF CLAIMS—LIMITATION OF ENTRY—CONFLICTING CLAIMS—EXISTING RIGHTS.

§ 243. Entry of coal lands.
§ 244. Pre-emption of coal lands.
§ 245. When claims are to be presented.
§ 246. Only one entry allowed.
§ 247. Conflicting claims.
§ 248. Existing rights.
§ 249. Departmental regulations and instructions.
§ 250. Restrictions as to purchase.
§ 251. School sections containing coal.
§ 252. Coal lands and town sites.
§ 253. Actual possession of coal mines upon railroad sections.
§ 254. Coal lands in Minnesota, Wisconsin, and Michigan.

CHAPTER XVIII.

MISCELLANEOUS PROVISIONS.

§ 255. Power of the President as to appointments.
§ 256. Pending applications—Existing rights.
§ 257. Possessory actions relative to mines.
§ 258. Practice before the Land Department—Hearings, contests, and appeals—Witnesses and testimony.
§ 259. Appeals, exceptions, evidence.
§ 260. Fees of registers and receivers.
§ 261. Payment pending contest.
§ 262. Decisions of the Land Department—Their authority.
§ 263. Right of inspection of mine.
§ 264. Mining claims in river beds.
§ 265. Timber on mineral lands—Railroad companies.
§ 266. Claims not within any mining district.
§ 267. Removal of machinery.
§ 268. Criminal offenses.
§ 269. Various provisions

CHAPTER I.

INTRODUCTORY—THE FIRST MINING ACT.

§ 1. The Act of 1866—The repealed sections.—It had been a well-known policy of the Government of the United States, from the time of its foundation, to reserve from sale all lands containing minerals, or " known mines." This policy was not in any degree disturbed until the passage of the Act of 1866, known as the first Congressional mining act. In 1850, the policy of the nation's selling the mines for the purpose of obtaining public revenue began to be discussed in the National Legislature. But, after much controversy, the arguments in favor of leaving the mines free and open for exploration and development prevailed, and adverse measures were defeated or abandoned. From that time until 1866, non-action was the policy of the Government. At this time, the necessity of Congressional action, long before felt, came to be directly recognized. In the annual report of the Secretary of the Treasury for the year 1865, the substitution of an absolute title in fee for the indefinite possessory rights or claims under which the mines were held by private parties, was earnestly recommended. It was urged that the right to obtain a " fee-simple title " would invite to the mineral districts men of character and enterprise, and would give permanency to settlements by the stimulus which ownership always produces. Under the then existing condition of things, constant fear was felt, by those who were engaged in mining pursuits, that some disturbance and interference with their rights of property, such as they had, would occur. This fear was not groundless.[1] Measures for the sale of the mines

---

[1] Valuable coal-fields had been discovered on the public lands of California, and large quantities of coal were being taken therefrom by intruders on the public lands. The Act of March 3d, 1807, provided (2 U. S. Stats. 445) that if any person or persons should take possession of, or make a settlement on, the public lands of the United States, which lands should not have been previously sold, ceded, or leased by the United States, or the claim to which by such

and for the taxation of the miners, as a class, had from time to time been proposed; and, besides, the Government had the undoubted *legal* right to treat every miner upon the public domain as a naked trespasser.

The passage of the Act of 1866, with all its defects, marked a change in the Governmental policy, and introduced a new era in the history of mining enterprise.

§ 2. Section one—License without title.—The first section (now repealed) provided: "Sec. 1. That the mineral lands of the public domain, both surveyed and unsurveyed, are hereby declared to be free and open to exploration and occupation by all citizens of the United States and those who have declared their intention to become citizens, subject to such regulations as may be prescribed by law, and subject also to the local customs or rules of miners in the several mining districts, so far as the same may not be in conflict with the laws of the United States."[1]

This section declared the freedom of the mines by opening the whole public domain to exploration or prospecting in search of mines and minerals, and to the occupation and use of such mines as were unoccupied, or which might be discovered, to be worked for the use and benefit of individuals, partnerships,

---

person or persons should not have been previously recognized and confirmed by the United States, or if any person or persons should cause such lands to be thus occupied, taken possession of, or settled, or should survey, or attempt to survey or designate, any boundaries on such lands, such person or persons should forfeit any right to such lands, and the President of the United States might direct the marshal of the district to remove from such lands any such person or persons, and to employ such military force as might be necessary for that purpose. And the persons on such lands in violation of the provisions of the act, were liable to fine and imprisonment in the manner declared in the act. This act was considered by the then Attorney-General of the United States, in 1852, as providing ample means of protecting the lands in question from the intrusion complained of until Congress should establish some method of bringing the lands into the market. The Attorney-General then advised that the President should issue instructions to the marshal of the district to remove all "intruders" from the lands, and, in such manner as might be most effective, prevent the spoliations of which complaint was made, and for this purpose, if necessary, to authorize him to call to his assistance any military force which might be near at hand. (Opinion of Atty.-Gen. U. S. February 11th, 1852; 10 Op. Atty.-Gen. 184.) See, also, United States *v.* Parrott, 1 McA. C. C. 271; Blanchard & Weeks' Leading Cases on Mines and Mining Water Rights, 91.

[1] Act of July 26th, 1866; 14 U. S. Stats. 251.

corporations, or companies. These privileges were limited to citizens of the United States, and to those who might declare their intention to become such. Exploration and occupation were subject to such regulations as might be prescribed by law : it was not said whether by the law of Congress alone, or by the laws of the respective States and Territories in which the mines might be situated. But the fifth section [1] provided, as a condition of sale to be expressed in the patent, that "in the absence of necessary legislation by Congress, the local legislatures may provide rules for working mines, involving easements, drainage, and other necessary means to their complete development.". The laws referred to, then, must have been those of the local legislature, in the absence of regulations by Congress.

The other restriction was that of the local customs or rules of miners in the several mining districts, not in conflict with the laws of the United States. These regulations were fully recognized throughout the act. But they also were required to be consistent with the laws of the States and Territories.

The first section conferred no *title* to mining claims, other than possessory rights, and was entirely distinct in many respects from the residue of the act, relating to the acquisition of title by patent to veins of the four metals named. It conferred a right of occupation and appropriation, without charge, of all minerals, ores, and metals dug from the placers or beds of ore, or raised by vein-mining. It constituted a mere license by the Government to go upon the public domain and search for minerals—metalliferous or non-metalliferous—and appropriate them. In effect, it conferred a right to commit an act on the public land, which, before the law, was clearly a trespass. And no other title than the possessory claim conferred by the license was provided for, under the law, to any class of claims but to gold, silver, copper, and cinnabar in place, as provided in section two. Placer claims, river diggings, gravel and cement beds, as valuable as some of them were, seemed totally without the purview of the law, so far as title was concerned.[2] It became important, therefore, to know what the extent or effect of this license was. In the first place, it might have been revoked at any time by

---

[1] . Stat. Sec. 2328.
[2] But see Decision of Comr. Aug. 27th, 1868; Zabriskie's Land Laws, 218.

Congress by a simple repeal of the law, and the right would end. Unlike a pre-emption claim, which gives precedence to purchase, no equity attached upon an entry with an intention to improve and purchase.[1]

The Government could set up no right to the minerals extracted from a claim while the law remained in force, or after its repeal, because ownership attached to them. It was by the express sanction of the Government that the miner entered upon the claim and worked it.

No estate was granted by the act in the land, or minerals in the land, by the license, of any certain or determinate tenure, as in the case of a lease. The property in the minerals was granted only when they became severed from the soil, and liable to be recovered in an action of trover. At common law, the grantee of a license could not bring ejectment, because the grantor was supposed to continue in possession, and this feature was the test in distinguishing between a lease and a license.[2] But the grantee of such a license from the Government could maintain such an action for the possession of his claim, both by State laws and at least by implication under the Act of Congress of the 27th of February, 1865, relating to the Courts of Nevada.[3] In other respects, the common-law rules applicable to licenses to dig ore governed those claims, and the act was to be interpreted as such an authority as excused a trespass, and which regarded rather the act itself than its connection with the land, or that merely amounted to a bare permission or dispensation to do or suffer certain temporary acts, and gave no real beneficial interest in the land.[4] The Supreme Court of the United States, in the case of Gratiot, gave the legal definition of a lease, as distinguished from a mining license, upon a contract made by the authority of the President under the Act of 1807, authorizing him to lease lead-mines.

The Court say: " The contract purports to be a license for smelting lead-ore, and it is objected that it is not a lease within the act of Congress. The legal understanding of a lease for

---

[1] See People v. Shearer, 30 Cal. 645; Yale's Mining Claims, 355, 356.
[2] Bainbridge on Mines, 246.
[3] 13 Stats. at L. 441.
[4] Bainbridge on Mines, 252; Blanchard & Weeks' Leading Cases on Mines and Mining Water Rights, Chaps. 14, 15.

years is a contract for the possession and profits of land for a determinate period, with the recompense of rent. The contract in question is strictly within the definition."[1] The license was in the nature of a tenancy at will, revocable at pleasure, and until an entry and purchase of a lode-claim, was applicable to all the privileges granted under the Act of 1866.[2]

There was nothing obligatory on claimants to proceed under the Act of 1866, and where they failed to do so, there being no adverse interest, they held the same relations to the premises they worked as before the passage of the act, with the additional guarantee that they possessed the right of occupancy under the statute.[3]

§ 3. **Duties of registers and receivers.**—It became the duty of registers and receivers, upon the passage of the Act of 1866, to acquaint themselves with the local mining customs and usages. In acting upon individual claims, a perfect record thereof was required to be taken and preserved by the register and receiver, and accompanied by a diagram or plat fixing the out-boundaries of the district in which such customs and usages existed.[4] As Sec. 1 of the Act of 1866 did not relate to title, the surveyors, receivers, and registers had no duty to discharge under it. The instructions were only applicable to the other sections.[5]

The diagram or plat fixing the out-boundaries of the district in which the customs and usages existed, was not in practice found easy of execution. The names of the districts were accidentally given, governed by no rule; the boundaries uncertain and undefined, except when controlled by well-known, natural objects; the districts frequently changed or divided, and perhaps never made the subject of actual survey—a record of the customs and regulations was an easier matter.[6]

§ 4. **Title and patent.**—**The second section** (also repealed) read: " Sec. 2. That whenever any person or association of

---

[1] U. S. v. Gratiot, 14 Pet. U. S. 526.
[2] Yale's Mining Claims, 355, 356, 357.
[3] Instructions Jan. 14th, 1867; Zabriskie's Land Laws, 200; Copp's U. S. Mining Decisions, 239; Gold Hill Quartz Mining Co. v. Ish, 5 Oregon, 104.
[4] Instructions Jan. 14th, 1867; Zabriskie's Land Laws, 200; Copp's U. S. Mining Decisions, 239.
[5] Yale's Mining Claims and Water Rights, 357, 358.
[6] Ibid. 359.

persons, claim a vein or lode of quartz, or other rock in place, bearing gold, silver, cinnabar, or copper, having previously occupied and improved the same according to the local customs or rules of miners in the district where the same is situated, and having expended in actual labor and improvements thereon an amount of not less than one thousand dollars, and in regard to whose possession there is no controversy or opposing claim, it shall and may be lawful for said claimant or association of claimants to file in the local land office a diagram of the same, so extended, laterally or otherwise, as to conform to the local laws, customs, and rules of miners, and to enter such tract and receive a patent therefor, granting such mine, together with the right to follow such vein or lode, with its dips, angles, and variations, to any depth, although it may enter the land adjoining, which land adjoining shall be sold subject to this condition." [1]

§ 5. **Limitation of the right to obtain patents** under the Act of 1866. This section limited the right to apply for and receive patents for mining claims to persons: 1st. Who had occupied and improved their claims according to the local customs or rules of miners. 2d. Who had, by themselves or their grantors, held and worked their claims for a period equal to the time prescribed by the Statute of Limitations for mining claims of the State or Territory where the same might be situated. 3d. Who had expended, in actual labor and improvements upon their respective claims, an amount of not less than one thousand dollars. 4th. In regard to whose possession there was no controversy or opposing claim.[2]

§ 6. **Applicants for patent** must have had the local possessory rights. The mining act authorized applications for patents by persons having previously occupied and improved their claims according to the local customs and rules of miners, and who had expended in actual labor and improvements thereon an amount not less than $1,000 on each claim, and such claimants were authorized to include in their applications only those

---

[1] Act of July 26th, 1866; 14 U. S. Stat. 251.
[2] Instructions Aug. 8th, 1870; Copp's U. S. Mining Decisions, 255.

claims to which they had possessory titles, under and by virtue of such local customs, and they had no right to include premises to which no such possessory rights had attached.

Persons having no possessory rights according to the local mining laws and regulations, and who had not made the improvements required by the mining act, were not authorized to apply for patents, and the attempt to do so was held to be a fraud, not only against the rightful owners, but against the policy of the act itself.[1]

The evidence was required to show that the proper notice and diagram were posted upon the premises, and identify the claims alleged in the petition and advertisement. Proof of citizenship was required, and the amount of land could not exceed that authorized by law.

The vein or lode of quartz, or other rock in place, bearing gold, silver, cinnabar, or copper, to which a patent could be obtained, was one which had been previously occupied and improved according to local customs, and on which not less than $1,000 had been expended in actual labor and improvements, and also one " in regard to whose possession there was no controversy or opposing claim."[2]

---

[1] Decision Commissioner, January 28th, 1869; Copp's U. S. Mining Decisions, 20.

[2] *Evidence required*—In a given case, a company, being the applicants, presented the following documents to substantiate their claim:

1st. The written application of the company; 2d. Copy of the original location; 3d. Copy of three sections of the mining customs of the district; 4th. Affidavits as to posting the notice and diagram on the claim; 5th. A copy of the notice and diagram; 6th. The register's certificate of application to enter the land; 7. Copy of published notice; 8th. Receiver's receipt for the price of the land; 9th. The register's certificate of entry and payment of purchase-money; 10th. Receipt of newspaper publisher; 11th. The certificate of the Surveyor-General of payment of fees for surveying and office work. In addition to this evidence, the company was required to present: 12th. Evidence under the State law that the company was incorporated as stated, and that the applicants were entitled to represent the same as trustees; 13th. Evidence of the character of the vein exposed, the evidence to be furnished by the Surveyor-General by indorsing it on the plat; 14th. Evidence that not less than $1,000 had been expended on the claim in actual labor and improvements. This fact, in regard to the actual labor and improvements, was also to be certified by the Surveyor-General, by indorsement on the plat, in addition to which there was required an affidavit, or the verbal testimony reduced to writing by the local land officers, of two or more reliable persons cognizant of the facts of such improvements, who would state particularly of what the improvements consisted, when they were made, and by what claimants, and estimate the value of the

Congress had the power to make such qualifications in granting mineral lands as it saw fit. It chose to say that no such

same specifically. 15th. The notice of location was required to name the lode. Some proof was required of holding the possessory title, and that the copy of the original location transmitted referred to the lode claimed. 16th. The land office recognized the mode of transfer from the original locator to the applicant authorized by the State law. The proof as to how title was acquired by applicants was required to be such as to enable the commissioner to act understandingly. If parties held by deed or bill of sale, a properly certified copy of the same was to be transmitted; and where a law sanctioned a verbal sale of a mining claim accompanied by immediate transfer, proof of the vendee's title, actual possession at the time of the sale, accompanied by immediate transfer of the same to the vendees, were also ordered to be furnished, together with a certificate from the county recorder that no adverse conveyance appeared on his records for the premises claimed; 17th. Proof was required (a certificate from the officers being sufficient) that a diagram of the claim had been filed in the local land office, and a notice of the application posted in the register's office; 18th. A certified copy of a portion of the mining regulations was deemed insufficient. A certified copy of all the customs complete, as they existed at the date of the location, was required; 19th. The survey and plat of the claim was also ordered to be transmitted to the General Land Office, together with a copy of the field-notes, with an approved plat, having on it the indorsements required by the third section of the Act of 1866, and representing the claim in relation to the township and standard lines of the public surveys; 20th. An affidavit was required of reliable person, or persons, acquainted with the premises, to the effect that the claim, as surveyed and platted, contained but one *known* vein or lode. When applicants presented, as proof of important facts, the affidavits of absent persons, the local officers were ordered to have the characters for truth of such absent deponents vouched for by some responsible officer to whom they were known, and if this were not done in good faith, and the officers satisfied of the credibility of the deponents, the affidavits were not to be received. All transmittals were to be accompanied with a letter of advice, the testimony in the case, and the joint opinion of the register and receiver on the claims. (Decision of Com. June 6th, 1868; In re Kelsey Lode, Zabriskie's Land Laws, 212.)

In still another case the papers were found satisfactory on the following points: 1st. The character of the vein exposed; 2d. The expenditure in labor and improvements; 3d. That diagram and notice were filed and posted in the register's office for ninety days, and that the required notice was published in a newspaper for the same period; 4th. That the claim was surveyed and platted, and survey and plat approved and indorsed as required by statute; 5th. That expenses of survey, plat, and notice, and the price of the land, had been paid by the claimants; 6th. As to the citizenship of the claimants; 7th. As to certain transfers of interest, the proof was found insufficient, and further proof required; 8th. That the notice and diagram were posted in a conspicuous place on the claim, as required by the statute; 9th. That the premises claimed and surveyed contained but one vein or lode; 10th. In locations of so large a size as 1,000 feet, the printed copy of the mining laws transmitted was considered insufficient proof, and was required to be supported by the corroborative testimony of at least two intelligent practical miners of the district, familiar with its mining customs and regulations. Proof was allowed to be furnished either orally in the office, to be reduced to writing, or by affidavits; but in case the latter

lands should be patented unless they were those that were free from all questions relating to the possession.

Not content with saying that the possession should be free from controversy, which might imply active assertion of right by proceedings in Court, or otherwise, it did say that the mines should be free from all opposing claims. That is to say, no patent shall issue for any mineral lands about which any one, other than the petitioner, asserts any right of possession, and all controversy must relate to possession, for title is in the Government, and therefore cannot be in question.

A controversy and opposing claim was, therefore, sufficient, irrespective of its merits, to prevent the issuance of a patent until the claim was decided.[1]

Section two was the most important, and the controlling section of the act. The occupation, either before or after the act, preceded the right of entry. The expenditure in actual labor or improvements on the claim, in a sum equal to that named in the act under the district laws, was absolutely essential as the condition of development. This provision was not a condition subsequent, merely directory, to be dispensed with at discretion, but one to be performed in good faith. Such is the rule of interpretation under the Spanish code, which applies the principle of *strictissimi juris* to the necessary work. If a controversy occurred about the title to the claim, the entry could not be made till it was judicially decided. But the existence of a controversy did not prevent the filing of the application for an entry of the claim as a jurisdictional fact, necessary to be averred negatively, as the right to make the application could not be defeated by any loose or indefinite assertion of title, but

---

method was adopted, the credibility of the deponents was to be properly vouched for by a responsible officer, to whom they were either personally known, or who, upon information obtained from competent and reliable sources, would feel himself justified in certifying to their characters for truth and veracity, and, in matters requiring judgment and discrimination, to their intelligence also. The papers transmitted were required to be accompanied with an opinion as to the character and intelligence of all the witnesses and the good faith of the whole proceeding. (In re Clear Creek Quicksilver Mine, Decision of Commissioner, May 15th, 1868; Zabriskie's Land Laws, 216.

[1] New Idria Case, Opinion of Assistant Atty.-Gen. U. S. July 21st, 1871. Decision Acting Secretary of Interior, August 4th, 1871; Copp's U. S. Mining Decisions, 47.

was to be presented under the act in form, in order to be recognized as a controversy.

This section did not explicitly require a written application for the title or patent to the claim or mine. The claimant was only required "to file in the local office a diagram of the same." The third section spoke of the notice of an "intention to apply for a patent." In no other part of the act was reference made to written applications for a patent. The first subdivision of the instructions under the section referred to "the application filed as aforesaid," but the language was not to be found in the section. It was, however, necessarily implied, and the implication had all the force of positive expression. The written application should have possessed all the requisites stated by the law as conditions precedent, and shown upon its face, by a plain and succinct statement, that the applicant brought himself within the terms of the law.

The form should have possessed the substantial requisites of a pleading; as the source of title it became the original muniment in the claim of title, followed by the final paper—the patent.[1]

§ 7. Citizenship required.—There was a difference between the restricted language of the first section, giving the freedom of the mines to citizens of the United States and those who declared their intention of becoming citizens; and the unqualified language of the second section: "Any person, or association of persons" "who claim a vein or lode," etc. There was a question whether an alien in possession of a lode, who had expended the necessary sum and conformed to the local rules and customs of miners, could make an application alone, or with his associates, for a patent. The instructions were at first silent.

Ordinarily, an alien might be entitled to make application. The general language of the section was not restrained by the first section, which had no relation to titles. That aliens were in possession of such claims, alone and associated with citizens, was a well-known fact, and that foreign capital had assisted to a large extent in developing our mines was also well-known. There was no provision of the law to deprive them of their

[1] Yale's Mining Claims, 362.

claims by confiscation, or providing for a proceeding in the nature of office found, the subject will be more fully discussed hereafter in considering applications under the Act of 1872 and the Revised Statutes.[1]

§ 8. Entry and diagram.—Under the Departmental Instructions, it was held that mining claims might be entered at any district land office in the United States under this law, by any person or association of persons, corporate or incorporate. In making the entry, however, such a description of the tract was required to be filed as would indicate the vein or lode, or part or portion thereof, claimed, together with a diagram representing, by reference to some natural or artificial monument, the position and location of the claim, and the boundaries thereof, so far as such boundaries could be ascertained. In all cases, the number of feet in length claimed, on the vein or lode, was to be stated in the application filed, and the lines limiting the length of the claim were required in all cases to be exhibited on the diagram, and the course or direction of such end-lines, when not fixed by agreement with the adjoining claimants, nor by the the local customs or rules of the miners of the district, were ordered to be drawn at right angles to the ascertained or apparent general course of the vein or lode.[2]

Where, by the local laws, customs, or rules of miners of the district, no surface ground was permitted to be occupied for mining purposes, except the surface of the vein or lode, and the walls of such vein or lode were unascertained and the lateral extent of such vein or lode unknown, it was sufficient, after giving the description and diagram, to state the fact that the extent of such vein or lode could not be ascertained by actual measurement, but that the vein or lode was bounded on each side by the wall of the same, and to estimate the amount of ground contained between the given end-lines and the unascertained walls of the vein or lode; and in such case the patent issued for all the land contained between such end-lines and side-walls, with the right to follow such vein or lode, with all its dips, angles, and varia-

---

[1] Yale's Mining Claims and Water Rights, 361, 362, 363.
[2] Instructions Jan. 14th, 1867; Zabriskie's Land Laws, 200; Copp's U. S. Mining Decisions, 209.

tions, to any depth, although it might enter the land adjoining: provided, the estimated quantity should be equal to a horizontal plane bounded by the given end-lines, and the walls on the sides of such vein or lode.[1] Where, by the local laws, customs, or rules of miners of the district, a given quantity of surface-ground was fixed for the purpose of mining or milling the ore, the diagram and description in the entry were required to correspond with and include so much of the surface as was allowed by such laws, customs, or rules for that purpose.

But where, by such customs and rules, no surface-ground was permitted to be occupied for mining purposes, except the surface of the vein or lode, and the walls of such vein or lode were ascertained and well-known, such wall was required to be named in the description and marked in the diagram in connection with the end-lines of such claims. In the absence of uniform rules in any mining district, limiting the amount of surface to be used for mining purposes, actual and peaceable use and occupation for mining or milling purposes were to be regarded as evidence of a custom of miners authorizing the same. And the ground so occupied and used in connection with the vein or lode, and being adjacent thereto, might be included within the entry, and the diagram was to embrace the same as appurtenant to the mine.[2]

Where the claimant or claimants desired to include within their entry and diagram any surface-ground beyond the surface of the vein, it was necessary, upon filing the application, to furnish the register of the land office with proof of the usage, law, or custom under which he or they claimed such surface-ground, and such evidence might consist either of the written rules of the miners of the district, or the testimony of two credible witnesses to the uniform custom, or the actual use and occupation; which testimony was required to be reduced to writing by the register and receiver, and filed in the register's office with the application, a record thereof to be made.[3] Where the diagram showed and the application stated that "no surface-ground is claimed

[1] Instructions Jan. 14th, 1867; Zabriskie's Land Laws, 200; Copp's U. S. Mining Decisions, 230.
[2] Ibid.
[3] Ibid.

along the line of the lode," this, it was held, failed to comply with the requirement of a diagram " so extended, laterally or otherwise, as to conform to the local laws, customs, and rules of miners." Such an application was rejected.[1]

§ 9. Defects in the instructions.—The greater part of the instructions related to the mode of the surveys, descriptive of the land to be patented. The law did not change the system of surveys applicable to the public lands, except in ordering the survey of the claims upon the land not " hitherto surveyed by townships, ranges, and sections," from established base and meridian lines. The deviation from the rectangular system, authorized by this law, exists in certain cases in the general laws, where rectangular lines are impracticable and inexpedient, as remarked by Mr. Yale,[2] the *proviso* in the instructions under the second section, that the estimated quantity of the land surveyed should be equal to a horizontal plane bounded by the given end-lines, and the *walls* on the sides of each vein or lode, and the other parts of the instructions under this section, making the walls of the lode the lateral boundaries, whether ascertained or unascertained, where the local laws did not give surface-ground beyond the walls, were found inadequate to the purposes of the law. Such local laws were not generally found in force in any important mining camp, as surface-ground beyond the known, probable position of the walls of the veins is nearly always absolutely required for working purposes, independent of drifting, when necessary. Nor was the instruction consistent with the geological formations of metalliferous veins, whether we adopt the theory of the filling of a fissure vein by an expansive force from below, as applicable to true fissure lodes, or an injection, as it is sometimes called, or by infiltration from above, as applicable to another class of lodes. " Each theory," says Yale, " is correct in given cases, according to received opinions. In the first case, where the walls of the vein are rough, an increase of mineral is expected with an increase of depth and frequent irregularities of width in the vein; and in the second case, where the vein

---

[1] In re Gould and Condo Lodes, and McKibben Lode. Decision of Comr. March 24th, 1873; Copp's U. S. Mining Decisions, 165.
[2] Yale's Mining Claims, 363.

is wide at top, with smooth walls of the same material on both sides, we are justified in assuming that the vein is wedge-shaped, thinning gradually as the walls converge. A large proportion of metalliferous veins have their opposite walls nearly parallel, and Lyell gives an example in the celebrated vein of Audrensburg, in the Hartz, which has been worked at a depth of 500 yards perpendicularly and 200 horizontally, retaining almost the entire length a uniform width of three feet. But many lodes are extremely variable in size, being only one or two inches in one part and eight or ten feet in others, and again narrowing as before. Such alternate swelling and contracting is characteristic of these lodes, and is fully explained."

"De la Beche observes that the walls of fissures in general are rarely perfect planes throughout their entire course; nor could we well expect them to be so, since they commonly pass through rocks of unequal hardness and of different mineral composition. If, therefore, the opposite sides of such irregular fissures slide upon each other, or if there be a fault, as in the case of so many lodes, the parallelism of the opposite walls is at once entirely destroyed. The great mother-vein of Mariposa and Tuolumne, according to Prof. Whitney's report, is very irregular in width, varying from two feet to several rods. A surface width, upon a horizontal plane, bounded by the ascertained walls of the lode at the surface, did not satisfy the law by giving the entire lode within the length to the claimant, including its increased width as ascertained in descending.[1] The law itself gives the right to follow the lode with its dips, angles, and variations. The instructions were better adapted to the Comstock lode, which is very wide near the surface, than to lodes in general."

§ 10. **The application**, under the Acts of 1866 and 1870, was required by the Land Department to be in writing, and filed in the office of the register and receiver of the land district in which the claim lay. It stated the name of the applicant, and whether the claim was applied for by an individual, an association, or an incorporation; the name and extent of the claim; the character of the ore; the mining district, county, and State; the date of its original location, according to the mining customs;

---

[1] Yale's Mining Claims and Water Rights, 361-364.

where the same was recorded; whether the applicant claimed as a locator or purchaser; gave a description of the premises claimed, and the nature of the improvements made or labor performed; and finally, that the claimant had posted a "diagram" of the claim in a conspicuous place thereon, together with notice of his intention to apply for a patent, giving the date of the posting.

With the application the claimant filed a copy of the "diagram" posted on the claim, which "diagram" was required to represent the boundaries of the premises, as fixed by the local laws, customs, or rules of miners; and when the claim lay upon surveyed land, it showed its relation to the public surveys.

Diagrams of placer claims upon surveyed lands represented the subdivision which the claimant desired to enter, as the act required such entries in their exterior limits to conform to the legal subdivisions. With the diagram, it was necessary to file a copy of the "notice" posted upon the claim. This notice stated the name of the claimant; described the claim; gave the names of the adjoining claims, or if none adjoined, the names of the nearest claims; stated whether it was a placer or rock claim: if the former, the approximate area; if the latter, the estimated extent of the surface-ground, and the number of feet claimed on the course of the vein, distinctly stating the name of the lode, and the character of the vein exposed; the mining district, county, and State in which it lay; whether upon surveyed or unsurveyed lands: if the former, in what section, township, and range; if the latter, the location of the claim relatively to some well-known natural object or landmark in the vicinity; and finally, the notice stated that it was the intention of the claimant to apply for a patent for the premises designated, and upon which it was posted.[1]

There was also to be filed with the application satisfactory evidence that the applicant had the possessory right to the claim, agreeably to the local laws or customs of miners. This consisted of a certified copy of the laws or customs of the miners of the district, in force at the date of the location of the claim, of a certificate under seal, of the county or mining recorder,

[1] Instructions Aug. 8th, 1870, Copp's M. D. 256; Instructions Jan. 14th, 1867, Ibid. 241; Zabriskie's Land Laws, 200.

giving a copy of the record of the original location of the claim, with the name or names of the locators; and, if the applicant claimed as a purchaser, an abstract of title was to be filed, tracing the right of possession from the original locators to the applicant. Where applicants furnished satisfactory evidence that they and their grantors had held and worked their claims for a period equal to the time prescribed by the Statute of Limitations of mining claims of the State or Territory where the same might be situated, such evidence being sufficient to establish a right to a patent for a claim so held and worked, upon compliance with the other provisions of the law and instructions, the proofs above enumerated were not required.[1]

§ 11. **Publication of the notice.**—Upon filing these papers, the register and receiver gave the same careful examination, and if found to be regular, the register ordered the publication of the "notice" for ninety days in a newspaper published nearest the location of the claim; but before ordering such publication, the register required the claimant to enter into an agreement with the publisher to the effect that no claim or demand should be made against the United States for the payment of such publication, until the filing of which agreement the register was required to decline to order the publication. The cost of the publication of notice was, therefore, not to be estimated by the Surveyor-General. The register also posted copies of the "notice" and "diagram" in his office for ninety days, and on forwarding the case to the General Land Office, certified that they were so posted. On the expiration of the ninety days, the claimant, or his duly authorized agent, filed with the register his own affidavit, supported by at least one other person, cognizant of the fact, that the "notice" and "diagram" were posted in a conspicuous place upon the claim for the period of ninety consecutive days, giving the date of the same. The affidavit of the publisher was also required to be filed, to the effect that the "notice," a printed copy of which was attached, was published in his newspaper for ninety days, giving the dates on which such publication commenced and ended, and that he had received payment in full for the same. These affidavits were to

---

[1] Instructions Aug. 8th, 1870; Copp's U. S. Mining Decisions, 257.

be taken before the register and receiver, or any officer authorized to administer oaths within their district; but if taken before a magistrate without an official seal, his official character was to be authenticated under seal by the county clerk, in the usual manner. If all the proof furnished was satisfactory to the register and receiver, and no adverse claim had been filed, those officers, at the end of the ninety days, so informed the applicant for patent, and the Surveyor-General, who made an estimate of the expense of surveying and platting the claim, except in the case of placer claims on surveyed land, where no further survey was required, and when the claimant deposited the amount so estimated with any assistant United States treasurer, or designated depository in favor of the United States Treasurer, to be passed to the credit of the fund created by "individual depositors for surveys of the public lands," and filed with the Surveyor-General one of the duplicate certificates of deposit, that officer ordered the claim to be surveyed and platted in accordance with the regulations, except in cases where the claimant had had a preliminary survey made by the United States deputy surveyor, for the purpose of perfecting the diagram and notice posted on the claim, in which case such preliminary survey might be platted and adopted by the Surveyor-General for the final survey. Copies of plat and field-notes of survey were to be sent to the register and receiver, and to the General Land Office, the latter accompanied by the certificate of deposit.[1]

The register and receiver examined the returns of survey, and, if satisfactory, allowed the entry to be completed at the rate of five dollars per acre, or fractional part of an acre, for lode claims, or two and one-half dollars per acre for placer claims, and transmitted all the papers on their files bearing upon the case to the General Land Office, together with their joint opinion thereon, so that a patent might be issued if the proceedings were found regular.[2]

§ 12. **The duties of claimants, registers, and receivers** under the Act of 1866, were abstracted by the Department as fol-

[1] Instructions August 8th, 1870; Copp's U. S. Mining Decisions, 258.
[2] Ibid.

lows : " Claimant to post a notice on the claim, giving information of his intention to apply for a patent; to file a diagram with the register, together with the evidence of the rules of miners in support of the claim and its extent. After the diagram and notice have been posted ninety days, and no adverse claim filed, the claimant to apply to the Surveyor-General for a survey of the claim, deposit the amount estimated by the Surveyor-General to cover the expenses of the survey, platting, and notice, with any Assistant United States Treasurer, or designated depository in favor of the United States Treasurer, to be passed to the credit of the fund created by " individual depositors for the surveys of public lands," taking duplicate certificate of deposit, filing one with the Surveyor-General, to be sent to the General Land Office, and retaining the other; and when the survey is approved, and diagram thereof, together with the Surveyor-General's certificate as to improvements, and character of the vein exposed, the claimant to pay to the receiver the price of the claim. The register and receiver to examine the testimony filed by the claimant, showing the applicability of miners' rules in reference to the extent of the claim, which testimony is to be reduced to writing, and filed with the claimant's application in the register's office; also to examine the returns of survey approved by the Surveyor-General, and filed by the claimant.

Receiver to receive from the claimant the price of the claim on his filing with the register and receiver the approved plat and certificate of the Surveyor-General, as to the value of improvements and character of vein exposed, based on the testimony of two reliable witnesses.

The register's diagram of the claim being filed by the claimant, the register shall publish a notice in a newspaper nearest the claim, naming the mine, claimant, adjoining claimant, district, and county, informing the public that application has been made for a patent. The register will post the notice in his office for ninety days, and on the publisher presenting his account to the register, immediately on the expiration of the ninety days he will transmit it to the Surveyor-General; and on the receipt from the claimants of the Surveyor-General's certificate of the improvements on the claim, together with plat and other evidence of the survey approved, also the receiver's receipt for the payment for the claim, the register will transmit the same, with proof indorsed by the register and receiver as satisfactory, to the Commissioner of the General Land Office for patent." [1]

"*Surveyor-General's* duty when no adverse claim is filed, proof furnished that the diagram and notice had been posted for ninety days, and on receiving also from the register the account of the publisher of the notice.

"The Surveyor-General, when applied to by the claimant for the survey of his claim, shall estimate the expense of the survey, platting, and notice, and when a certificate of deposit is filed with him by the claimant, he shall order the survey to be made and transmit the certificate of deposit to the General Land Office. When the returns of survey are made to the Surveyor-General's office, he will approve the same, hand the necessary evidence thereof to the claimant, to be filed by him in the register's and receiver's office, for examination and final preparation of patent certificate by the register for transmission to the Commissioner of the General Land Office. The Surveyor-General will also transmit returns of the survey to the commissioner, with the account of the surveyor, and that of the publishers of the notice for direct payment from the United States Treasury to the parties entitled, as in case of payments made out of the funds deposited under tenth section of the Act of Congress, approved May 30th, 1862, and joint resolution of June 1st, 1864." [2]

§ 13. What a patent conveyed.—Every patent issued under the act expressly conveyed to the patentee the surface-ground embraced by the exterior boundaries of the survey of his claim, together with the right to follow the vein or lode along the course to the number of feet expressed in the patent, with its dips, angles, and variations, to any depth, although the lode should, in its dip or course, leave the surface-ground patented, and enter the land adjoining. The restriction was to one vein or lode. None of the patentees' rights existing under this section were affected by its repeal.[3] In all applications, therefore,

---

[1] Instructions June 25th, 1867; Copp's U. S. Mining Decisions, 247.
[2] Ibid.
[3] Act of 1872, Secs. 9, 12, 16; 17 U. S. Stats. 92.

pending at the date of the passage of the Act of 1872, although the patents were not issued till afterward, they conveyed the surface-ground embraced by the interior boundaries of the survey, and the right to follow the vein as above indicated, and also all other veins, lodes, or ledges, throughout their entire depth, the top or apex of which lay inside of such surface-lines extended downward vertically, although such other veins, lodes, or ledges, might so far depart from a perpendicular in their course downward as to extend outside the vertical side-lines of the surface-location, *provided*, that their right of possession to such outside parts of such other veins, lodes, or ledges was confined to such portions thereof as lay between vertical planes drawn downward through the end-lines of their location, so continued in their direction that such planes would intersect such exterior parts of such veins, lodes, or ledges; no right being granted, however, to the claimant of a vein or lode which extended in its downward course beyond the vertical lines of his claim, to enter upon the surface of a claim owned or possessed by another.

The Act of 1872 enlarged those rights, and in the applications for patents pending at the date of its passage, May 10th, 1872, authorized the issuance of patents upon such applications, which patents, in addition to granting to the patentee the right to follow the particular vein or lode along its course, although it might enter the land adjoining, to the number of feet expressed in the patent along the course thereof, and to any depth, also gave such patentee the right to follow all other veins, lodes, or ledges, the top or apex of which should lie within the exterior boundaries, if the same were not adversely claimed on May 10th, 1872, only to such extent, however, along the course thereof as might be embraced by such external boundaries, but to any depth; and furthermore, the act granted the exclusive right of possession to the surface-ground embraced by the survey.[1]

§ 14. **Diagram, notice, survey, and patent.**—The third section, which was also repealed by the Act of 1872, read as

---

[1] In re Hercules Lode; Decision of Commissioner, Dec. 26th, 1872; Copp's U. S. Mining Decisions, 151.

follows: " Sec. 3. That upon the filing of the diagram as provided in the second section of this act, and posting the same in a conspicuous place on the claim, together with a notice of intention to apply for a patent, the register of the land office shall publish a notice of the same in a newspaper published nearest to the location of said claim, and shall also post such notice in his office for the period of ninety days; and, after the expiration of said period, if no adverse claim shall have been filed, it shall be the duty of the Surveyor-General, upon application of the party, to survey the premises and make a plat thereof, indorsed with his approval, designating the number and description of the location, the value of the labor and improvements, and the character of the vein exposed; and upon the payment to the proper officer of five dollars per acre, together with the cost of such survey, plat, and notice, and giving satisfactory evidence that said diagram and notice have been posted on the claim during said period of ninety days, the register of the land office shall transmit to the General Land Office said plat, survey, and description; and a patent shall issue for the same thereupon. But said plat, survey, or description shall in no case cover more than one vein or lode, and no patent shall issue for more than one vein or lode, which shall be expressed in the patent issued."[1]

Notice.—The notice required in the third section was required to state the name of the claimant, the name of the mine, the name of adjoining claimants on each end of the claim, the district and county in which the mine was situated, informing the public that application had been made for a patent for the same. If no adverse claim was filed, and satisfactory proof was produced that the diagram and notice had been posted in the manner and for the period stipulated in the statute, it became the duty of the Surveyor-General to proceed in the manner pointed out in the section.[2]

The register was to give the notice required for the period of ninety days, and adverse claimants had the entire ninety days in which to file their claims; and immediately upon the expiration of the ninety days, if there had been no adverse claim filed, the claimant had the right to apply to the Surveyor-General

---

[1] Act of July 26th, 1866, 14 U. S. Stat. 252.
[2] Instructions Jan. 14th, 1867; Zabriskie's L. L. 200; Copp's Decis. 239.

for a survey, and upon its being approved, and the land paid for and the proper papers forwarded to the Commissioner, he was entitled to his patent. Ninety days were given in which to file adverse claims. They were required to be filed within that period.[1]

Where there was no evidence that a proper notice or diagram was posted on the claim, and the affidavits that were filed did not describe the notice or diagram, and did not state when they were posted up, this was held not to be a compliance with the statute.

There was, besides, no proof that the published notice agreed with the description in the application, and the application was rejected.[2]

§ 15. **Survey.**—As preliminary to the survey, the Surveyor-General was required to estimate the expense of surveying and platting, and ascertain from the register the cost of the publication of notice, the amount of all of which was to be deposited by the applicant for survey with any assistant United States Treasurer, or designated depository in favor of the United States Treasurer, to be passed to the credit of the fund created by "individual depositors for the surveys of the public lands." Duplicate certificates of such deposits were to be filed with the Surveyor-General for transmission to the General Land Office, as in the case of deposits for surveys of public lands, under the tenth section of the Act of Congress approved May 30th, 1862, and joint resolution of July 1st, 1864.

After the survey thus paid for was duly executed, and the plat thereof approved by the Surveyor-General, designating the number and the description of the location, accompanied by his official certificate of the value of the labor and improvements, and character of the vein exposed, with the testimony of two or more reliable persons cognizant of the facts on which his certificate was founded as to the value of the labor and improvements, the party claiming filed the same with the register and receiver, and thereupon paid to the receiver five dollars per acre

---

[1] In re Flagstaff Lode, Decision of Secretary, March 14th, 1872; Copp's U. S. Mining Decisions, 72.

[2] In re New Idria Mining Company's Application; "McGarrahan's Case," Decision of Acting Secretary, Aug. 4th, 1871; Opinion of Assistant Attorney-General, July 21st, 1871; Copp's U. S. Mining Decisions, 47-59.

for the premises embraced in the survey, and filed with those officers a triplicate certificate of deposit, showing the payment of the cost of survey, plat, and notice, with satisfactory evidence, which was the testimony of at least two credible witnesses, that the diagram and notice were posted on the claim for a period of ninety days as required by law. Thereupon, it was the duty of the register to transmit to the General Land Office the plat, survey, and description, with the proof indorsed as satisfactory by the register and receiver, so that a patent might issue if the proceedings were found regular; but neither the plat, survey, description, nor patent was allowed to issue for more than one vein or lode.[1]

The unity of the surveying system was to be maintained by extending over the mining districts the rectangular method, at least so far as township lines were concerned. The contemplated surveys of the mineral lands were to be made by district deputies, under contracts, according to the mode adopted in the survey of the public lands and private land claims, embracing in them all such veins or lodes as might be called for by claimants entitled to have them surveyed.

In consideration of the very limited scope of surveying involved in each mining claim, the per mileage allowed by law was not considered adequate to secure the services of scientific surveyors, and hence the necessity of resorting to a per diem principle, it being thought the most equitable under the circumstances.

The Surveyor-General was, therefore, authorized to commission resident mineral surveyors for different districts, where, isolated from each other, and absolutely inconvenient for one surveyor promptly to attend to the several calls for surveying in such localities, the compensation not to exceed $10 per diem, including all expenses incident thereto. Bonds in the sum of $10,000 were required from such surveyors.[2]

§ 16. **Posting the notice of application to make the entry.**—The details of the instructions under this section, and of the section itself, were to be strictly attended to.

---

[1] Instructions January 14th, 1867; Zabriskie's L. L. 200; Copp's Decis. 239.
[2] Ibid.

The application to the Surveyor-General to make the survey after the register and receiver had acted, was to be made by the claimant in writing, the necessary proof made before him of the work and its value, and payment made of the money for the survey by the deposit. When the survey was approved by him, and his certificate given to the claimant, based upon the testimony of two witnesses, the certificate was to be filed with the register and receiver, into whose hands the case came for the second time. Evidence was then given before them of the posting of the notice and diagram for ninety days on the claim, by two witnesses. Five dollars per acre was then paid to the receiver for the quantity of land embraced in the survey. The two officers then transmitted to the General Land Office " the plat, survey, and description, with the proof indorsed as satisfactory." It was not stated whether the written application of the claimant was to be transmitted or not. The patent was then issued " if the proceedings were found regular " for the limited quantity of one vein or lode.[1]

§ 17. **Effect of irregularities—Notice of application—Requisites.**—The purpose of the diagram and notice was analogous to a legal summons, by which any and all parties are notified that unless within a given time they come forward and defend any rights or interest they may have in certain premises, their rights to do so shall become barred, and judgment rendered for claimant.

The diagram and notice should, therefore, have been carefully prepared: any deception in the notice might have been a cause for the rejection of the claim.[2] But immaterial discrepancies, not likely to deceive parties to be notified, have been disregarded.[3]

Informal and irregular applications were not countenanced. A case presented the following irregularities: the notice was published nearly a month prior to the date of the application, and for the same length of time before the notices and diagrams

---

[1] Yale's Mining Claims and Water Rights, 366, 367.
[2] In re Flagstaff Lode, Decision of Commissioner, December 8th, 1871; Copp's Decis. 75.
[3] In re Flagstaff Lode, Decision of Secretary, Nov. 24th, 1871; Ibid. 71.

were posted on the claim and in the office of the register, and not during the ninety days of posting notices; the description and location of the premises as given in the notices and diagrams were meager and incorrect; the evidence submitted by the applicant showed that he had the record title to 240 linear feet only, whereas the application was for 1,200 feet. The office declined to issue a patent under these circumstances, and rejected the application. The applicant subsequently made a motion for a rehearing, and filed an abstract of title showing that at the time of the motion he had the record title to 1,200 linear feet of the lode, but as the applicant had not become sole owner until eleven months after the application for patent, the motion for rehearing was overruled.[1]

If the proceedings were irregular by a defect in the application, either in the allegations on substantial points in the failure to prove the necessary work and expenditure, the posting and advertisement, the proof of the local laws, the compliance with them, the diagram and survey, and the payment for the land and expenses, the application for the patent was rejected, and the applicant left without evidence of title. These were the equities to be maintained before the Government parted with the fee.[2]

§ 18. **Fees of surveyors.**—The per diem allowance to deputy surveyors, including all expenses of assistants for surveys of mineral claims, (see Instructions Jan. 14th, 1867) having been found inadequate, and in consequence, parties, in order to induce deputies to make the surveys, found it necessary to pay additional sums as on private account, the surveyors-general were authorized to increase the maximum per diem allowance according to the difficulty of the service, taking care, however, to have the work performed on the most economical scale by skillful and responsible surveyors, and in no case to exceed a maximum of twenty dollars per day. In each case where over ten dollars per day were allowed, the reasons showing the necessity thereof

---

[1] In re Red Warrior Lode, Decision Acting Commissioner, June 18th, 1873; Copp's U. S Mining Decisions, 201, 203; Decision of Commissioner, October 8th, 1873; Ibid.
[2] Yale's Mining Claims, 366, 367.

were required to be stated in the contract and then reported to the General Land Office; and no extra compensation was to be exacted or received by the deputy under penalty of forfeiting the contract, and exclusion from the public surveying service.[1]

§ 19. **Size of locations—Adjustment of surveys.**—The fourth section provided: "Sec. 4. That when such location and entry of a mine shall be upon unsurveyed lands, it shall and may be lawful, after the extension thereto of the public surveys, to adjust the surveys to the limits of the premises, according to the location and possession and plat aforesaid, and the Surveyor-General may, in extending the surveys, vary the same from a rectangular form to suit the circumstances of the country, and the local rules, laws, and customs of miners: *Provided*, That no location hereafter made shall exceed two hundred feet in length along the vein for each locator, with an additional claim for discovery to the discoverer of the lode, with the right to follow such vein to any depth, with all its dips, variations, and angles, together with a reasonable quantity of surface for the convenient working of the same as fixed by local rules: *And provided further*, That no person may make more than one location on the same lode, and not more than three thousand feet shall be taken in any one claim by any association of persons."[2]

§ 20. **Duties of deputy-surveyors.**—The deputy-surveyors were required to be scientific men, capable of examining and reporting fully on every lode they surveyed, and to bring in duplicate specimens of the ore, one of which was ordered to be sent to the General Land Office, and the other the Surveyor-General was authorized to keep, to be ultimately turned over with the surveying archives to the State authorities.

The surveyors of mineral claims, whether on surveyed or unsurveyed lands, were ordered to designate those claims by a progressive series of numbers, beginning with No. 37, so as to avoid interference in that respect with the regular sectional series of numbers in each township; and were to designate the four corners of each claim, where the side-lines of the same

---

[1] Instructions Aug. 8th, 1870; Copp's U. S. Mining Decisions, 253.
[2] Act of July 26th, 1866, 14 U. S. Stat. 252.

were known, so that such corners could be given by either trees, if any were found standing in place, or any corner-rocks existing in place, or posts might be set diagonally, and deeply imbedded, with four sides facing adjoining claims, sufficiently flattened to admit of inscriptions thereon; but where the corners were unknown, it was sufficient to place a well-built, solid mound at each end of the claim. The beginning corner of the claim nearest to any corners of the public surveys was to be connected by course and distance, so as to ascertain the relative position of each claim in reference to township and range when the same had been surveyed; but in those parts of the surveying district where no such lines had been extended, it was the duty of surveyors-general to have the same surveyed and marked, at least so far as standard and township lines were concerned, at the per mileage allowed, so as to embrace the mineral region, and to connect the nearest corners of the mineral claims with the corners of the public surveys. If found impracticable to establish independent base and meridian lines, or to extend township lines over the region containing mineral claims required to be surveyed under the law, then there was to be surveyed, in the first instance, such a claim, the initial point of which would start either from a confluence of waters, or such natural and permanent objects as would unmistakably identify the point of the beginning of the survey of the claim, upon which other surveys would depend.[1]

§ 21. Following the vein to any depth.—An applicant for a patent under the Act of 1866 might include surface-ground lying on either or both sides of the vein, as part of his claim, or apply for a patent for the vein alone. His rights upon the vein and in working into it were precisely the same, whatever might be the form of his surface-ground, or whether he had any or none. His end-lines and the distance between them were the same at all depths as upon the surface, no matter whether the position of the vein was vertical, or whether it dipped at a less or greater angle. This resulted directly from the right granted to the miner by all the local mining customs, as well as by the national

---

[1] Instructions Jan. 14th, 1867; Zabriskie's L. L. 200; Copp's Decis. 239.

mining act, of following the vein with all its dips, angles, and variations.

The Congressional enactment adopted in this respect the provisions of the mining customs, subordinating the rights of a patentee in respect to the surface-ground to the more important rights in respect to the vein, granting the right to follow the latter, with all its dips, angles, and variations, although it might enter the land adjoining, and requiring the adjoining land to be sold subject to this condition.

If a vein descends vertically into the earth, no controversy will arise. The measurement at the bottom is the same as at the surface. Suppose the cavity to have been made, and by a convulsion of nature the vein is swung from a vertical position to that of an angle, the first cavity in its last position represents the rights of a miner where the vein dips or inclines.[1]

§ 22. **Mode of survey—Quantity and restriction to one claim.**—The first instructions under this section were explicit except on three points:

1st. The number and quantity allowed each person or association before the passage of the act.

2d. The number of persons who constituted an association under the act, to take up 3,000 feet.

3d. Whether an indefinite number of claims could be taken up by the same individual by complying with the act, on different lodes.

The first omission was partly supplied by the law itself, as the quantity was referred to the local laws of the district, which governed before the law was passed.

The second by the subsequent interpretation which the Commissioner put upon the act.

The third point was left to the General Law, as the act contained no restriction as to the number of claims by an individual or association, except that not more than one should be taken up on the same lode. The mining laws limited the number of claims by location to one in the district, with an additional quantity to the discoverer; by the General Law, the number by

---

[1] In re Mountjoy Lode, Decision of Commissioner, Jan. 7th, 1870; Copp's U. S. Mining Decisions, 27.

purchase could not be limited by the district laws of the mines. The pre-emption laws limit the right to one location, and the declaration of the claimant made under oath is stringent in this respect. This act followed none of the analogies of the pre-emption laws in these particulars.

The third section having provided that the plat, survey, and patent should in no case cover more than one vein or lode, and that this should be expressed in the patent, this section again declared that no person could take more than one location on the *same* lode.

The reasonable quantity of surface for the convenient working of the claim was fixed by local rules, and should always have included sufficient ground for mills, workshops, dwellings, stables, drifting, and structures for manipulating the ores. Points for convenient drainage should always have been located with the claim, in anticipation of the necessity which might arise. The incident of timber was secured with the claim, but unless the surface-survey included a sufficient quantity, and there was timbered land in the vicinity, a pre-emption claim should have been located in connection with the mining claim, as there was nothing in the pre-emption acts prohibiting an entry to a lode under the act, in addition to the pre-emption claim.

The limitation of quantity under this act superseded all district laws, allowing a greater quantity than 200 feet to an individual location. The district laws generally limited the quantity fixed upon as so much for each man, and partners or corporations could only take up a number of feet corresponding to the number of the original locators.

§ 23. Deviation from the rectangular form of survey.—

The phrase, "circumstances of the country," undoubtedly meant the physical conformation of certain localities not admitting of lines at right angles in the location and description of mining claims. There were no local rules to comply with respecting the forms of the claims to which the surveys were to be adapted, but the extent and location of claims were regulated in many instances by the accidental conformation of the diggings and claims marked out with reference to the cardinal points or the form of the lines comprising the boundaries. In lode-claims located on the sides of hills or in ravines, the regulation is indispensable, and the form of the claim must take the shape of the country.[1]

§ 24. Number of feet located.—There was, after the passage of the Act of 1866, no authority of law for the location of more than 1,200 feet by five persons, provided they were discoverers, or 1,000 feet if claimed simply as locators.

Where certain applicants had applied for more, a patent was refused, but they, claiming as discoverers, were allowed to take 1,200 feet along the line of the lode, in which event they were instructed to have their monuments moved by a United States deputy-surveyor, and the plat and field-notes amended accordingly, a re-survey of the premises being held not necessary. They were also given the option of making re-locations under the Act of 1872, in which case they were required to commence *de novo*, after filing notice of location with the proper local officer, the proceedings being the same as if no previous application had been made. In this event, the Surveyor-General was allowed to adopt the field-notes of survey already made, with the necessary amendments as to distances along the vein and corner monuments, thus saving the applicants the expense of a re-survey.[2]

The act fixed a limit for claims on all veins or lodes from and after its passage, which limit could not be exceeded, no matter what the local regulations allowed, the Congressional maximum being 200 feet along the course of the lode to each locator, with an additional claim of 200 feet for discovery, and fixed 3,000 feet as the utmost extent that could be located or claimed upon the same by any association of persons after the 26th of July, 1866. After this date no individual in any district could " locate " or " claim " more than 200 feet on the course of any lode discovered thereafter unless he was the discoverer, when he could take an extra claim of 200 feet, and not more than 3,000 feet could thereafter be located or claimed upon any one vein by any association of persons, and to locate 3,000 feet of such

---

[1] Yale's Mining Claims and Water Rights in California, 369, 370.
[2] In re San Xavier Mine, Decision Commissioner, July 10th, 1873; Copp's U. S. Mining Decisions, 209.

lode required not less than fourteen bona fide locators to be associated together, each taking a claim of 200 feet, with 200 feet additional to the discoverer, or fifteen locators where they claimed without regard to the discovery right. In making these locations, the miners had the option of taking up and recording their claims either as segregated individual locations of 200 feet each, and working or disposing of them as such, or they could associate together and locate a number of these claims in common, provided the legal maximum of 3,000 feet was not exceeded, after the 26th day of July, 1866. This statute did not fix any amount of work or expenditure as necessary to *hold a claim*, but left that to be regulated by the miners themselves. It did, however, prescribe that an amount of not less than $1,000 should be expended on the claim as one of the conditions precedent to *obtaining a patent*.[1]

Where an applicant filed affidavit that the lode was discovered prior to the passage of the Act of 1866, and the record evidence showed it was not located till after the passage, the record evidence was held to control, and the act governed as to amount of location; and where 3,000 feet had been located the monuments were ordered removed by the deputy United States surveyor and placed at the four corners of 1,000 feet, the grantors having claimed by virtue of discovery, and the locators being only four in number, and the plat and field-notes were ordered to be amended accordingly.[2]

The construction first placed by the Land Office upon the provisos was that the limitation of claims in the aggregate to 3,000 feet on a lode to any person or association was wholly prospective, and related entirely to claims taken up after the date of said act, leaving the parties who held the possessory rights to claims previously located, although in excess of that maximum, at liberty to apply for and receive patents therefor, but the Assistant Attorney-General, in the New Idria Case,[3] had advised,

---

[1] In re Helmick Silver Mining Co. Decision of Commissioner, August 27th, 1872; Decision Acting Secretary, September 4th, 1872; Copp's U. S. Mining Decisions, 136, 139.

[2] In re Dunkirk Lode, Decision of Commissioner, Sept. 17th and Oct. 11th, 1873; Copp's U. S. Mining Decisions, 224.

[3] Decision of Assistant Atty.-Gen. July 21st, 1871; Copp's U. S. Mining Decisions, 57.

and the Acting Secretary of the Interior had followed the advice,[1] and held, reversing the decision of the Commissioner, that although the local law allowed in mining for cinnabar an appropriation of 160 acres, the statute limiting the claim to 3,000 feet controlled, and that Congress did not intend to provide that all new claims originating after the passage of the act should be limited to the 3,000 feet for each association, *and* at the same time provide that claims originating before its passage should be entitled to more.

Following this view, the office, on March 27th, 1872,[2] refused and declined to issue patents conveying more than 3,000 feet along the vein or lode, whether the location was made before or after the date of the Act of 1866.

The last proviso of this section limited the quantity of land that could be appropriated by any one association to 3,000 feet. Local rules when conflicting must give way to statutory provisions.[3]

---

[1] Decision of Acting Secretary, August 4th, 1871; Copp's U. S. Mining Decisions, 47.

[2] Decision of Commissioner, March 27th, 1872; Copp's U. S. Mining Decisions, 83.

[3] Many controversies arose in regard to the proper construction of the fourth section, some contending that under it a company formed merely for mining purposes and locating claims could take 3,000 feet on the vein, although such company or association might be composed of less than fourteen individuals. It was held by the Land Office that the manner of making locations, and the number of feet that could be taken on the same vein or lode by an individual or an association, depended upon the rules and customs of miners of the respective districts, the Act of 1866 in no respect superseding or modifying those customs, except where they authorized the location of more than 200 feet on the same lode by any one person, or more than 3,000 feet by any association of persons. In such cases, the statute restricted and reduced locations made after the act to the above named quantities respectively, as the maximum in each case. And this was the only difference existing between the local mining regulations and the controlling act of Congress.

An individual could not, therefore, locate more than 200 feet on the same lode, nor an association more than 3,000 feet, no matter how many persons might be associated together, or what the local customs prescribed. Whether a company or association could take as much as 3,000 feet depended upon the mining regulations of the particular district, and the number of persons associated in such company. Individuals could not, by forming themselves into companies, locate a greater number of feet to each person than could be done by each acting separately. They might locate as a company or an association at the rate of 200 feet to each individual embraced in it, with an additional 200 feet to the discoverer, if the local customs permitted that much to be taken, until 3,000 feet were located, after which no additional quantity could be claimed on the same lode

§ 25. **Adverse claims and contests.**—Section 6 (also expressly repealed) provided as follows: "Sec. 6. That when any adverse claimants to any mine, located and claimed as aforesaid, shall appear before the approval of the survey, as provided in the third section of this act, all proceedings shall be stayed until a final settlement and adjudication in the Courts of competent jurisdiction of the rights of possession to such claim, when a patent may issue as in other cases."

This provision was intended to protect the rights of third parties against the claim of the applicant, and authorized a contest before the proper tribunals, so that the right to the mine might be determined between the claimant and the contestant, and the patent issue to the proper party.

Under our land system, the registers and receivers of the Land Office have authority to determine contests between two or more pre-emptors claiming the same quarter-section of land, subject to the decision of the Commissioner, and an appeal from him lies to the Secretary of the Interior. So far, the question belongs to the Executive Department, and the decision is ministerial, although involving judicial questions. These decisions, however, may be revised by the judiciary in proper cases, and are not always conclusive. This provision, transferring the con-

---

by the same company, whatever might be the number of its members. In districts where the mining regulations limited locations to less than 200 feet to each individual, or less than 3,000 feet to any association of persons, claimants were restricted accordingly, such regulations remaining in full force, being unaffected by the act of Congress.

These remarks, however, applied wholly to original locations, made in pursuance of the rules and regulations of miners in the several mining districts. They had no application to claims in the hands of purchasers, and a mining claim of 3,000 feet might be owned and controlled by an association of less than fourteen persons, where possession was obtained by bona fide purchases for valuable consideration, or partly by purchase and partly by location, there being nothing in the act to prevent an association composed of any number of individuals from holding such claim, and, upon proper application and proof, obtaining a patent for it.

But as to associations or companies, formed for the purpose of *locating* claims, they were subject to the limitations of the fourth section, and the restriction of 200 feet to each locator could not be evaded by forming an association. The restriction to 3,000 feet was applied, whether the location was made prior or subsequent to the date of the act. (Decision of Commissioner, July, 1869; Zabriskie's L. L. 224; Land Office Reports, 1868-9, Report of Secretary of Interior, 144; Decision Commissioner, March 27th, 1872; Copp's M. D. 83; New Idria Case, Opinion of Assistant Att'y-Gen. U. S. July 21st, 1871; Copp's M. D. 57.)

test to the judiciary, was an innovation in the public land system. The provision was defective in many particulars.

Under the third section, the notice of the filing of the claim was to be given for the period of ninety days, and after the expiration of that period, if no adverse claim was filed, the Surveyor-General made the survey under the provision of this section, the adverse claimant might appear at any time before the approval of the survey, as provided in the third section. This extended the period indefinitely until the approval of the survey, and consequently, the contest might not be limited to the ninety days previously specified. As the surveyor did not get possession of the case till the lapse of the ninety days after the proof had been made in the register's office, this additional time was necessarily indefinite, as no time was limited within which the survey was to be made, other than reasonable convenience demanded.

This section did not state the mode of contesting the application before the land officers. No form of contest was indicated, whether written or oral. But the adverse claim was to be by necessary implication in writing, setting forth substantially the grounds of the contest by claiming the mine as against the applicant and all others. All proceedings were then stayed, both by the surveyor and the land-owners, "until a final settlement and adjudication in the Courts of competent jurisdiction of the *rights of possession* to such claim."

What were the Courts of competent jurisdiction was a matter left to be determined by the existing laws. No Court was designated as the competent tribunal, no attempt was made to confer jurisdiction upon Federal Courts, as Congress might have done, nor was an attempt made to give jurisdiction to State Courts, which already possessed it. The question to be determined was distinctly stated, viz: the *right of possession*. No other right could be involved, as no title had yet passed from the United States. The Federal Courts had already jurisdiction to determine such actions, in cases brought directly before them, under the Nevada Judicial Act of 1865, general in its provisions to all Courts. But under that act the jurisdiction was over the *subject;* and the jurisdiction over the *parties* must have been derived from existing laws other than the Act of 1866, where the paramount title was in the United States.

Where the applicant claimed under a derivative title from the locators, or those claiming under the locators, and the adverse claimant derived his title from the same original source, it was a proper case of contest, and the Court should have awarded the patent to the proper party. There was no limitation in the *act* within which the contest must have been commenced in the Courts after filing the adverse claim, or by which a party could be coerced to a speedy trial.[1]

§ 26. Proceedings on adverse claims.—Should a party appear as an "adverse claimant," as contemplated by the sixth section of the act, the register was ordered to require such person to show by proof the claim or interest he might have in the mine, and if satisfactory to the register, all proceedings were to be stayed until a final settlement and adjudication should be had in the Courts. But in case the adverse claimant, after proceedings had been stayed, failed to institute action in the Courts, either pending or at their next ensuing session, with a view to a final adjustment of the claims, the register was ordered to proceed with the case as if no objection had been filed.

The sufficiency of the adverse claim was a matter expressly referred to the local Courts by the statute, but the land officers were to be satisfied that the opposing claim was such as was contemplated by the sixth section. They were not to suffer the forms of law to be fraudulently used by pretended claimants, having in fact no rights worthy of investigation in the Courts. For instance, if it appeared that the adverse claim relied upon, related to a settlement claimed under the Pre-emption or Homestead Laws of the United States, it would have been decided not to be such a claim as was to be referred to the judicial tribunals for determination, and upon the filing of which the proceedings were to be stayed, and the case suspended to await trial in Court, these tribunals having no jurisdiction of claims arising under the Pre-emption and Homestead Laws.

The adverse claim must have been one arising under the local customs and rules of miners. The claimant was required to file an affidavit stating fully the nature of his claim, and if the facts

---

[1] Yale's Mining Claims and Water Rights, 372, 373, 379.

disclosed present opposing interests under these regulations, or the local laws of the State or Territory, the proceedings were to be stayed; after which, it became the duty of the party out of possession to carry the case into the Courts, and have his rights judicially determined. The language of the second section, " having previously occupied and improved the same," did not refer to an occupancy at some remote period. It meant an occupancy continuing up to the date of the application for a patent, otherwise the mine could not be said to be one " in regard to whose possession there was no controversy or opposing claim." The very fact, therefore, of the applicant being out of possession, and an adverse party in possession, showed the claim to be one for adjudication in the Courts before it could be disposed of in the Land Office. Hence, it was the duty of all applicants under the Mining Act to state in their applications whether they were occupying the premises for which a patent was asked; and if not, whether an adverse party was in possession. If the latter was the case, the party was notified that an application for a patent was made, in order that he might file an affidavit of his claim,[1] and the case was then suspended for action in the Courts.

§ 27. **Miscellaneous.**—The register was ordered to enter claims under the act in separate tract-books from those used for agricultural lands—dividing the books into townships and ranges, allowing about eight pages to each township. A new series of numbers was ordered to be commenced—beginning with No. 1, and continued in regular order. As no special fee was provided for, registers and receivers were allowed one per cent. each on amount of purchase-money, as in cash sales. The money received was to be accounted for in the receiver's returns as cash received from sale of mineral claims.[2]

Where the rules of miners did not permit ground to be occupied, except the surface of the vein or lode, the claims presented might contain less than an acre of ground. In such cases, the Land Office does not deal with a fraction, and the price of five

---

[1] Instructions July, 1869; Zabriskie's L. L. 239.
[2] Instructions June 25th, 1867; Copp's U. S. Mining Decisions, 245; Instructions July, 1869; Zabriskie's L. L. 239.

dollars was to be paid for the same. If the area exceeded that quantity, ten dollars; if more than two acres, fifteen dollars, and so on. In applications for mineral claims it was necessary, where a claim contained less than one acre, that the agreement expressed should be to pay five dollars for the claim.[1]

[1] Instructions June 25th, 1867; Copp's U. S. Mining Decisions, 245.

CHAPTER II.

RESERVATIONS AND EXCEPTIONS OF MINERAL LANDS IN GRANTS BY THE GOVERNMENT.

§ 28. Mineral lands reserved.
§ 29. Mineral lands in certain States not excepted.
§ 30. Exception from certain grants.
§ 31. The policy of the Government in reserving or excepting mineral lands.
§ 32. Excepting clause in placer and agricultural patents.
§ 33. Saline lands.
§ 34. School lands containing mineral.
§ 35. School lands in Nevada.
§ 36. Mineral lands in railroad grants.

§ 28. Mineral lands reserved.—Sec. 2318 of the Revised Statutes of the United States provides as follows: "In all cases, lands valuable for minerals shall be reserved from sale, except as otherwise expressly directed by law."[1]

And by Sec. 2258 of the Revised Statutes it is provided that: "The following classes of lands, unless otherwise specially provided for by law, shall not be subject to the rights of pre-emption, to wit: First. Lands included in any reservation by any treaty, law, or proclamation of the President, for any purpose. Second. Lands included within the limits of any incorporated town, or selected as the site of a city or town. Third. Lands actually settled and occupied for purposes of trade and business, and not for agriculture. Fourth. Lands on which are situated any known salines or mines."[2]

Minerals in the Indian Territory are said to be not reserved by the United States, and the Land Office has no control over such lands in such Territory.[3]

---

[1] See Act of July 4th, 1866, 14 U. S. Stats. 86.

[2] Rev. Stat. 2258, Sec. 10, Act Sept. 4th, 1841; 5 U. S. Stat. 455. See Wilcox v. Jackson, 13 Pet. 498; Josephs v. U. S. 1 N. & H. 197; Turner v. Am. B. Union, 5 McLean, 344; U. S. v. R. R. Bridge Co. 6 McLean, 517; Russell v. Beebe, Hemps. 704.

[3] Decision of Commissioner, June 26th, 1873, Copp's U. S. Mining Decisions, 208.

§ 29. **Mineral lands in certain States not excepted.**—Sec. 2345 of the Revised Statutes reads: "The provisions of the preceding sections of this chapter shall not apply to the mineral lands situated in the States of Michigan, Wisconsin, and Minnesota, which are declared free and open to exploration and purchase, according to legal subdivisions, in like manner as before the tenth day of May, one thousand eight hundred and seventy-two. And any bona fide entries of such lands within the States named, since the tenth day of May, one thousand eight hundred and seventy-two, may be patented without reference to any of the foregoing provisions of this chapter. Such lands shall be offered for public sale in the same manner, at the same minimum price, and under the same rights of pre-emption as other public lands."[1]

Exceptions from the operation of the act.—By an act to exclude the States of Missouri and Kansas from the provisions of the Act of May 10th, 1872, it is provided that, within the States of Missouri and Kansas, deposits of coal, iron, lead, or *other mineral*, are excluded from the operation of the Act of 1872, and all lands in those States are declared subject to disposal as agricultural lands.[2] Non-mineral affidavits, therefore, are not required from parties who desire to secure title to land within those States.[3]

§ 30. **Certain grants not to include mineral lands.**—"No act passed at the first session of the thirty-eighth Congress, granting lands to States or corporations to aid in the construction of roads or for other purposes, or to extend the time of grants made prior to the thirtieth day of January, one thousand eight hundred and sixty-five, shall be so construed as to embrace mineral lands, which in all cases are reserved exclusively to the United States, unless otherwise specially provided in the act or acts making the grant."[4]

§ 31. **The policy of the Government in reserving or excepting mineral lands.**—It has already been stated that it

---

[1] Rev. Stat. 2345; Act Feb. 18th, 1873; 17 U. S. Stat. 465.
[2] Act approved May 5th, 1876.
[3] Decision of Commissioner, July 21st, 1876, 3 Copp's Land-owner, 132.
[4] Rev. St. 2346; Act Jan. 30th, 1865; 13 U. S. Stat. 567.

had always been the policy of the Government of the United States to reserve from sale and pre-emption entry lands containing minerals or " known mines." [1] The truth of this statement will be demonstrated by reference to various acts of Congress.

"In 1785, the Continental Congress reserved one-third part of gold, silver, lead, and copper mines; but this principle was afterward abandoned. Salt-springs and lead mines were reserved by subsequent laws, and leased by the Government. The former were generally given to the new States on their admission; but under restrictions. They could not be sold, nor leased for a period exceeding ten years. By Acts passed in 1846, (9 Stat. at L. 37) and 1847, (Id. 181) the States were authorized to dispose of their salt springs; and lead and copper mines in the northwest were thrown open to settlers, and made subject to pre-emption. The Acts of July 1st, 1864, (13 Stat. at L. 343) and March 3d, 1865. (Id. 529) threw open coal lands to entry, but fixed the minimum price at twenty dollars per acre, instead of allowing pre-emption at the ordinary rate." [2]

The ordinance of the Revolutionary Congress, continuing until the Constitutional Congress of 1789, for ascertaining the mode of disposing of the lands in the western territory, passed May 20th, 1785—and which is the basis of our present land system—reserved "one-third part of all gold, silver, lead, and copper mines, to be sold, or otherwise disposed of, as Congress shall hereafter direct"; and in the form of grant, or patent prescribed by the act, the language is, "excepting and reserving one-third part of all gold, silver, lead, and copper mines within the same, for future sale or disposition." [3]

In numerous instances, from 1807, where lands were authorized to be sold in particular sections of the country, lead mines were reserved from sale; and by an Act of the 3d March, 1807, (2 Stat. at L. 445) the President was authorized to lease the lead mines for a period not exceeding three years. The Supreme Court of the United States held that power was given to Con-

---

[1] Ante, Sec. 1. Gold Hill Quartz M. Co. v. Ish. 5 Oregon, 107.
[2] Am. Law Review, Vol. 2, p. 388.
[3] American State Papers, Public Lands, Part 1, 13, 11; Yale's Mining Claims. 325.

gress by the Constitution to dispose of the public lands, and that power included the power to lease as well as to sell.[1]

By the eighth section of the Act of the 3d of March, 1849, (9 Stat. at L. 396) the powers of the Secretary of the Treasury over lead and other mines, relating to their supervision and lease, and the accounts with agents, were transferred to the Department of the Interior created by that act.

The tenth section of the General Pre-emption Law of 1841 (5 Stat. at L. 453) expressly excepted from the operation of the law "all lands on which are situated any known salines or mines."

The grants to railroad companies contain similar reservations of mineral lands; and even the Sutro Tunnel Act, while granting all the minerals discovered in the construction of the tunnel, and a pre-emption to two sections of land near the mouth of the tunnel for the use of the same, excepts the minerals contained in the *land*, and declares that the land shall not be selected from mineral land.[2]

By the Act of March 3d, 1853, (10 Stats. at L. 244) to provide for the survey of public lands in California, and to grant pre-emption rights thereon, the Act of September 4th, 1841, is extended to California, and the Surveyor-General prohibited from running other than township lines on mineral lands. The right of pre-emption on unsurveyed lands was limited to the period of one year, and mineral lands were excluded from its operation. The inhibition is repeated throughout the act. Mineral lands are excepted by the sixth section, and by the seventh section, which provides that no person shall obtain the benefits of the act by a settlement or location on *mineral lands*. (10 Stats. 244.)

The Act of July 23d, 1866, to quiet land titles in California, which required an approval of the selection under a school-land warrant issued by the State authorities, also, by Sec. 1, excepts mineral lands from the selections. (14 U. S. Stat. at L. 218.)

Reservations in railroad grants.—Sec. 10 of the Act of July 25th, 1866, (14 Stats. 239) granting lands to aid in the construction of a railroad from the line of the Central Pacific Rail-

---

[1] United States *v*. Gratiot, 14 Pet. 526.
[2] Sec. 2, Act of July 25th, 1866; 14 U. S. Stats. 242.

road, in California, to Portland, Oregon, limits the grant to alternate sections of land *not* mineral. The third section of the Act of 1862, (12 Stats. 489) to aid in the construction of a railroad from the Missouri River to the Pacific Ocean, reserves the mineral land, but grants the timber on it.

By Sec. 2 of the Act of July 13th, 1866, (14 Stats. 94) to aid in the construction of a railroad from Folsom to Placerville, in California, ten alternate sections per mile, designated by odd numbers, on each side of the line, " not containing gold or silver," were granted.

The proviso is added that the word " mineral " shall not be held to include iron or coal.

On the 30th of January, 1865, a joint resolution, reserving mineral lands from the operation of all acts of the first session of the Thirty-eighth Congress, was passed. (13 Stats. 567.)

But Fletcher v. Peck, 6 Cranch, 87, decides that where a right vests by legislative grant, even in the case where a fraud was committed by the party interested in obtaining it, a repeal of the act cannot divest the right; and certainly a legislative declaration could not effect it. While, therefore, such a resolution might be important as showing the *intention* of Congress, if an act had *actually granted* mineral land, and there was no room for construction, it would probably be inoperative.[1]

The Government never parted with the right to the mines.—It was said, in U. S. v. Parrott, 1 McAllister, C. C. 271, (which was a branch of the celebrated Castillero litigation, relative to the New Almaden quicksilver mine) that neither the policy nor legislation of a State could deprive the United States of any legal right they had to the mines.

The Act of 1850, (9 U. S. Stats. 452) admitting California into the Union, expressly provided that the people of that State, through their legislature, or otherwise, should never interfere with the primary disposal of the public lands within its limits, and should pass no law and do no act whereby the title of the United States to, and right to dispose of, said lands should be impaired or questioned. Congress had never parted with the

[1] Yale's Mining Claims, 329.

mines nor the right to dispose of them, but had a right at any moment to dispose of them.[1]

Reservations in grants to the States.—The Pre-emption Act of 1841, granting to certain States therein named, and to all new States afterward admitted, 500,000 acres of land, excluded reserved lands, salines, and known mines from sale. It was held, in California, that the question as to whether mineral or agricultural land prevailed, having been ascertained by the officers of the State, and the selection approved by the United States, a State patent would hold the land.[2]

By the seventy-two section grant for the use of a seminary of learning, and by the ten-section grant for public buildings, (see Secs. 12 and 13, Act of March 3d, 1853, 10 Stats. 248) mineral lands and reserved lands were reserved from location.

So, also, with the Act of May 30th, 1862, extending the pre-emption laws, Sec. 7 (12 Stats. 410).

And the Act of July 2d, 1862, (12 Stats. 503) for mining colleges.

The Illinois lead case.—Digging for minerals on the public domain of the United States, before the passage of the Mining Act of 1866, was a trespass, entitling the Government to damages in an action at law, and was such waste as might have been restrained by injunction.[3]

By an Act of the 3d of March, 1807, 2 U. S. Stats. 445, the lead mines of Indiana Territory, and afterward the State of Illinois, were reserved from sale, and the President authorized to lease them for a period not exceeding three years, and a grant of land, containing a lead mine discovered before the sale, was declared to be fraudulent and void; and, in U. S. *v.* Gear, 3 How. 120, the defendant in a civil action was held guilty of trespass, in mining for lead upon land in the State of Illinois, and an injunction was granted restraining him from the commission of waste.

By the fifth section of the act, the lead mines were reserved from sale. By the Act of June 26th, 1834, to create additional land districts in the State of Illinois and elsewhere, in connection with the Pre-emption Acts of 1830 and 1832, all the lands described in the Act of 1834 became the subject of sale and pre-

---

[1] U. S. *v.* Parrott. 1 McAll. C. C. 271.   [2] Ah Yew *v* Choate, 24 Cal. 562.
[3] U. S. *v.* Gear. 3 How. U. S. 120.

emption; and the question in Gear's case was, whether the Act of 1834 repealed the reservation contained in the Act of 1807, and subjected the land in question, containing lead ore, to pre-emption and sale. The Court held that there was no repeal, and that the land was not open to settlement by reason of the reservation; besides, the right of the Government, as owner of the land, to restrain the trespass, was complete and perfect.[1]

Implied license.—That there was ever an implied license from the Government to mine for the precious metals upon the public land, by reason of its indulgence, if not the direct encouragement extended to the mining population, was always denied in the Courts of the United States. The defendant, in the case of U. S. *v.* Parrott, claimed this license, and through it, immunity from damages for waste, but the claim was denied.[2]

The sale of lead mines.—By an Act of July 11th, 1846, (9 Stats. 37) Congress authorized the sale of the reserved lead mines in the States of Illinois and Arkansas, and the then Territories of Wisconsin and Iowa, at an increased rate of $1.25 per acre, as the minimum at public sale, but still excepting the lead mines from the operation of the pre-emption laws, until after they had been offered at public sale.

By the Act of March 3d, 1847, (9 Stats. 179) to create an additional land district in Wisconsin, pre-emption was granted to parties in the possession of lead mines by occupation, through discovery, or lease, under the United States.

By the Act of March 1st, 1847, (9 Stats. 146) the copper mines of Michigan were ordered to be sold, giving certain preferences to lessees under the Government, and persons in possession.

But by the Act of Sept. 26th, 1850, (9 Stats. 472) both of these were repealed, and this placed the mineral lands within these districts upon the same footing, as to sale, private entry, and pre-emption, as other public lands of the United States, saving certain vested rights.[3]

---

[1] See Cotton *v.* U. S. 11 How. U. S. 229; Yale's Mining Claims, 331.
[2] U. S. *v.* Parrott, 1 McAll. C. C. Rep. 271; U. S. *v.* Castillero, 2 Black. U. S. 17; Yale's Mining Claims, p. 333.
[3] See Cooper *v.* Roberts, 18 How. 173; Higgins *v.* Houghton, 25 Cal. 252.

Sale and pre-emption of coal lands.—Coal lands having been reserved under the General Pre-emption Act of 1841, an Act of July 1st, 1864, (13 Stats. 343) for the disposal of coal lands and town property on the public domain, authorized tracts embracing coal fields, or coal beds—and which by that act and past legislation were excluded, as mines, from ordinary private entry—to be sold.

By an amendatory Act of March 3d, 1865, (13 Stats. 529) citizens of the United States, who at that date were actually engaged in bona fide coal mining, on public lands not reserved for purposes of commerce, had a pre-emption to 160 acres of land, and might enter the same.

Grants from Indian Tribes in America.—The character of the grant of mines from Indian tribes came up for discussion in the important case of Chouteau v. Moloney, (16 How. U. S. 203) which was an action in the nature of ejectment for a large tract of land, including the whole city of Dubuque, Iowa, and which tract was formerly a part of the Louisiana territory acquired by the United States from France under the Treaty of Paris of 1803. It was admitted that the defendant had all the title that the United States possessed under the treaty. But the plaintiff claimed that he had acquired title to the land before the treaty; and, as private property was excepted from the cession to the United States by the terms of the treaty, the Government of the United States never had any title to give, nor any interest that could pass by its patent.

It is to be remembered that the French had retained possession of Louisiana up to 1762, when they ceded it to Spain; but in 1800, Bonaparte, then First Consul, induced the Spanish Government to retrocede it to France, and it remained French territory until the cession to the United States in 1803.

During the interval of time when it was the property of Spain—viz, in 1788—the tribe of Indians called the Foxes sold to the person under whom plaintiff claimed, a permit to work at a certain mine as long as he pleased, and also sold and abandoned to him all the adjacent coast and contents of the mine. In 1796, the grantee or licensee presented his petition to the governor of the territory, under the Spanish rule, for a confirmation of the sale, and a grant of the mine and land, and the governor made confirmation and grant accordingly. The question, therefore, was, did the title then pass?

The question was to be decided under the Spanish laws and regulations then in force, and under those laws and regulations it was held, in the first place, that the Indian tribes never had any right to interfere with or dispose of the mines within the dominion of Spain. The right of the Indians, as recognized by the latter country, extended to occupancy, but never to sale of the territory.

In that case, it was also held that the words of the grant from the Indians did not show an intent to part with anything more than a mining privilege, and that the governor, in his subsequent grants, only intended to confirm such rights as the Indians had previously given; and, further, that the proceedings to obtain the grant for the lands were irregular under the Spanish laws, and no complete title passed. The title confirmed and granted was good as a permission to dig and work the mines, but nothing more.[1]

By Article 4 of the Treaty, proclaimed January 17th, 1865, concluded on the 12th of October, 1863, between the United States and the Shoshone-Goship bands of Indians at Tuilla Valley, Utah, (now a part of Montana) it was provided that the country of the tribe might be explored or prospected for gold and silver, and other minerals and metals, and when mines were discovered that they might be worked, etc. (13 Stats. 682.) This is the first declaration to be found in the laws of the United States, authorizing, with the consent of Indian chiefs, the digging of gold upon public land—if such a provision can be called an authorization.

Further acts of Congress.—By the Act of February 27th, 1865, it was provided that no possessory action between individuals in any of the Courts of the United States, for the recovery of any mining title, or for damages to any such title, should be affected by the fact that the paramount title to the land on which the mines were was in the United States, but each case was to be adjudged by the law of possession. (13 Stats. 441.) This section is general in its terms, and applies to all Federal Courts.

---

[1] Chouteau *v.* Moloney, 16 How. 203.

The Act of May 5th, 1866, Section 2, (14 Stats. 43) concerning the boundaries of the State of Nevada, recognized and protected possessory rights to mining claims in Nevada, but proceeded further to state that the act was not to be construed as granting a title in fee to any mineral lands held by possessory titles in the mining States and Territories.

Next came the Act of July 25th, 1866, (14 Stats. 242) granting the right of way, and other privileges, to Adolph Sutro and his assigns, and commonly known as the Sutro Tunnel Act. This was the first act of Congress which, in express terms, granted a mining privilege on public land to any individual, or the public at large.

§ 32. Excepting clauses in placer and agricultural patents.

—A patent for a placer claim conveys " all valuable mineral and other deposits within the boundaries thereof," if no veins or lodes are claimed or known to exist within the exterior limits of the claim patented at the date of patent. In cases arising under the eleventh section,[1] an excepting clause is inserted as follows: "That, should any other vein, or lode, of quartz, or other rock in place, bearing gold, silver, cinnabar, lead, tin, copper, or other valuable deposit, be claimed or known to exist within the above described premises at the date hereof, the same is expressly excepted and excluded from these presents."

In all agricultural land patents the following clause is inserted, viz: "Subject to any vested and accrued water-rights for mining, agricultural, manufacturing, or other purposes, and rights to ditches and reservoirs used in connection with such water-rights as may be recognized and acknowledged by the local customs, laws, and decisions of Courts, and also subject to the right of the proprietor of a vein or lode to extract and remove his ore therefrom should the same be found to penetrate or intersect the premises hereby granted, as provided by law."

No title to a mining claim can be secured under an agricultural land patent. (See Sec. 2258, Rev. Stat.) [2]

---

[1] Rev. Stat. U. S. Sec. 2333.
[2] Decision of Commissioner, July 29th, 1875, 2 Copp's Land-owner, 82.

§ 33. Saline lands.—These lands are not subject to homestead or pre-emption entry. (See Secs. 2258 and 2289, Rev. Stat. U. S.) The policy of the Government has been uniform since the inauguration of the land system, to reserve from sale salt-springs.

The Act of May 18th, 1796, (1 U. S. Stat. 466) requires every surveyor to note in his field-book the true situation of all mines, salt-licks, and salt-springs, and reserves for future disposition by the United States every salt-spring which may be discovered, together with the section of one mile square which includes it.

The Act of May 10th, 1800, (2 U. S. Stat. 73) continued these reservations, and authorized sales to be made of the public lands by the Register and Receiver, excluding the sections reserved by the above mentioned act.

The Act of March 26th, 1804, (2 U. S. Stat. 277) providing for the disposal of the public lands in the Indiana Territory, declares that "the several salt-springs in the said Territory, together with as many contiguous sections to each as shall be deemed necessary by the President of the United States, shall be reserved for the future disposal of the United States."

It has been the policy of the Government to reserve these salt-springs and lands from sale, as is evidenced by the text of the different acts regulating the disposal of the public lands.

The Act of April 30th, 1802, (2 U. S. Stat. 173) admitting the State of Ohio, granted to the State certain salt-springs.

The Act of April 18th, 1818, (3 U. S. Stat. 429) authorizing the admission of the State of Illinois, grants all the salt-springs and the lands reserved for the use of the same to the State.

The Act of March 6th, 1820, (3 U. S. Stat. 545) authorizing the people of Missouri to form a State government and for the admission of the State, provides "that all salt-springs, not exceeding twelve in number, with six sections of land adjoining to each, shall be granted to the said State for the use of said State. * * * Provided, that no salt-spring, the right whereof now is, or hereafter shall be, confirmed or adjudged to any individual or individuals, shall by this section be granted to the said State."

The same provision is made in the acts providing for the admission of the following named States, as was provided in case of Missouri, viz: Arkansas, 5 U. S. Stat. 58; Michigan, 5 U. S. Stat. 59; Florida, 5 U. S. Stat. 789; Iowa, 5 U. S. Stat. 789; Wisconsin, 9 U. S. Stat. 58; Minnesota, 11 U. S. Stat. 166; Oregon, 11 U. S. Stat. 383; Kansas, 11 U. S. Stat. 269; Nebraska, 13 U. S. Stat. 47.

The Act approved March 3d, 1875, (18 U. S. Stat. 474) enabling the people of Colorado to form a State government, and for the admission of the State into the Union, has the same provisions in regard to salt-springs as those contained in the Missouri act.

The Supreme Court of the United States, in the case of Morton *v.* Nebraska, 21 Wall. 660, construed the proviso in the grant to Nebraska of salt lands. This proviso reads the same in the Nebraska and Colorado acts, viz: "Provided, that no salt-spring or lands, the right whereof is now vested in any individual or individuals, or which hereafter shall be confirmed or adjudged to any individual or individuals, shall by this act be granted to said State."

The State of Nebraska is within the limits of the Louisiana purchase. That part of Colorado which embraced the salt-springs in controversy lies within the boundaries of the territory ceded by Mexico to the United States.

The Court held that "the purpose Congress had in view is to be found in the unbroken line of policy in reference to saline reservations, from 1796 to the date of this act. To perpetuate this policy, and apply it equally to all the lands of the three Territories, (Kansas, Nebraska, and New Mexico) was the controlling consideration for the incorporation of the section (Sec. 4, July 22d, 1854, 10 U. S. Stat. 308); and although the words of the section are loose and general, their meaning is plain enough when taken in connection with the previous legislation on the subject of salines. It cannot be supposed, without an express declaration to that effect, that Congress intended to permit the sale of salines in Territories soon to be organized into States, and thus subvert a long established policy, by which it had been governed in similar cases."

W. C.—4.

Where it is not shown that any valuable deposit of salt is found upon the land, but the lands appear to be valuable only on account of salt-springs, the office has no authority to dispose of the tracts, either as agricultural or mineral lands. Accordingly, certain salt-springs in Colorado, "with six sections adjoining, and as contiguous as may be to each," were reserved, in order that Colorado might be placed on an equal footing with other States in the matter of salt-spring reservations. Filings thereon were rejected.[1]

The status of saline lands and salt-springs was fully considered by the Supreme Court of the United States in the case of Morton v. Green et al. and the State of Nebraska, already adverted to.[2] The action was ejectment, plaintiff's title being based upon locations of certain warrants. The real defendant, the State of Nebraska, insisted that the locations were without authority of law, because the lands on which the warrants were laid were saline lands, and, therefore, not subject to entry. Justice Davis, delivering the opinion of the Court, said: "The policy of the Government, since the acquisition of the northwest territory and the inauguration of our land system, to reserve salt-springs from sale, has been uniform. The Act of May 18th, 1796, (1 Stats. 464) the first to authorize a sale of the domain ceded by Virginia, is the basis of our present rectangular system of surveys. That act required every surveyor to note in his field-book the true situation of all mines, salt-licks, and salt-springs; and reserves for the future disposal of the United States a well-known salt-spring on the Scioto River, and every other salt-spring which should be discovered.

"These reservations were continued by the Act of May 10th, 1800, (2 Stats. 73) which created land districts in Ohio, with registers and receivers, and authorized sales by them; the preceding act having recognized the governor of the northwest territory and the Secretary of the Treasury as the agents for the sale of the lands. And the same policy was observed when provision was made in 1804 for the disposal of the lands in the Indiana Territory—embracing what are now Illinois and Indi-

---

[1] Hall v. Litchfield, Decision of Acting Commissioner, March 2d. 1876, 2 Copp's Land-owner, 179.
[2] 21 Wall. U. S. 660.

ana. (2 Stats. 277.) It was then declared 'that the several salt-springs within said Territory, with as many contiguous sections to each as shall be deemed necessary by the President, shall be reserved for the further disposal of the United States.' Without referring particularly to the different acts of Congress on the subject, it is enough to say that all the salines in the Virginia cession were reserved from sale, and afterward granted to the several States embraced in the ceded territory. Congress, in the disposition of the public lands in the Mississippi Territory, (2 Stats. 548; 3 Stats. 489) and in the Louisiana purchase, preserved the policy which it had applied to the country obtained from Virginia. Over all the territory acquired from France the general land system was extended. The same rules which were prescribed by law for the survey and sale of lands east of the Mississippi River were transferred to this new acquisition. (2 Stats. 324.) At the first sale of lands in this region which the President was authorized to make, salt-springs, and lands contiguous thereto, were excepted. (2 Stats. 391.) And this exception was continued when, in 1811, a new land district was created. Prior to this time, no portion of the country north of the State of Louisiana had been brought into market. The Act of March 3d, 1811, authorized this to be done, but the President, in offering the lands for sale, was directed to except salt-springs, lead mines, and lands contiguous thereto, which were reserved for the future disposal of the States to be carved out of this immense territory, which included the present State of Nebraska. (3 Stats. 665, Sec. 10.) And so particular was Congress not to depart from this policy, that in giving lands, in 1815, to the sufferers by the New Madrid earthquake, every lead mine and salt-spring were excluded from location. Indeed, in all the acts creating new land districts, in the territory now occupied by the States of Arkansas and Missouri, the manner of selling the public lands is not changed, nor is a sale of salines in any instance authorized. On the contrary, they incorporate the same reservations and exceptions which are contained in the Act of March 3d, 1811. In all of them, the Act of 18th May, 1796, is the rule of conduct for all Surveyors-General and their deputies, as the Act of 10th May, 1800, is the rule for all registers, requiring them to exclude from sale all salt springs, with the sections containing them

"In this state of the law of saline reservations, the Act of 22d July, 1854, was passed. It is by no means certain that the Act of March 3d, 1811, did not work the reservation of every saline in the Louisiana purchase; but, without discussing this point, it is enough to say that the Act of 1854 leaves no doubt of the intention of Congress to extend to the territory embraced by the States of Kansas and Nebraska, the same system that had been applied to the rest of the Louisiana purchase. There was certainly no reason why a long established policy, which had permeated the land system of the country, should be abandoned. On the contrary, there was every inducement to continue, for the benefit of the States thereafter to be organized, the policy which had prevailed since the first settlement of the north-western territory. In the admission of Ohio and other States, Congress had made liberal grants of land, including the salt-springs. This it was enabled to do by reserving these springs from sale. Without this reservation, it is plain to be seen there would have been no springs to give away, for every valuable saline deposit would have been purchased as soon as it was offered for sale. An intention to abandon a policy which had secured to the States admitted before 1854 donations of great value, cannot be imputed to Congress, unless the law on the subject admits of no other construction.

"But the Law of 1854, (10 Stats. 308) instead of manifesting an intention to abandon this policy, shows a purpose to continue it. It was the first law under which lands were surveyed in Nebraska, offered at public sale, and so made subject to private sale by entry. By it, Surveyors-General for New Mexico, and for Kansas and Nebraska, were appointed, with the usual powers and duties of such officers. And, although there are provisions relating to New Mexico applicable to that Territory alone, yet the leading purpose of this act was to bring into market, as soon as practicable, the lands of the United States in all of these Territories. In New Mexico this could not be done as soon as in Kansas or Nebraska, on account of the policy adopted of donations to actual settlers who should remove there before the first of January, 1858, and because of the necessity of segregating the

Spanish and Mexican claims from the mass of the public domain. For this reason, doubtless, local land offices were not created in New Mexico, but they were in Kansas and Nebraska, and registers and receivers appointed, with the powers and duties of similar officers in other land offices of the United States. And the President was authorized to cause the lands, when surveyed, to be exposed to sale from time to time, in the same manner and upon the same terms and conditions as the other public lands of the United States. If there were no other provisions in the law than we have enumerated, we should hesitate to say, in view of the limitation on sales prescribed by law wherever public lands had been offered for sale, that they did not of themselves work a reservation of the land in controversy. In conducting the public sales the register always reserved salines, as it was his duty to do, when marked on the plats, and this was never omitted, except by the neglect of the Surveyors-General or their deputies. But the fourth section of the act removes all doubt upon that subject. That section declares that none of the provisions of this act shall extend to mineral or school lands, *salines*, military or other reservations, or lands settled on or occupied for purposes of trade and commerce.

"It is contended that this section applies to the donations, conceded in the preceding sections, to actual settlers in New Mexico. But why make this restriction? To do it would require the importation of the word 'foregoing,' so that the section would read: None of the (foregoing) provisions shall extend to salines or mineral lands. There is no authority to make this importation, and in this way subtract from the general words of the section. The language of the section is imperative, and leaves no room for construction. Besides, why should an intention be imputed to Congress to exclude actual settlers from saline lands, but leave them open to private entry by speculators? The legislation upon the subject of public lands has always favored the actual settlers, but the construction contended for would discriminate against them, and in favor of a class of persons whose interests Congress has never been swift to promote.

"Apart from this, however, the purpose Congress had in view is to be found in the unbroken line of policy in reference to saline reservations, from 1796 to the date of this act. To perpetuate this policy, and apply it equally to all the lands of the three Territories, was the controlling consideration for the incorporation of the section, and although the words of the section are loose and general, their meaning is plain enough when taken in connection with the previous legislation on the subject of salines. It cannot be supposed, without an express declaration to that effect, that Congress intended to permit the sale of salines in Territories soon to be organized into States, and thus subvert a long-established policy by which it had been governed in similar cases. If anything were needed to show that the fourth section did reserve saline from sales, it can be found in the Act of the 3d of March, 1857, (11 Stats. 186) rearranging the land districts in Nebraska. This act excepts from sale such lands 'as may have been reserved.' This is a declaration that lands had been reserved, and obviously it is a legislative construction of the fourth section of the Act of 1854, for nowhere else, except by implication, had there been reservations of any sort in the Territory of Nebraska.

"Besides this, the Nebraska Enabling Act of April 10th, 1864, (13 Stats. 47) affords still further evidence that the Act of 1854 was intended to reserve salines. The purpose of reserving them was to preserve them for the use of the future States, and no State had been organized without a grant of salt-springs. In some of the States, the grant was of all within their boundaries, but on the admission of Missouri, and since, the number was limited to twelve. This number, with a certain quantity of contiguous lands, were granted to Nebraska on her admission. In doing this, Congress must have assumed that the springs had been reserved from sale, for if this had not been done, the presumption is, there would have been nothing for the grant to operate upon. It may be true that lands only fit for agriculture will remain a long time unentered, but this would never be the case with lands whose surface was covered over with salt. It would be an idle thing to make a grant of such lands, if there had been a previous right of entry conceded to individuals. This was in the mind of Congress, and induced the reservation in the Act of 1854, by means of which Nebraska could be placed on an equal footing with other States in like situation.

"But it is said the locations in question are ratified by the proviso to the section granting the salt-springs. This proviso was as follows: 'Provided, that no salt-spring or lands, *the right whereof is now vested in any individual* or individuals, or which hereafter shall be confirmed or adjudged to any individual or individuals, shall by this act be granted to said State.' This provision, with an unimportant change in phraseology, was first introduced into the Enabling Act for Missouri, (3 U. S. Stats. 547, Sec. 6) and exactly similar provisions with the one in question were inserted in the acts relating to Arkansas and Kansas. (5 U. S. Stats. 58; 12 Id. 126.) The real purpose of the proviso is to be found in the situation of the country embraced in the Louisiana purchase. The Treaty of Paris of April 30th, 1803, by which the 'Province of Louisiana' was acquired, stipulated for the protection of private property. This comprehended titles which were complete as well as those awaiting completion, (Soulard *v.* United States, 4 Peters, 511) and Congress adopted the appropriate means for ascertaining and confirming them. They were numerous and of various grades, and covered town sites and every species of lands. In Missouri, as the records of this Court show, they were quite extensive; and when she was admitted into the Union, many of these titles were perfect, and still a large number imperfect. In this condition of things, Congress thought proper, in granting the salt-springs to the State, to say that no salt-springs, *the right whereof now is* or shall be confirmed or adjudged to any individual, shall pass under the grant to the State. Whether this legislation was necessary to save salt-springs claimed under the French treaty, it is not important to determine, but manifestly it had this purpose in view and nothing more. It could not refer to salt-springs not thus claimed, because all entry upon them was unlawful, on account of previous reservation. It speaks of confirmations which had been made and those which were awaiting Governmental action, and in this condition were all the titles the United States were bound to protect.

"Although the words employed in the first division of the proviso to the saline grant to Nebraska are not the same as those used in the Missouri grant, they mean the same thing. There can be no difference between a right which has been confirmed and one which is now vested. Both are perfect in themselves, and refer to completed claims, while the last division in each proviso has reference to claims in course of completion, but not finally passed upon. This proviso can have little significance in the Enabling Act of Nebraska, nor, indeed, in many other enabling acts, but Congress doubtless thought proper to introduce it out of the superabundance of caution, as there could be no certainty that in purchased or conquered territory, however remote from settlement, there might not be private claims protected by treaty stipulations to which it would be applicable. It cannot be invoked, however, for the protection of these plaintiffs. When a vested right is spoken of in a statute, it means a right lawfully vested, and this excludes the locations in question, for they were made on lands reserved from sale or entry. If Congress had intended to ratify invalid entries like these, they would have used the language of ratification. Instead of doing this, the language actually employed negatives any idea that Congress intended to give validity to any unauthorized location on the public lands.

"The pre-emption act of the 4th of September, 1841, (5 Stats. 456) declares that 'no lands on which are situated *any known* salines or mines shall be liable to entry,' differing in this respect from the Acts of 1796 and 1854, which reserve every 'salt-spring' and 'salines.' The salines in this case were not hidden, as mines often are, but were so incrusted with salt that they resembled 'snow-covered lakes,' and were consequently not subject to pre-emption. Can it be supposed that a privilege denied to pre-emptors in Nebraska was conceded in the Act of 1864 to persons less meritorious?

"It appears, by the record, that on the survey of the Nebraska country, the salines in question were noted on the field-books, but these notes were not transmitted to the registers' general plats, and it is argued that the failure to do this gave a right of entry. But not so, for the words of the statute are general, and reserve from sale or location *all* salines, whether marked on the plats or not.

"What effect the statute might have on salines hidden in the earth, not known to the surveyor or the locator, but discovered after entry, may become a question in another case. It does not arise in this. Here the salines were not only noted on the field-books, but were palpable to the eye. Besides this, the locators of the warrants, before they made their entries, were told of the character of the lands. Indeed, it is quite clear that the lands were entered solely on account of the rich deposits of salt which they were supposed to contain.

"It does not strengthen the case of the plaintiffs that they obtained certificates of entry, and that patents were subsequently issued on these certificates. It has been repeatedly decided by this Court that patents for lands which have been previously granted, reserved from sale, or appropriated, are void. (Polk v. Wendell, 9 Cranch, 99; Minter v. Cromelin, 18 Howard, 88; Reichart v. Felps, 6 Wallace, 160.) The executive officers had no authority to issue a patent for the lands in controversy, because they were not subject to entry, having been previously reserved, and this want of power may be proved by a defendant in an action at law. (Minter v. Cromelin, Supra.)"

The judgment of the Supreme Court of Nebraska was affirmed.

The fact that a salt-spring exists upon a quarter-section withdraws the tract from the operation of the Homestead and Preemption Laws. A hearing to prove that the land is agricultural was not permitted where the township plats showed the existence of the salt-spring, and it was not alleged that the Surveyor-General's return was incorrect in regard to the location of the springs.[1]

---

[1] See Sec. 2258, 2289, Rev. St.; Decision of Commissioner, Nov. 5th, 1875; 2 Copp's Land-owner, 131. Since the above decisions were rendered, an act of Congress has been passed providing for the sale of saline lands. The text is as follows: "An Act providing for the sale of saline lands. *Be it enacted by the Senate and House of Representatives of the United States of America in Congress assembled,* That whenever it shall be made appear to the register and receiver of any land office of the United States that any lands within their district are saline in character, it shall be the duty of said register and said receiver, under the regulations of the General Land Office, to take testimony in reference to such lands to ascertain their true character, and to report the same to the General Land Office; and if, upon such testimony, the Commissioner of the General Land Office shall find that such lands are saline and incapable of being purchased under any of the laws of the United States relative to the public domain, then, and in such case, such lands shall be offered for sale by public auction at the local land office of the district in which the same shall be situated, under such regulations as shall be prescribed by the Commissioner of the General Land

§ 34. School lands containing minerals.—The question as to the ownership (State or National) of school lands has recently been passed upon by the Supreme Court of the United States in a very important case,[1] that of Sherman v. Buick, in which the plaintiff in error, who was plaintiff in the action, brought his suit in the proper Court of the State of California to recover possession of a piece of land of the defendant in error. On the trial, the plaintiff asserted title under a patent from the United States, of the date of May 15th, 1869, and the defendant under a patent from the State of California, of the date of January 6th, 1869. The land in question was a part of Sec. 36, township 5 south, range 1 east, Mt. Diablo Meridian, and the title of the State was supposed to rest on the Act of Congress of March 3d, 1853, (10 U. S. Stats. 246) granting to said State for school purposes, with certain limitations, every sixteenth and thirty-sixth section, according to the surveys thereafter to be made of the public lands.

The plaintiff, in aid of his patent and to defeat the title of the State under the Act of 1853, offered to prove that he settled upon the land described in his patent as early as December 20th, 1862, and had ever since resided on it; that the land was not surveyed until August 11th, 1866, and that he filed and proved his pre-emption claim to it, November 6th, 1866, and paid for and received a patent certificate, on which his patent was duly issued.

Office, and sold to the highest bidder for cash, at a price not less than one dollar and twenty-five cents per acre ; and in case said lands fail to sell when so offered, then the same shall be subject to private sale, at such land office, for cash, at a price not less than one dollar and twenty-five cents per acre, in the same manner as other lands of the United States are sold ; *Provided*, That the foregoing enactments shall not apply to any State or Territory which has not had a grant of salines by act of Congress, nor to any State which may have had such a grant, until either the grant has been fully satisfied, or the right of selection thereunder has expired by efflux of time. But nothing in this act shall authorize the sale or conveyance of any title other than such as the United States has, and the patents issued shall be in the form of a release and quit-claim of all title of the United States in such lands.

"Sec. 2. That all executive proclamations relating to the sales of Public Lands shall be published in only one newspaper, the same to be printed and published in the State or Territory where the lands are situated, and to be designated by the Secretary of the Interior.

"Approved, January 12, 1877."

[1] No. 45, Oct. 31, 1876, 3 Copp's Land-owner, 135, to be reported in about the 3d or 4th vol. of Otto, 93 or 94 U. S. R.

The Court excluded this evidence and gave judgment for defendant, and the Supreme Court of California affirmed that judgment.

The contest in the case was between a patent of the United States and a patent of the State of California, and the decision required a construction of the Act of 1853, so far as to determine which of these patents conveyed the real title under the facts offered in evidence.

The statute is entitled "An Act to provide for the survey of the public lands in California, the granting of pre-emption rights therein, and for other purposes."[1]

It is the first act of Congress which extends the land system of the United States over the newly acquired territory of that State. It provided for surveys, for sales, for the protection of the rights of settlers, miners, and others; and among the other purposes mentioned in the caption, for the donations to the State of lands for schools and for public buildings.

The importance of the subject is such that the following full extract is given from the opinion of Mr. Justice Miller, who delivered the opinion of the Court:

"The sixth and seventh sections of the act are of chief importance in the matter under consideration; the preceding sections having provided for surveying all the lands. The clause of the sixth section, in which the grant to the State of the sixteenth and thirty-sixth sections for school purposes is found, reads as follows: 'All the public lands in the State of California, whether surveyed or unsurveyed, with the exception of sections sixteen and thirty-six, which shall be, and hereby are, granted to the State for the purposes of public schools in each township; and with the exception of lands appropriated under this act, or reserved by competent authority, and excepting, also, the lands claimed under any foreign grant or title, and the mineral lands, shall be subject to the pre-emption laws of fourth of September, eighteen hundred and forty-one, with all the exceptions, conditions and limitations therein, except as is herein otherwise provided; and shall, after the plats thereof are returned to the office of the register, be offered for sale, after six

---

[1] 10 U. S. Stats. 244. See, also, Huff v. Doyle, No. 661, Oct. Term, 1876, to be reported in 93 or 94 U. S. R. Supreme Court of the United States.

months' public notice in the State of the time and place of sale under the laws, rules, and regulations now governing such sales, or such as may be hereafter prescribed.' Then come several provisos which we will consider hereafter, but we pause here to note the effect of this granting and excepting clause on the lands which should, by the future surveys of the Government, be found to be sections sixteen and thirty-six.

"It is obviously the main purpose of the section to declare, that after the lands are surveyed they shall be subject to sale according to the general land system of the Government; and secondly, to subject them to the right of pre-emption as defined by the Act of 1841, and to extend that right to lands unsurveyed as well as to those surveyed. But here it seemed to occur to the framer of the act, that California, like other States in which public lands lay, ought to have the sixteenth and thirty-sixth sections of each township for school purposes, and that they should not be liable to the *general pre-emption* law as other public lands of the Government would be. He accordingly injected into the sentence the grant of these lands to the State and the exception of them from the operation of the Pre-emption Law of 1841, together with other lands which in like manner were neither to be sold nor made subject to pre-emption. These were lands appropriated under the authority of that act, or reserved by competent authority; lands claimed under any foreign grant or title, (*i. e.*, Mexican grants) and mineral lands; all these were by this clause exempted from sale and from the general operation of the pre-emption laws.

"But the experience of the operation of our land system in other States suggested that it might be ten or twenty and in some instances thirty years before all the surveys would be completed and the precise location of each school section known. In the meantime, the State was rapidly filling up by actual settlers on these lands, whose necessities required improvements, and that those improvements, when found to be located on a school section, should have some protection. What this protection should be, and how the rights of the State should be also protected, and the relative rights of the settler and of the State under these circumstances, is a subject of a distinct section of the act—the one succeeding that we have just considered.

"That section (Sec. 7) provides: 'That when any settlement, by the erection of a dwelling-house, or the cultivation of any portion of the land, shall be made upon the sixteenth and thirty-sixth sections *before the same shall be surveyed*, or when such sections may be reserved for public uses, or taken by private claims, other lands shall be selected by the proper authorities of the State in lieu thereof.' That it was the purpose of this section to provide a rule for the exercise of the right of pre-emption to the school lands granted by the previous section cannot be doubted. The reason for this is equally clear, namely, that these lands were not only granted away by the preceding section and inchoate rights conferred on the State, but they were, with other classes of lands, by express terms excepted out of the operation of the pre-emption laws which it was a principal object of that section to extend to the public lands of California generally.

"Whether a settler on these school lands must have all the qualifications required by the Act of 1841, as being the head of a family, a citizen of the United States, etc., or whether the settlement, occupation, and cultivation must be precisely the same as required by that act, we need not stop to inquire. It is very plain that by the seventh section, so far as related to the dates of the settlement, it was sufficient if it was found to exist at the time the surveys were made which determined its locality; and as to its nature, that it was sufficient if it was by the erection of a dwelling-house, or by the cultivation of any portion of the land. These things being found to exist when the survey ascertained their location on a school section, the claim of the State to that particular piece of land was at an end; and being shown in the proper mode to the proper officer of the United States, the right of the State to that land was gone, and in lieu of it she had acquired the right to select other land agreeably to the Act of 1826, subject to the approval of the Secretary of the Interior.

"But it is said that the right of pre-emption thus granted by the seventh section was subject to the limitation prescribed by the third proviso to the sixth section, namely, 'that nothing in this act shall be construed to authorize any settlement to be made on any public lands not surveyed, unless the same be made within one year from the passage of this act; nor shall any right of such settler be recognized by virtue of any settlement or improvement made of such unsurveyed lands subsequent to that day.' And such was the opinion of the Supreme Court of California. And that Court, assuming this to be true, further held that the grant made by the act of the school sections was a present grant, vesting the title in the State to the sixteenth and thirty-sixth sections absolutely, as fast as the townships were surveyed and sectionized. (Higgins *v.* Houghton, 25 Cal. 252.) As a deduction from these premises, it held that the right to pre-emption on these lands expired with the lapse of the year from the passage of the act, and that no subsequent act of Congress could revive or extend it, even if it was so intended.

"But we are of opinion that the first of this series of propositions is untenable.

"The terms of the proviso to the sixth section and those of the seventh section, if to be applied to the same class of lands, are in conflict with each other. The one says that if settlement be made on land *before the survey,* which by that survey is found to be on the sixteenth or thirty-sixth section, the settlement shall be protected. The other says that no settlement shall be protected unless made within one year after the passage of the act. In view of the well-known fact that none of these surveys would be completed under several years, the provision of the seventh section was a useless and barren concession to the settler, if to be exercised within a year; and in the history of land titles in that State would have amounted to nothing. This apparent conflict is reconciled by holding to the natural construction of the language and the reasonable purpose of Congress, by which the limitation of one year to the right of pre-emption in the sixth section, is applicable alone to the general body of the public lands not granted away, and not excepted out of the operation of the Pre-emption Law of 1841, as the school lands were by the very terms of the previous part of the section; while section seven is left to control the right of pre-emption to the school sections, as it purports to do.

"In this view of the matter, the very learned argument of counsel on the question of the character of the grant, as to the time when the title vests in the State, and the copious reference to the acts of Congress and of the State as authorizing pre-emption after the expiration of one year from the date of the statute, are immaterial to the issue. Actual settlement before survey made accompanied the grant as a qualifying limitation of the right of the State, which she was bound to recognize when it was found to exist, and for which she was authorized to seek indemnity in another quarter. There is, therefore, no necessity for any additional legislation by Congress to secure the pre-emption right as to school sections, and no question as to whether it has so legislated, or whether such legislation would be valid, and we do not enter on those questions.

"No question is made in the argument here, none seems to have been made in the Supreme Court of the State, and none is to be found in its opinion in the case, as to the admissibility of the rejected testimony, if the fact which it sought to establish could be recognized by the Court. Nor do we think such objection, if made, is sustainable. The testimony offered does not go to impeach or contradict the patent of the United States, or vary its meaning. Its object was to show that the State of California, when she made her conveyance of the land to defendant, had no title to it. That she never had, and that by the terms of the act of Congress under which she claimed, the only right she ever had in regard to this tract was to seek other land in lieu of it. The effect of the evidence was to show that the title set up by defendant under the State was void—not merely voidable, but void *ab initio*. For this purpose it was competent and it was sufficient, for it showed that when the survey was actually made, and the land in question was found to be part of section thirty-six, plaintiff had made a settlement on it, within the meaning of the seventh section of the Act of 1853, and the State could do nothing but seek indemnity in other land.

"It has always been held that an absolute want of power to issue a patent could be shown in a Court of Law to defeat a title set up under it, though where it is merely voidable the party may be compelled to resort to a Court of Equity to have it so declared. (Stodard *v.* Chambers, 2 How. 317; Easton *v.* Salisbury, 21 How. 426; Reichart *v.* Felps, 6 Wall. 160.)"

The judgment was reversed and case remanded to the Supreme

Court of California, with direction to order a new trial in conformity to the principles of the opinion.[1]

[1] The Supreme Court of California held (45 Cal. 656) that the title to each sixteenth and thirty-sixth section upon its being surveyed vested absolutely in the State of California; that Congress had no power, after the passage of that act, to impair the grant or prevent the title to those sections, upon their being surveyed, from vesting in the State, and that therefore the Act of Congress of May 30th, 1862, (12 U. S. Stats. 409) did not have the effect to extend the right of preemption over those sections.

In arriving at this conclusion, the Court followed the doctrine laid down in Higgins v. Houghton, 25 Cal. 252, to the effect that the Act of Congress of March 3d, 1853, vested in this State the title to the sixteenth and thirty-sixth sections in each township; that the power of locating the land granted by means of a survey of the public lands was reserved to the General Government, and "as fast as townships thereafter were surveyed and sectionized, that the State became the owner of the sixteenth and thirty-sixth sections absolutely, not only as to quality, but as to position also"; and "that by the grant of the sixteenth and thirty-sixth sections to the State in full property, they were effectually withdrawn from the operation of the acts relating to pre-emptions." In the latter case the lands were *mineral lands*. The grants, so far as respects the *location* of the lands granted, were held to be subject to the exception of lands reserved by competent authority, and lands to which a valid right of pre-emption should attach, under the provisions of the act, prior to the survey, that is to say, the lands to which a valid right of pre-emption might be acquired by means of a settlement which had already been made, or which might be made within one year after the passage of the act. (See Higgins v. Houghton, 25 Cal. 252; Doll v. Meador, 16 Cal. 296; Van Valkenburg v. McCloud, 21 Cal. 330; Foley v. Harrison, 15 How. 447.)

The words of the Swamp Land Act of Congress of the 28th September, 1850, are that the lands "shall be and the same are hereby granted to said State." And these words were held to constitute a grant *in presenti* in the following cases: Summers v. Dickinson, 9 Cal. 554; Owen v. Jackson, Ibid. 322; Keeran v. Griffith, 27 Cal. 87; Robinson v. Forrest, 29 Cal. 317.

The case of Sherman v. Buick, Supra, is followed in the still later case of Morrow v. Kingsbury, Oct. 7th, 1875, Supreme Court of California, No. 4504, not reported. (See Finney v. Berger, 50 Cal. 248.)

The ruling upon this important subject in the General Land Office of the United States should be referred to.

The question arose in the Keystone Case, which involved the right of the State of California to school sections which are mineral in character.

See Keystone Case, Decision of Commissioner of Land Office, June 18th, 1872; Report of Secretary of Interior, 1873, 24; Copp's U. S. Mining Decisions, p. 105; Decision of Secretary of Interior affirming the decision of the Commissioner, April 28th, 1873; Report of Secretary of Interior, 1873, 24; Copp's U. S. Mining Decisions, p. 109.

The controversy arose between certain mining claimants, the town of Amador, Amador County, Cal., and the grantee of the State of California, of a portion of a thirty-sixth section. The question was as to the right of the State of California, under the grant of March 3d, 1853, to lands found upon survey to be numerically designated under the public land system as Secs. 16 and 36, where such lands were, at the date of such survey and designation, in the bona fide possession of parties properly qualified, who claimed the right of having the

In the construction of the Act of 1853, therefore, it must be considered as settled that the school sections, sixteen and thirty-

mining and town-site laws of the United States executed in their favor. There had been a decision in the case of Cooper v. Roberts, 18 How. 173, affirming the right of the State of Michigan to certain copper-bearing lands in School Sec. 16, in that State.

The Commissioner drew a distinction between the acts applicable to Michigan— the Act of June 23d, 1836, (5 U. S. Stats. at L. 59) and the act applicable to California—the Act of 1853, Supra.

He referred to the inhibition in the latter, Sec. 6, against the survey of any other than township lines, "where the lands are mineral," and remarked that this inhibition was not repealed by Congress until the passage of the laws of July 26th, 1866, and July 9th, 1870, commonly known as the "mining acts," which provided for extending the United States surveys to mineral lands. He said it was not easily understood how the sixth section of the Act of 1853 could be construed to be a present grant of Secs. 16 and 36 of lands which were, by the second section of the same act, expressly excluded from survey, as mineral. The Commissioner reviewed the seventh section of the Act of 1853, and the Act of July, 1866—the first "mining act"—and ruled that prior to the 7th of October, 1870, (the date of filing the plat of the township) the land in controversy formed a part of the unsurveyed mineral lands of the public domain, and that parties who were in the actual occupancy and possession of mining claims, under local regulations, in said subdivision at the date of the filing of the township plat, were in such occupancy and possession under authority of the Statute of July 26th, 1866, and that they or their grantees, upon compliance with the mining laws of Congress, would be entitled to patents for their respective claims, the same as if they were upon unsurveyed lands, or within sections other than sixteen or thirty-six.

In affirming this ruling, the Secretary of the Interior said that it was conceded, upon the facts, that each of the mining companies was entitled to a patent, unless the title to the half-section was vested in the State of California or its grantee, and assumed that in every valid grant there must be a grantor capable of making the grant, a grantee capable of taking it, and a thing granted capable of identification with reasonable certainty ; that all grants made by the General Government to individuals, corporations, or States are to be construed strictly against the grantee, and that nothing passes by implication ; that the intent of the law-makers is to govern, and that such intent is to be gathered from the entire act. He considered these three questions :

1st. When does title vest in the State to Secs. 16 and 36 under the Act of 1853 ?

2d. Does the seventh section except from the grant land upon which settlement has been made prior to survey, for other purposes than pre-emption appropriation ?

3d. Does the grant include mineral lands in Secs. 16 and 36 ?

In answer to the first query, and construing the sixth section of the act, he held that it was a grant to the State *in presenti*, in the nature of a float, taking effect upon specific tracts when the same are surveyed by the United States, and not before. The grant is in words *de presenti;* but, until survey, there are no tracts or parcels of land in existence answering to the calls of the grant.

The grant was held to be in its nature the same as that usually made by Congress to railroad companies, to aid in the construction of their roads. These grants are generally for a certain number of sections, designated by odd numbers on each side of the road, with a provision for indemnity selection, in case any of

W. C.—5.

six, granted by section six to the State, are also excepted from the operation of the Pre-emption Law governing the public

such sections shall have been sold, or otherwise disposed of, prior to the definite location of the line.

The cases of Railroad v. Smith, 9 Wallace, 99, and Railroad v. Fremont Co. 9 Wallace, 90, decided that these grants did not vest any right in the companies to specific sections until the line of the road was definitely fixed on the face of the earth. The title to specific tracts vested only on the happening of a contingency—the definite location of the road. The same rule was applied by the Secretary to the grant to the State. The title only vested upon the happening of the contingency that made the grant certain as to location, viz., the survey. As sustaining this position the following cases were cited : Gaines v. Nicholson, 9 How. 365; Cooper v. Roberts, 18 How. 173 ; Kissell v. St. Louis Public Schools, 18 How. 19; Terry v. Megerle, 24 Cal. 624; Grayson v. Knight, 27 Cal. 507; Middleton v. Lowe, 30 Cal. 596; West v. Cochran, 17 How. 413. And the following cases, relied upon in opposition, reviewed : Higgins v. Houghton, Supra; Rutherford v. Green, 2 Wheat. 196; Lessieur v. Price, 12 How. 59; How v. Missouri, 12 How. 126; Veeder v. Guffey, 3 Wis. 520 ; Sherman v. Buick, 45 Cal. Supra; Van Valkenburg v. McCloud, 21 Cal. 330.

In answering the second question, and construing the seventh section of the Act of 1853, the Secretary was of opinion that it excepts from the grant to the State lands upon Secs. 16 and 36, upon which any settlement by the erection of a building or buildings, or the cultivation of any portion of the land, has been made prior to survey.

And as regards the third proposition, he was of opinion that Congress, by the Act of 1853, did not intend to grant and did not grant to the State any mineral lands that, by survey, are shown to be in Secs. 16 and 36—on the contrary, the intention to reserve those lands was considered apparent from the act itself. (Secs. 6, 8, 12, 13.)

The Mining Act of July 26th, 1866, was regarded as providing an exclusive method for appropriating the mineral lands of the United States. It was the first act passed by Congress undertaking to dispose of its mineral lands, and it made no exceptions in favor of school or other grants.

If the State could obtain two sections in every mineral township, it was urged that it might establish a mineral system for itself, and one in conflict with that of the General Government. It was held that such was not the intention of Congress, and no mineral lands passed by the grant.

The Keystone case was approved in Delaney v. Thomas, Decision of Commissioner, June 25th, 1875 ; 2 Copp's Land-owner, 50.

The legislature of California, by the Act of Feb. 3d, 1876, Stats. 1875-6, p. 20, and the Act of March 28th, 1874, Stats. 1873-4, p. 765, did indeed provide for the "sale of the mineral lands belonging to the State," and undertook to provide machinery regulating the same. It provided for the affidavit for purchase, as to who should be preferred purchasers; for contests and actions, manner of sale, vested rights, patents, and payments.

In the General Land Office a case arose in relation to the right of the State of Nevada to Secs. 16 and 36 of each township, for school purposes, when such sections are found to contain mines.

The seventh section of the Enabling Act of the 21st of March, 1864, passed at the first session of the Thirty-eighth Congress, grants to said State said sections, unless sold or otherwise disposed of by any act of Congress.

Joint resolution of the 30th of January, 1855, (13 Stats. 557) declares that no

lands generally. That under the seventh section providing a rule by which the right of pre-emption on the school sections is governed; a settlement is protected, if the surveys ascertain its location to be on a school section when those surveys are made. The only right then conferred on the State is to select other land in lieu of that occupied.

The proviso in the sixth section forbidding pre-emption on unsurveyed lands, after one year from the date of passage of the act, is limited to the lands not excepted out of that section, and has no application to the school sections excepted.

§ 35. **School lands containing minerals, in Nevada.**—The question whether the grant of school lands to the State of Nevada was one *in præsenti* or *in futuro*, was presented for decision in the case of Heydenfelt v. Daney Gold and Silver Mining Co.[1] in the Supreme Court of Nevada.

The Court, however, assumed, for the purposes of the decision, that the grant took effect absolutely upon the admission of the State into the Union.

The case was an action of ejectment to recover a portion of the west half of the southwest quarter of section sixteen, township sixteen, range twenty-one east, Mount Diablo base and meridian. The plaintiff claimed title under a patent issued to his grantors and predecessors in interest by the State of Nevada, on the 14th day of July, 1868, under and by virtue of the statute authorizing the conveyance of lands granted to the State by the seventh section of the Enabling Act of Congress, entitled "An Act to enable the people of Nevada to form a constitution and

---

act passed at the first session of the Thirty-eighth Congress, granting lands to States or corporations, to aid in the construction of roads, or *for other purposes*, or to extend the time of grants heretofore made, shall be so construed as to embrace mineral lands, which in all cases shall be and are reserved exclusively to the United States, unless "otherwise specially provided" in the act making the grant.

In view of this legislation, and of the considerations set forth, it was held to be clear that an executive officer must regard a section of land No. 16 or 36, situated in Nevada, and "rich in minerals," as the property of the United States, and not as passing to the State under the act. (Decision of Secretary of the Interior, May 20th, 1870; Copp's U. S. Mining Decisions, pp. 30, 31.)

The State Register was allowed to select other lands as indemnity when school sections Nos. 16 and 36 should be found to be mineral. (Decision of Commissioner of General Land Office, May 24th, 1870; Copp's U. S. Mining Decisions, pp. 30, 31.)

[1] 10 Nevada, 290.

State government, and for the admission of such State into the Union on an equal footing with the original States," approved March 21st, 1864, which reads as follows: "That sections numbers sixteen and thirty-six in every township, and where such sections have been sold or otherwise disposed of by any act of Congress, other lands equivalent thereto, in legal subdivisions of not less than one-quarter section, and as contiguous as may be, shall be and are hereby granted to said State for the support of common schools." (13 U. S. Stat. 32; Stat. 1864–5, 37.) The defendant claimed title under a patent issued to it by the United States, on the 7th day of March, 1874, under and by virtue of the Act of Congress entitled "An Act granting the right of way to ditch and canal-owners over the public lands, and for other purposes," approved July 26th, 1866, (14 U. S. Stat. 251) the act amendatory thereof, approved July 9th, 1870, (16 U. S. Stat. 217) and the act entitled "An Act to promote the development of the mining resources of the United States," approved May 10th, 1872 (17 U. S. Stat. 91).

The land in controversy was mineral land, and the defendant was in possession of the same, and was engaged in conducting and carrying on the business of mining thereon, and had erected improvements thereon, for mining purposes, of the value of over $80,000. In the year 1867, prior to the date of the survey or approval of the survey of the land by the Government of the United States, the grantors and predecessors in interest of defendant entered upon the land for mining purposes, and claimed and occupied the same in conformity with the laws, customs, and usages of miners in the locality and mining districts in which the land was situate, and were so possessed and engaged in mining thereon when the land was first surveyed, and when the State issued its patent to the grantors and predecessors in interest of plaintiff.

Two leading questions were presented for consideration in determining the legal rights of the respective parties: 1st. When does the title vest in the State to the sixteenth and thirty-sixth sections granted by the seventh section of the Enabling Act? 2d. Did the patent issued by the State include mineral lands?

The Court did not deem it necessary to decide whether the grant was one *in præsenti* or *in futuro*, assuming, for the sake of the argument, that the proper construction to be given to the seventh section of the Enabling Act is, that the grant took effect absolutely upon the admission of this State into the Union, and that the title to the lands then vested in the State, although subsequent proceedings might, as was said in Schulenberg v. Harriman, 21 Wall. 62, "be required to give precision to that title and attach it to specific tracts," and likewise assuming that Congress had no power, after the admission of the State into the Union, to impair the grant, without the consent of the State. The Court said: "Still, we think it must be admitted that Congress could thereafter, with the consent of this State, prior to the disposal by the State of any of the lands embraced in said sections, and at any time prior to the survey, change the terms of the grant, and we are of opinion that, by the subsequent act of Congress and the act of acceptance by the legislature of this State, the mineral lands were reserved from sale by the government of the United States, with the consent of this State, and that the patent issued by this State did not, upon the admitted facts of this case, include the mineral lands in controversy. If we accept the definition announced by text-writers, 'that a grant is a contract; executed, it is true, but still a contract,' (3 Parsons on Contracts, 527) and it was so decided in Fletcher v. Peck, 6 Cranch, 87, it would be within the power of both parties, by mutual consent, to modify or change the terms of the contract after its execution; and if we adopt the rule as stated by Field, J., in Schulenberg v. Harriman, Supra, that 'a legislative grant operates as a law as well as a transfer of the property, and has such force as the intent of the legislature requires,' the same principle follows, and the law could be changed or modified at any time by the consent of both parties before the rights of others attached, certainly this must be true, unless there is some constitutional provision against such acts of legislation.

"In Higgins v. Houghton, 25 Cal. 255, where it was held that the State of California, by virtue of the grant of March 3d, 1853, which in some respects is similar to the grant under consideration, 'became the owner of the sixteenth and thirty-sixth sections absolutely, not only as to quantity, but as to position also,' the Court impliedly recognized the fact that it was within the power of Congress and the State by mutual agreement to change the provisions of the grant. After stating that there had been no legislation by Congress prior to the grant which would interfere with the conclusions reached in said case, the Court said: 'And if there has been any legislation since the grant that conflicts with the conclusion, it must be null and void, *unless, indeed, it has been acceded to by the grantee.*' Here such subsequent legislation was had by Congress, *and it was acceded to by the grantee.*

"After the sixteenth and thirty-sixth sections had been granted, and after this State had been admitted into the Union, Congress passed an act entitled 'An Act concerning certain lands granted to the State of Nevada,' approved July 4th, 1866. After confirming the appropriation made by the constitution of this State, to educational purposes, of the land granted to this State by the law of September 4th, 1841, and providing for the appointment of a 'Surveyor-General for Nevada,' who was to perform certain duties therein prescribed under the direction of the Secretary of the Interior, it was further enacted: 'That in extending the surveys of the public lands in the State of Nevada, the Secretary of the Interior may, in his discretion, vary the lines of the subdivisions from a rectangular form to suit the circumstances of the country; but in all cases lands valuable for mines of gold, silver, quicksilver, or copper shall be reserved from sale.' (14 U. S. Stat. 85–6, Sec. 5.) This State, in accepting the grant, unequivocally consented to the reservation by Congress of the mineral lands, and accepted the grant with all the conditions and reservations mentioned in said section. The act passed by the legislature of this State, entitled 'An Act in relation to and accepting the lands granted to the State of Nevada by the Government of the United States,' approved February 13th, 1867, is explicit upon this point. It reads as follows:

"'Sec. 1. The State of Nevada hereby accepts the grants of lands made by the Government of the United States to this State, in the following acts of Congress, to wit: "An Act donating public lands to the several States and Territories which may provide colleges for the benefit of agriculture and the mechanic arts," approved July 2d, 1862, as amended and approved

April 14th, 1864, and as extended July 4th, 1866, by an act entitled " An Act concerning certain lands granted to the State of Nevada," upon the terms and conditions in said acts expressed, and agrees to comply therewith.

"' Sec. 2. The State of Nevada hereby accepts the grants of lands made by the Government of the United States to this State, in the Act of Congress entitled " An Act concerning certain lands granted to the State of Nevada," approved July 4th, 1866, upon the terms and conditions in said act expressed, and agrees to comply therewith.

"' Sec. 3. The State of Nevada hereby accepts *all grants* of public lands *heretofore made* by the Government of the United States to this State, *upon the terms and conditions so granted, as modified in the Act of July 4th*, 1866, *above in this act referred to.*' (Stat. 1867, 57.)

"This act was passed prior to the survey, by the United States, of the land in controversy, which, from the record in this case, is shown to have been made in August, 1867.

"This State, by its act of acceptance of the grant as modified by the Act of Congress of July 4th, 1866, was estopped from thereafter claiming title to any lands valuable for mines of gold, silver, quicksilver, or copper, for such lands were, by said act, expressly reserved from sale. It is evident that when Congress passed the Act of July 4th, 1866, it thought that, by the effect of the grant and the law of the event, that this State would not acquire an absolute ownership in the lands until the surveys were made; but even if it was mistaken as to the legal effect of the grant, its action received the sanction and approval of this State before the title of the State, under any rule of construction, absolutely attached *to any specific tract of land*. If it be conceded that the State had a vested title to the mineral lands contained in the sixteenth and thirty-sixth sections, prior to the Act of February 13th, 1867, it is certain that by said act it relinquished its rights thereto, and thereby agreed to accept other lands in lieu thereof.

"The passage of said act was a recognition by the legislature of this State of the validity of the claim made by the Government of the United States to the mineral lands.

" Whatever might, therefore, be the construction of the language of the Enabling Act, as interpreted from the act itself, we think it is controlled by the subsequent legislation we have referred to, and that the title of the State to the land conveyed to appellant's grantors was, at the time of the survey thereof, subject to the terms and conditions imposed by the Act of Congress of July 4th, 1866; and as the portion of said land in controversy in this action was then 'rich in minerals,' and occupied and claimed by respondent's grantors for mining purposes, the grantors of appellant acquired no title thereto by virtue of the patent issued by this State.

"Against the views we have above expressed, counsel for appellant make three objections: 1st. It is first argued that the Act of July 4th, 1866, is prospective in its terms, and that it only applies to future acts or grants. We think that the act, when read entire, is susceptible of but one construction. It refers to lands granted prior to, and at the time of, the passage of the act. The title of the act clearly indicates that it was the intention of Congress to make the act apply to lands already granted —'An Act concerning lands *granted* to the State of Nevada '— not lands to be thereafter granted, but lands granted by that and other prior acts of Congress. The construction we have placed upon this act must certainly be correct, if it be true, as was argued by appellant's counsel upon another branch of this case, and held to be the law in Whitney v. Whitney, 14 Mass. 92, that we should not be encouraged to direct our conduct, in arriving at the intentions of the legislature, 'by the crooked cord of discretion, but by the golden meteward of the law'; that we are not to construe statutes by equity, but to collect the sense of the legislature by a sound interpretation of its language, according to reason and grammatical correctness. But we do not think there is any room for argument as to its meaning. It applies to all grants made by Congress to the State of Nevada, where the lands granted had not been surveyed by the Government of the United States, and included the grant mentioned in the Enabling Act, and such was the evident understanding of the legislature of this State when it passed the act of acceptance, approved February 13th, 1867.

"2d. It is argued that the Act of Congress applies only to the public lands then belonging to the United States, and it is claimed that, inasmuch as the sixteenth and thirty-sixth sections had already been granted, the act did not affect the title to them, as they were no longer public lands. An examination of the various acts of Congress relative to the surveying of the public lands, has convinced us that the word 'public' is applied by Congress to all the unsurveyed lands, whether the same or any portion thereof had been previously granted or not.

"All lands are public within the meaning of that word, as used in the act referred to, until the survey is made. This is necessarily so, because, until the surveys are made, the rights of the grantee to any specific tract of land could not be ascertained; hence, it is that the word 'public' is used to distinguish the unsurveyed from the surveyed and segregated lands where the rights of private proprietorship had attached.

"3d. The last objection argued by appellant's counsel is, that the Act of February 13th, 1867, is in violation of the third section of Article 11 of the constitution of this State. It is claimed that, by the provisions of said section, the sixteenth and thirty-sixth sections are set apart and dedicated to the public schools, and that it was not, therefore, within the power of the legislature to relinquish the title of the State to these sections. Section 3 provides that 'all lands, including the sixteenth and thirty-sixth sections in every township, donated for the benefit of public schools, in the Act of the Thirty-eighth Congress, to enable the people of Nevada Territory to form a State government, * * * shall be and the same are hereby solemnly pledged for educational purposes, and shall not be transferred to any other fund for any other uses,' etc.

"The plain object of this provision of the constitution was to prevent the legislature from passing any law that would appropriate the proceeds received by the State from the sale of such lands to any other than educational purposes. The title to said sections is vested in the State, not in the schools. The lands are solemnly pledged to educational purposes, and when sold by the State the proceeds arising therefrom must, under the provisions of the constitution, be paid into the school fund, and only be used for educational purposes, 'and shall not be transferred to any other fund for any other uses.' The same disposition must also be made of the proceeds derived by the State from the sale of lands selected in lieu of the sixteenth and thirty-sixth sections. There is nothing in the act which attempts to make any disposition of said lands for any other than educational purposes. The school fund is fully protected, and, in our opinion, this provision of the constitution has not been violated.

"We have not, in this opinion, considered the legal effect of the joint resolution of Congress, approved January 30th, 1865, which provides that no act passed at the same session of Congress as the Enabling Act, 'shall be so construed as to embrace mineral lands, which in all cases shall be and are reserved exclusively to the United States, unless otherwise specially provided in the act or acts making the grant,' and which was construed by the Secretary of the Interior to exclude from the operation of the Enabling Act all mineral lands (Copp's U. S. Mining Decisions, 31); nor have we deemed it necessary to discuss many other points that were urgently pressed by counsel, as the results we have reached, upon the points decided, are in our judgment conclusive of this case." [1]

In the Land Office it has been held that Sections 16 and 36, when mineral, did not pass to the State of Nevada, under the Act of Congress of March 24th, 1864, in view of Joint Resolution of January 30th, 1865, but remained the property of the United States. [2]

---

[1] Heydenfelt v. Daney G. & S. M. Co. 10 Nevada, 290.

The author is informed that this case has been recently affirmed (March, 1877) by the Supreme Court of the United States. The opinion of the Court, delivered by Davis, J., is said to hold: First—That the Act of Congress of March 21st, 1864, authorizing the people of Nevada to frame a constitution, under which act Nevada selected and conveyed the land in controversy to the grantees of Heydenfelt, did not constitute a grant *in presenti* of the premises, but the grant remained inchoate and incomplete until the land was surveyed by the United States authorities, and the survey properly approved. Second—That the survey and approval not having been made prior to the entry by the company's predecessors in interest for mining purposes, the land was not, by act of Congress, or in any other manner, ever granted by the United States to Nevada. Third—That, under the entry, the company's grantors, and their right thereto, having become established prior to the survey of Section 16 by the United States, the land was not included in, nor did it pass to Nevada by, the granting clause of 1864, but, on the contrary, was excluded therefrom because previously possessed and occupied by defendants' grantors for mining purposes in conformity with the mining laws, rules, and customs of the miners in the locality where it is situated, and in conformity with the Mining Act of Congress, approved July 26th, 1866.

The opinion is not, at the date of writing, accessible to the author. See Appendix.

[2] Decision of Secretary, May 20th, 1870, Copp's Mining Decisions, 31.

§ 36. **Mineral lands in railroad grants.**—Two cases in California have been decided, wherein the ownership by railroads of mineral lands within the boundaries of their grants, has been considered. In McLaughlin v. Powell, 50 Cal. 64, the action was ejectment. The defendant, in his answer, set up that the plaintiff claimed the land under a grant made to the Central Pacific Railroad Company of California, and that the land was mineral land, and was by the express terms of the grant excepted from the operation of the same. On the trial, the plaintiff offered in evidence a patent from the United States to the Western Pacific Railroad Company of California, dated May 31st, 1870, conveying the demanded premises as a portion of the land granted by Congress, to aid in the construction of a railroad, by the Act of July 1st, 1862, and the act amendatory thereof, passed July 2d, 1864. By the terms of said acts, the grant was limited to public land which was not mineral land, and which was not sold, reserved, or otherwise disposed of by the United States, and to which a pre-emption or homestead claim should not have attached at the time the line of the road should be definitely fixed.

The objections to the admission of the patent being overruled, the plaintiff then deraigned title by mesne conveyances from the company, and rested, and the defendant then offered to prove that the land was mineral land, containing large quantities of cinnabar and quicksilver, and that he had held the land as a mining claim since October, 1866, under the rules and regulations and customs of miners in the district where the land was situated. The objections to this testimony were sustained.

It was assumed in the decision that lands valuable because of cinnabar or quicksilver ores are "mineral lands" within the meaning of the act of Congress.

The defendants' objection to the patent, that it was "irrelevant," was held properly overruled. It was held not necessary to decide whether it was for the plaintiff, who relied on the patent, to prove that the land in controversy was not one of the excepted tracts, because no motion for nonsuit was made, and it was held that if the plaintiff was not required to prove that the land was not within the exception, the defendant was entitled affirmatively to establish that it was within it.

The exception contained in the patent was part of the description, and was equivalent to an exception of all the subdivisions of land mentioned, which were "mineral" lands.

The patent granted all of the tracts named in it which were not mineral lands. If all were mineral lands, it was suggested that the exception might be void; but as the fact could not be assumed, it was held that the defendant should have been allowed to prove that the demanded premises were mineral lands.[1]

In Alford v. Barnum, 45 Cal. 482, an action to abate a ditch as a nuisance, the defense was that the land upon which the ditch was dug was the public mineral land of the United States, and that the defendants were mining thereon for gold.

The land was within the grant to the Central Pacific Railroad Company, and the company, prior to the excavation of the ditch, had received a patent for it, which patent excepted from its operation all "mineral lands." The plaintiff, at the time the ditch was dug, was in possession of the land under a contract of purchase from the railroad company, who claimed under grant and patent from the United States.

No license from plaintiff was pleaded, nor attempted to be pleaded in time. It was found that the ditch was injurious to the premises, and interfered with the plaintiff's full and free enjoyment of the land.

It was virtually found below that the character of the land was not mineral, but the appellate Court further remarked that the mere fact that portions of the land contained particles of gold, or veins of gold-bearing quartz rock, would not necessarily impress it with the character of mineral land within the meaning of the acts of Congress reserving mineral lands from the grant, nor within the reservations of the patent which followed the terms of the granting acts.

It should be shown that the land contains metals in quantities sufficient to render it available and valuable for mining purposes. Any narrower construction, it was thought, would

---

[1] McLaughlin v. Powell, 50 Cal. 64. See, also, Railroad v. Smith, 9 Wallace, 98; People v. Stratton, 25 Cal. 242; Kernan v. Griffith, 27 Cal. 87; Robinson v. Forrest, 29 Cal. 317; Read v. Caruthers, 47 Cal. 181; Patterson v. Lynch, Circuit Court of California. Decision of Mr. Justice Sawyer.

operate to reserve from the uses of agriculture large tracts of land which are practically useless for any other purpose.

The land, therefore, was held not within the exception, and the plaintiff had judgment.[1]

---

[1] Alford v. Barnum, 45 Cal. 482; 12 U. S. Stat. at L. 489; 13 Ibid. 356. See Decision of Commissioner, March 14th, 1871, Copp's U. S. Mining Decisions, 40.

Upon the general subject of reservations in grants of mines, see Blanchard & Weeks' Leading Cases on Mines and Mining Water Rights, Chap. X.

CHAPTER III.

RIGHT OF EXPLORATION AND PURCHASE OF VALUABLE MINERAL DEPOSITS, AND THE OCCUPATION AND PURCHASE OF MINERAL LANDS—CITIZENSHIP AND PROOF THEREOF.

§ 37. Right to purchase.
§ 38. Valuable deposits.
§ 39. The general rule stated.
§ 40. Borax deposits.
§ 41. Mineral deposits.
§ 42. What is a mineral vein?
§ 43. Mineral veins, classifications.
§ 44. Definitions of terms in common use.
§ 45. Who may acquire patents.
§ 46. Application by aliens.
§ 47. Citizenship.
§ 48. Proof of citizenship.
§ 49. Affidavit of citizenship.
§ 50. Foreign corporation.
§ 51. Restriction as to proof.

§ 37. Right to purchase.—Sec. 2319 of the Revised Statutes is as follows: " All valuable mineral deposits in lands belonging to the United States, both surveyed and unsurveyed, are hereby declared to be free and open to exploration and purchase, and the lands in which they are found, to occupation and purchase by citizens of the United States, and those who have declared their intention to become such, under regulations prescribed by law, and according to the local customs or rules of miners in the several mining districts, so far as the same are applicable and not inconsistent with the laws of the United States." [1]

---

[1] Rev. Stats. Sec. 2319, same as Sec. 1, Act 1872, 17 U. S. Stats. 91.
Sec. 1, Act of 1866, 14 U. S. Stats. 251, read : " That the mineral lands of the public domain, both surveyed and unsurveyed, are hereby declared to be free and open to exploration and occupation by all citizens of the United States, and those who have declared their intention to become citizens, subject to such regulations as may be prescribed by law, and subject also to the local customs or rules of miners in the several mining districts, so far as the same may not be in conflict with the laws of the United States." (See Sec. 2329 Rev. Stat.) See Ante, § 2.

§ 38. Valuable deposits.—The word "deposit" has always been construed by the Land Office to be a general term, embracing veins, lodes, ledges, placers, and all other forms in which valuable metals have ever been discovered. In the sense in which the term "mineral" was used by Congress, it seems difficult to find a definition that will embrace what mineralogists agree should be included. Borax, nitrate and carbonate of soda, sulphur, alum, and asphalt, are generally classified and discussed as minerals.[1] Whatever is recognized as a mineral by standard authorities, where the same is found in quality and quantity sufficient to render the land sought to to be patented more valuable on this account than for purposes of agriculture, is treated by the land office as coming within the act. Lands, therefore, valuable on account of borax, carbonate of soda, nitrate of soda, sulphur, alum, and asphalt, it is held, may be applied for and patented.[2]

The first section of the Act of 1872 says: "All valuable mineral deposits."[3] The sixth section uses the term "valuable deposits."[4]

Diamond-producing lands are "valuable mineral deposits" under the act, and the provisions are as applicable as to lands containing gold or silver.[5] Deposits of fire-clay may be patented under the act, and so may iron deposits, which may be patented as vein or placer claims.[6]

Lands, more valuable on account of deposits of limestone, marble, kaoline, and mica, than for purposes of agriculture, may be patented as mineral land.[7]

Where valuable deposits of roofing slate had been discovered,

---

[1] Phillips' Mineralogy; Webster's Dictionary.

[2] Decision of Commissioner, July 15th, 1873, Copp's U. S. Mining Decisions, 316; Decision of Acting Comr. Oct. 23d, 1874, 1 Copp's Land-owner, 132; Report Comr. Genl. Land Office, 1873, p. 17.

[3] Rev. Stats. Sec. 2319; 17 U. S. Stats. 91.

[4] Rev. Stats. Sec. 2325; 17 U. S. Stats. 92.

[5] Decision of Acting Secretary, Aug. 31st, 1872, Sept. 3d, 1872, Copp's U. S. Mining Decisions, 140; Report Comr. Genl. Land Office, 1873, p. 16.

[6] Decision of Comr. July 10th, 1873 and July 26th, 1873, Copp's U. S. Mining Decisions, 209-214; 1 Copp's Land-owner 34; Decision of Comr. Jan. 30th, 1875, 1 Copp's Land-owner, 179.

[7] Decision of Comr. June 28th, 1875, 2 Copp's Land-owner, 66; Decision of Comr. Dec. 3d, 1875, 2 Copp's Land-owner, 131.

and large amounts spent in their development, the applicants were allowed to proceed to obtain patent.[1]

But under the Act of 1866, the office did not regard sulphur-springs as mineral so as to come within the inhibition of the statutes excluding mineral and saline lands from pre-emption entry or scrip location.[2]

Auriferous cement claims found in ancient river-beds, and usually worked by hydraulic process, do not come within the definition of "rock in place," but are patented as placers.[3] Petroleum claims may be patented under the Act of 1872.[4] Lands containing valuable deposits of umber may be patented as placer claims at $2.50 per acre, if not found in veins or "rock in place." If they are so found, then they may be patented at the rate of $5 per acre, like other lode-claims.[5]

§ 39. The general rule stated.—The rule may be stated in general terms, that where valuable mineral deposits are found in such quantity and quality as to render the land sought to be patented more valuable on this account than for purposes of agriculture, the tracts containing such valuable mineral deposits may be patented under the mining acts. But if this is not the case, they cannot be patented under the act except in the case of mill sites, which must be non-mineral in character. If parties have the possession and right of possession to salt-springs, and the deposit of salt renders the land more valuable on this account than for agricultural purposes, a patent may be secured upon full compliance with the laws and instructions.[6]

§ 40. Borax deposits cannot be entered under the Agricultural Laws of Congress, but may be under the Mining Acts, upon full compliance with the laws, as they provide for the patent-

---

[1] Decision of Acting Commissioner, Oct. 23d, 1874; 1 Copp's Land-owner, 132.
[2] Decision of Commissioner, Aug. 25th, 1869, Copp's U. S. Mining Decisions, 22.
[3] Decision of Commissioner, Feb. 12th, 1872, Copp's U. S. Mining Decisions, 78
[4] Decision of Commissioner. Jan. 30th, 1875, 1 Copp's Land-owner, 179.
[5] Ibid.
[6] Decision of Acting Commissioner, April 27th, 1874; 1 Copp's Land-owner, 19, reversing Decision Commissioner, July 28th, 1873, Copp's U. S. Mining Decisions, 211, which was to the effect that there was no general law under which salt-springs could be patented, and that they could only be disposed of by special act of Congress

ing of lands claimed and located for valuable deposits. The proceedings required are the same as in applications for placer mines.[1]

§ 41. **Mineral deposits.**—The useful minerals are found in deposits, which are classified into superficial, stratified, and unstratified deposits. Superficial deposits are those in which the materials are yet unconsolidated, and have been washed down from cliffs and mountain slopes, composed of rocks that contain metals, ores, and gems, either in veins or irregularly disseminated. The "placers" of California are familiar instances of this kind of deposit.

2d. Stratified deposits—where the minerals form entire strata, such as beds of coal and iron ore.

3d. Unstratified deposits, which are subdivided into:

Eruptive masses—Composed of the ingredients of volcanic rocks.

Minerals disseminated through eruptive rocks.—These, as a rule, are neither numerous nor valuable.

Contact deposits.—Metals or ores accumulated in the plane of junction between two rocks of different kinds, such as igneous and sedimentary rocks.

Impregnations—Which are accumulations of metalliferous minerals found diffused irregularly through rocky masses, the deposits of ore having no definite boundaries, or any regularity of structure, and appearing as though the rock had soaked up or absorbed the minerals as water saturates a sponge. Deposits of mercury exhibit this characteristic.

Fahlbands—A name given to a peculiar kind of deposit, where the ore is sparingly diffused through certain layers which are apt to disintegrate, and are more *fahl* (*i. e.*, foul or rotten) than the associated strata.

Stock work—Where the masses of metalliferous rocks are penetrated in every direction by threads or strings of ore, so that the whole must be taken out together.[2]

And mineral veins.

[1] Decision of Commissioner, April 18th, 1873, Copp's U. S. Mining Decisions, 194.
[2] Am. Cyclop. Art. Mineral Deposits, by Prof. Newberry.

§ 42. What is a mineral vein?—A mineral vein, as commonly understood, is a collection of mineral matters which have been slowly brought together and consolidated in elongated cracks or fissures in the rocks. Dikes are collections formed of molten rock, as lava, which has suddenly flowed into fissures and cooled. Among the earthy minerals which form the gangue or vein-stones are often found metallic ores, and it is from this source that the chief supplies of the useful metals are obtained. Veins worked for these are called by the miners "lodes."

Veins are met with in almost all rocks, are traced for miles in length, and penetrate the crust of the earth deeper than man has ever been able to follow them.

§ 43. Mineral veins—Classifications.—Mineral veins are usually sheets of mineral matter, of greater or less lateral and vertical extent. They have been divided into three principal varieties, which are generally well marked, but which sometimes blend in such a way as not to be easily separated. These varieties of mineral veins are known as gash veins, segregated veins, and fissure veins.

Gash veins—Are such as are confined to a single *stratum or formation*, and hence are of limited extent, both laterally and vertically. They may be vertical at right angles with the stratum, or horizontal and parallel with it.

Segregated veins—Are usually lenticular sheets of ore-bearing mineral, which are conformable to the bedding of the associated rocks, *i. e.*, are interposed between the layers of such rocks. They always occur in metamorphic rocks, and are usually inclined at a high angle with the horizon. They are called segregated veins because they are supposed to have been formed in the process of metamorphism, by the separation or withdrawal of the materials which compose them from the adjacent strata, and their concentration along certain lines. Segregated veins are limited, both laterally and vertically. They rarely exhibit anything of the banded structure which characterizes fissure veins, are chiefly composed of quartz, and form the great repositories of gold. Though segregated veins have usually no great lateral or vertical extent, they sometimes attain a thickness of twenty and thirty feet, and have a length on the surface of a mile or more.

Fissure veins.—These are of indefinite extent, laterally and vertically. They have been formed by volcanic or earthquake action, by which the rocks have been fractured and displaced. In all cases where an important crack or fissure is made by subterranean upheaval, either by the slipping in of wedges of rock or by the shifting of the sides of the fissure, so that their irregularities fail to match, the walls are prevented from returning to their original positions, and an irregular, open crevice is produced. When subsequently filled by foreign matter containing metals or ores, such a fissure becomes a fissure vein. In some instances the fracture of the rocks has considerable regularity, and the fissure may be of uniform width for several hundred feet in either direction. More generally, and especially where a fracture is attended with displacement, the fissure is of very unequal width, the vein-matter has in places a thickness of many feet, while at other points where the projecting walls approach or come in contact, the vein becomes very thin, and may be quite "pinched out." From their mode of formation, fissure veins are without definite limits, horizontally or vertically, They may frequently be traced for miles upon the surface, and their limits in depth are rarely reached. They therefore hold more extensive and continuous deposits of ore than any other kind of mineral veins, and constitute the most trustworthy bases for mining operations. Fissure veins frequently present a banded structure in the materials which compose them, and this forms one of their most striking characteristics. This feature is produced by the deposition on their walls of successive layers of different minerals. These layers often correspond on either side of the central line, showing that the deposition of the different sheets took place simultaneously on both walls. Sometimes a fissure vein exhibits a double or triple series of bands, showing that after being filled with ores it was again opened and a new fissure formed, and then this was filled in the same way as the first. The quartz, which constitutes a large part of the material composing fissure veins, frequently shows a "comby" structure, due to the formation of crystals, which shoot out from the walls, and interlock where they meet.

Another common feature in fissure veins is the "fluccan" or "selvege," a sheet of clay which lines either wall, and causes the vein-matter to cleave off readily. This fluccan seems to be due partly to the attrition of the sides when moved with immense force upon each other, and partly to the action on the walls of chemical solutions filling the fissure. The sides—and sometimes the interior—of fissure veins generally show polished and vertically striated surfaces ("slickensides"). These are produced by the friction of the walls on each other, or on the material composing the vein. Fissure veins cut indiscriminately through all kinds of rock. They frequently traverse stratified rocks across their lines of deposition and outcrop, and are then called cross-cut veins, to distinguish them from those that are more or less accordant with the stratification.[1]

The origin of fissures is more readily understood than the source of the materials that fill them. The forces which produce cracks in clays by their shrinkage, and in other substances by change of temperature, also operate to rend apart the solid strata, and fissures in these are also produced by earthquakes and volcanic action. Such openings are naturally found very irregular in their dimensions, and in districts where earthquake movements have been frequent, interrupted in their continuity, crossed by other fissures of later formation, and ramifying into side openings, some of which may prove as extensive as the main fissure.[2] The fissure is sometimes seen still open, containing only loose earth and stones that have fallen in from above, and sometimes it is partially filled with vein-stones, or ores, open spaces still remaining unfilled, and forming caverns on the line of the vein. The fissure again may be quite filled with mineral substances, which may be closely attached to the walls, as if all were originally formed at the same time, or, as is more commonly the case, a parting seam may be found on one or both sides, separating the vein-stone from the wall rock, and the faces of each are then often seen presenting a smooth surface, as if they

---

[1] In this description of mineral deposits and mineral veins, the author has given a synopsis of an article in the American Cyclopædia, Edition of 1875, contributed by Prof. J. S. Newberry, LL. D., Columbia College, New York. See further, as to contents of veins, "Filling of Veins," Theory of Injection, of Aqueous Deposition, of Lateral Secretion, Sublimation, Chemical Precipitation, etc

[2] New Am. Cyclop. "Mineral Vein."

had been rubbed together. A thin layer of tough clay, called by the miners "fluccan," is commonly interposed in the seam between the veins and its walls. Veins usually occur in groups of several together, lying nearly parallel to each other, both in direction and inclination downward; but as they are followed in one or the other direction, along the surface, or on their slope down, which is called their "dip" or "under-lay," they are often found to run into each other. While their general line is straight, it is more or less waving in places, and their dip is more variable, often becoming steeper at great depths, and changing to greater or less steepness along their course. The position of veins in regard to the rocks which contain them is sometimes across their strata, and sometimes with them, and in the latter case the veins are often found both in dip and direction to pass across one stratum, and continue between different layers from those in which they were first seen, thus establishing their character as veins formed in fissures in contradistinction to beds. Along the line of contact of two rocks of different character, as granite, gneiss, or trap with sandstone or limestone, veins frequently occur, and branches lead off into the rock on one side or the other.

It appears as if fissures may have opened originally in such positions, and also between adjoining strata of the same rock, for the reason that the disrupting force, when not directed at right angles across the strata, found along these lines the least resistance. A vein which cuts through rocks of different kinds changes not merely as regards its contents, but also in respect to its dimensions with the rocks which include it, and it is almost universally the case that a vein which is productive in one rock ceases to be so as it is followed from this into another. The width of veins is very variable, as would naturally result from the form of the original fissures; and that of any single vein is subject to great irregularities, especially where its walls have been moved subsequently to their separating, so as no longer to present corresponding depressions and prominences opposite each other. In places, the fissures will thus be found nearly closed by the contact of the two walls, and in others, opening out into wide spaces by their separation. The common width of a vein is about six feet. They range from a width of a few inches to hundreds of feet. Veins are, however, not rich in proportion to their size, and some of the smallest are the most profitable.

Masses of the wall-rock are sometimes met with in the vein, so large that the dividing of the vein around them appears like the leading off of a branch, nor is this found to be a mistake until the divided portions meet again on the other side of the interposed mass. The miners call such a mass a "horse," probably from the vein going down each side of it like a saddle on the back of a horse. The ores occur in bunches, strings, and layers, very irregularly distributed, and usually of many varieties associated together. The ores of one metal commonly prevail either throughout the mine, or to a certain depth, below which others may be found more productive. They often lie in courses or parallel belts, which slope in one or the other direction on the line of the vein, and between such courses the workings are comparatively unproductive. Large developments of ore are looked for where branches drop into the main vein. Near the surface, veins are not often found so rich as at some depth below, at least beyond the reach of atmospheric influences; but when once in what the miners call "settled ground," no improvement need generally be expected as the result merely of greater depth.[1]

§ 44. Definitions of terms in common use.

Rock in place—As used in the Mining Acts of Congress, has always received the most liberal construction that the language will admit of, and every class of claims that, either according to scientific accuracy or popular usage, can be classed and applied for as a "vein or lode," may be patented under the law. The object of the law is to dispose of the mineral lands of the United States for money value, and it is a matter of indifference to the Government whether the metal occurs in the form of a true or false vein.

[1] New Am. Cyclop. "Mineral Vein," citing "Report on the Geology of Cornwall, Devon and West Somerset," by Henry T. De la Beche (London, 1839). *De la richesse minerale*, by A. M. H. de Villefosse (Paris, 1819). *Lehrbuch der Chemischen und physikalischen Geologie*, by Gustav Bischof (Bonn, 1851). Papers of the "Proceedings of the Geological Society of Cornwall," by Messrs. Robert W. Fox, Joseph Carne, John Hawkins, and others. Whitney's "Metallic Wealth of the United States." Cotta's "Contributions to the Knowledge of Mineral Veins."

A lode—In mining, is a vein of mineral substance; usually a vein of metallic ore. "A metallic vein, or any regular vein or course, whether metallic or not, commonly a metallic vein."[1]

A vein—In mining parlance, is usually applied to a small lode; in geology, any seam of rock material, intersecting strata crosswise.

"A seam or layer of any substance, more or less wide, intersecting a rock or stratum, and not corresponding with the stratification; often limited, in the language of miners, to such a layer or course of metal or ore."[2]

It often includes, as in the instance of the Pennsylvania coal formations, layers, or what are properly strata or stratifications.

A quartz ledge—In a particular case, was defined to be "a stratum of quartz rock, running in a seam of the bed-rock along the face of the hill, in some places showing above the surface of the bed-rock, in some places above the bed-rock, but covered with surface earth, and in other places dipping entirely beneath the bed-rock.'

A spur—As used among the quartz miners of California, means a lateral branch from the main lead, not returning to it, but losing itself in the surrounding soil, and diminishing the dimensions of the main ledge by its own breadth.

A feeder.—A small vein starting from some distant point, running into the main lead, and enlarging it to the extent of its own breadth.

Float ore—Means those isolated masses of ore or mineral which are separated from the regular leads, and corresponds with the term "masses" as found in the books. The primary condition of all minerals is in "leads," but the convulsions of the world have produced what are now termed "masses," "lodes," "nests," etc., all separate and distinct things.[4]

Where quartz rock was broken and parted from the original vein, but it was found, as a fact, that it was a portion of the same quartz lode or claim, it was held that, if this were the

---

[1] Webster's Dic.
[2] Ibid. See Supra.
[3] Brown v. Quartz M. Co. 15 Cal. 155
[4] Ibid.; Ure's Dic. Vol. 2, pp. 166, 167; Blanchard & Weeks' Cases on Mines, Minerals, and Mining Water Rights, 21, 22. See Ibid. Glossary, for further definitions.

case, it was immaterial whether it was separated from the original vein or not, whether it was upon the surface or beneath it, or in what condition the quartz was, the first locator of the lode was entitled to it; and was not confined simply to the solid quartz actually embodied in the bed-rock, but was entitled to the loose quartz rock and decomposed material which were once a part of the lode, and were detached so far as the general formation of the ledge could be traced. The right of the quartz miner came from his appropriation; and whenever his claim is defined, there is no reason, said the Court, " why the appropriation may not as well take effect upon quartz in a decomposed state as any other sort, or why the condition to which natural causes may have reduced the rock should give character to the title of the locator." Such quartz rock, therefore, would be included under the general term of a " quartz ledge."[1]

Silver-bearing ore.—In an indictment for grand larceny of silver-bearing ore, it was held that the words " silver-bearing ore," as used in the indictment, had reference to a portion of vein-matter which had been extracted from a lode and assorted, separated from the mass of waste rock and earth, and thrown aside for milling or smelting purposes, or taken away from the ledge; and that the language used in the indictment necessarily implied that the ore had been severed from the freehold prior to the time of its asportation.[2] Of course, had the ore been considered as annexed to the freehold, it could not have been the subject of larceny.[3]

Tailings.—" 'Tailings' may be defined to be the refuse part of stamped ore, thrown behind the tail of the buddle or washing apparatus, which is dressed over again to secure whatever metal may exist in it." Where land was of no value, except for the " tailings," and they were valuable only for the gold and silver which they contained, and neither party claimed the land for any purpose except that of securing such tailings, and for mining purposes only, it was held that, " although not a mining claim within the strict meaning of the expression, as generally used in this

---

[1] Blanchard & Weeks' Cases on Mines, Minerals, and Mining Water Rights, Chap. I.
[2] State v. Berryman, 8 Nev. 270.
[3] People v. Williams, 35 Cal. 673.

country, still it is so closely analogous to it that the propriety of subjecting the acquisition and maintenance of the possession of it to the rules governing the acquisition of the right of possession to a strict mining claim, at once suggests itself. The only value attached to the land results from the precious metals which may be obtained from it. What is the difference how these metals may have been deposited there, so far as a case of this kind is concerned? It is distributed through a certain stratum of earth, which must be dug up and put through a certain milling process, as in case of any ordinary metalliferous earth. If the land be valuable only for the metal which it may contain, and it is claimed by neither party for any other purpose, the acquisition of title to it should manifestly be governed by the rules ordinarily controlling the acquisition of title or the right of possession to mining claims. We do not pretend to hold the land here in question to be mineral land, but only that it is so clearly analogous thereto that the laws controlling the possession of one should govern the other."[1]

"Mined coal" and ore, completely severed from the land, is personal property, and so treated.[2]

§ 45. **Who may acquire patent.**—No one but a citizen, or a person who has declared his intention to become such, can have the privilege of locating a mine, or acquiring a patent therefor. The reason of the rule was to prevent foreigners who might be inimical to the well-being and prosperity of the Government from obtaining possession and control of the vast interests which grow out of the mineral lands of the United States.

In the case of the Kempton mine, the application, after setting out the location and transfer of the claim, alleged "that all the above named locators of said claim and their grantees *are* citizens of the United States."

This was the only allegation or proof on this point contained in the entire record. It was objected that there was no allegation or proof that the original locators of the Kempton mine were citizens of the United States, or that they had declared their intention to become such at the time the location was made.

[1] Rogers *v.* Cooney, 7 Nevada, 213. [2] Lykens *v.* Dock, 62 Penn. St. 232.

The Secretary, in deciding the case, said: "I do not wish to be understood as deciding that a person who is not a citizen, or has not declared his intention to become such, cannot make a location of a mine, or dispose of it, provided he afterward becomes a citizen before he disposes of the mine. 'Naturalization has a retroactive effect so as to be deemed a waiver of all liability to forfeiture.' (Osterman v. Baldwin, 6 Wall. 122.) An assignor can transfer no greater interest to his assignee than he himself possesses. While he is unnaturalized he has no right to locate a mine. If he does so, and disposes of it before naturalization, a subsequent naturalization would not, in my opinion, save his location. If, therefore, it appeared in this case that the original locators were not citizens, or had not declared their intention to become such at the time their location was made, and that they had not become citizens when they transferred the mine, I should have no hesitation in holding that the transfer was invalid, and the claim of the applicants was not good. But there is no such allegation or proof in this case, and I should not be justified in presuming a state of facts which would work a forfeiture of the claim. The allegations, or pleadings, (if I may be allowed the expression) in proceedings of this kind, should be construed liberally, as I have heretofore held, and not, as common law, most strongly against the pleader. Under this rule of construction, I find myself obliged to overrule the objection as to citizenship, which is accordingly done."[1]

§ 46. **Application by aliens.**—No title in mining claims can be held by aliens prior to the issuance of patents therefor. They can claim no title, and can convey none.

In case of the application for patent for the Kempton mine, the Secretary of the Interior held that "if it appeared in the case that the original locators were not citizens, or had not declared their intention to become such at the time their location was made, and that they had not become citizens when they transferred the mine, I should have no hesitation in holding that the transfer was invalid, and the claim of the applicants was not good." And accordingly, where a party alleged that he came

[1] In re Kempton Mine, Decision of Secretary Interior. Jan. 2d, 1875, 1 Copp's Land-owner. 178.

into possession by purchase from certain Chinese, and did not show any other or previous right, it was held that, as at the time the aliens claimed the premises, they could not, under the law, hold title to the same, and, having no authority of law for laying claim to the premises, they could transfer no title to the applicant. It was, however, said, that had the applicant, after the purchase, made a re-location of the mine, made the required improvements, and otherwise complied with the law, he would have been in condition to apply for a patent. But none of these points were shown or alleged, and the application was rejected.[1]

§ 47. **Citizenship—Who is a citizen.**—The only parties entitled to the benefits of the acts are citizens of the United States, and those who have dec'-~ed their intentions to become citizens. None others are entitled.

A citizen is sometimes said to be a person who, under the laws and Constitution of the United States, has a right to vote for representatives in Congress and other public officers, and is qualified to fill offices in the gift of the people.[2] This definition excludes women and children. A citizen of the United States is a native-born or naturalized person, of either sex, who owes allegiance to, and is entitled to protection from, the United States, or a person who is made a citizen by treaty, stipulation, or by constitutional or statutory law.

A corporation, being artificial, invisible, intangible, existing only in contemplation of law, and in theory immortal, it is apparent that, in general, the terms "citizen" and "corporation" are distinct, and have widely different significations.[3]

Still, there has grown up a rule which treats a corporation, for the purpose of suing and being sued, as a citizen. It is a rule of convenience. But the extent of the doctrine in the United States Courts at first appeared to be that if all the stockholders were citizens of the State where the suit was pending, the corporation might sue, and the rule seems to have been that a corporation *aggregate* could not in its corporate capacity be

---

[1] Beckner *v.* Coates, Decision of Acting Commissioner, April 24th, 1876, 3 Copp's Land-owner, 18; Affirmed by Secretary of Interior, Sept. 20th, 1876; Ibid. 104. Ante, Sec. 45.
[2] 1 Bouv. Dic. 231.
[3] Dartmouth College *v.* Woodward, 4 Wheat. 636

a citizen. The parties could only sue when considered as a company of individuals, who must be citizens.[1] This rule, however, was afterward greatly modified, and in Louisville R. R. Co. *v.* Letson, 2 How. 550, it was held that a citizen of one State could sue a corporation which had been created by and transacted its business in another State, where the suit was brought, although some of the members of the corporation were not citizens of the latter State.

Justice Wayne said that a corporation, created by a State to perform its functions under the authority of that State, and only suable there, though it may have members out of the State, seemed to him to be a person, though an artificial one, inhabiting and belonging to that State, and therefore entitled, for the purpose of suing and being sued, to be deemed a citizen of that State.

He further said, that he was unable to reconcile these qualities of a corporation—residence, habitancy, and individuality—with the doctrine that a corporation aggregate cannot be a citizen for the purposes of a suit in the Courts of the United States, unless in consequence of a residence of all the corporators being of the State in which the suit is brought.[2]

Notwithstanding that Justice Wayne, in L. R. Co. *v.* Letson, Supra, said that a corporation was to be deemed a citizen, for the purpose of suing or being sued, Chief Justice Taney, in a later case, said that he presumed that no one ever supposed that the artificial being, created by an act of incorporation, could be a citizen of a State, in the sense in which that word is used in the Constitution of the United States.[3] Now, inasmuch as the Constitution of the United States refers to the jurisdiction of the United States Courts, as extending to controversies between citizens of different States, etc., and therefore declares them to be capable of suing and being sued in the United States Courts, it would seem that some one had supposed that a corporation could be a citizen, in the sense used in the Constitution, and that the tribunal making the " supposi-

---

[1] Strawbridge *v.* Curtis, 3 Cranch, 266; U. S. *v.* Deveaux, 5 Cranch, 84; Vicksburg *v.* Slocum, 14 Pet. 60.
[2] See, also, Marshall *v.* B. & O. R. R. Co. 16 How. U. S. 327.
[3] Covington Drawbridge Co. *v.* Shepherd, 20 How. 233.

tion" was no less an authority than the same Supreme Court of the United States, in the Letson case.[1]

If any rule can be deduced from these fluctuating opinions of the highest Court of the country, it is probably this. That the quality of citizenship was given to a corporation from the necessity of the case, and for the purpose of allowing it to sue and be sued, and for that alone, and that a corporation is in no other sense a citizen within the meaning of that term as used in the Constitution, or in the general laws relating to public lands. Such was the construction given to the first section of the Mining Act of 1866, by the Assistant Attorney-General.[2]

The Act of 1872, however, was more explicit upon the rights of corporations to apply for patents. Sec. 6[3] conferred the right of application upon "any person, association, or *corporation* authorized to locate a claim," etc., and in regard to proof of citizenship, it is provided[4] that it may consist, in the case of a "corporation organized under the laws of the United States, or of any State or Territory thereof," of the filing of a certified copy of their charter or certificate of incorporation.

These provisions were doubtless intended to obviate the perplexities and difficulties arising from the ambiguous language of the Act of 1866, which referred to "any person or association of persons." But these provisions, in their turn, it is important to consider, as they are not so full as to have escaped the necessity of construction.[5]

§ 48. **Proof of citizenship.**—Sec. 2321 of the Revised Statutes is as follows: "Proof of citizenship, under this chapter, may consist, in the case of an individual, of his own affidavit thereof; in the case of an association of persons unincorporated, of the affidavit of their authorized agent, made on his own knowledge, or upon information and belief; and in the case of a corporation organized under the laws of the United States, or of

---

[1] See Decision of Commissioner, July, 1869, and September 1, 1868, Zab. L. L. 221-234.
[2] In re New Idria Mine, Opinion of Assistant Att'y-Gen. July 21st, 1871; Copp's U. S. Mining Decisions, 47, 56.
[3] Rev. St. Sec. 2325.
[4] Ibid. Sec. 2321.
[5] Sec. 2, 14 U. S. Stats. 252.

any State or Territory thereof, by the filing of a certified copy of their charter or certificate of incorporation."[1]

The proof, therefore, necessary to establish the citizenship of applicants for mining patents, may consist, in the case of an individual claimant, of his own affidavit of the fact; in the case of an association not incorporated, of the affidavit of their authorized agent, made on his own knowledge, or upon information and belief; and setting forth the residence of each person forming the association. This affidavit must be accompanied by a power of attorney from the parties forming such association, authorizing the person who makes the affidavit of citizenship to act for them in the matter of their application for patent; and in the case of an incorporated company, organized under the laws of the United States, or the laws of any State or Territory of the United States, by the filing of a certified copy of their charter or certificate of incorporation. These affidavits of citizenship may be taken before the register or receiver, or any other officer authorized to administer oaths.[2] In case of an individual or an association of individuals, who do not appear by their duly authorized agent, there is now required the affidavit of each applicant, showing whether he is a native or naturalized citizen, when and where born, and his residence. In case an applicant has declared his intention to become a citizen, or has been naturalized, his affidavit must show the date, place, and the Court before which he declared his intention, or from which his certificate of citizenship issued, and present residence.

The requirements in reference to proof of citizenship have not been uniform. At first, under the Act of 1866, in requiring proof of citizenship, where the applicant was a corporation, a copy of their charter, or certificate of incorporation, might be filed in lieu of evidence of citizenship. In case, however, the applicant was an individual or an association of persons unincorporated, affidavits of citizenship, or of having filed declara-

---

[1] R. S. 2321, Sec. 7, Act 72; 17 U. S. Stat. 91. See Sec. 2325 R. S. The words "and nothing herein contained shall be construed to prevent the alienation of the title conveyed by a patent for a mining claim to any person whatever," were added to Sec. 7, and are now in Sec. 2326, Rev. Stat. U. S.

[2] Subdivisions 93, 94; Instructions June 10th, 1872; Instructions Feb. 1st, 1877, Subdivisions 78–81; Land Office Report, 1872, page 44.

tions of intention to become citizens, were required to be filed.[1] But subsequently, this order was revoked, and in case the application was made by an association of persons, incorporated or unincorporated, satisfactory proof was required that each member of such association was a citizen of the United States, or had filed his declaration of intention to become a citizen.[2]

Where such application was made by persons claiming to be native-born citizens of the United States, there was required the affidavit of each person so claiming that he was a native-born citizen, stating the place of his birth, such affidavit to be taken before a notary public, officer of a Court of Record, or the register or receiver of the land district wherein the claim lay.

Where the application was made by a person claiming to have filed a declaration of intention to become a citizen, he was required to present a certified copy of such declaration, under seal of the Court in which it was made.

Where such application was made by a person claiming to be a naturalized citizen of the United States, he was required to present his naturalization certificate, or a certified copy thereof, under seal of the Court from which the original issued. Where the application was made by an incorporated company, it was again required that they furnish a certified copy of their certificate of incorporation, besides evidence of the citizenship of each member or stockholder of such company.

Where the application was made by an association of persons unincorporated, evidence was required of the citizenship of each person forming such association, as before stated.[3]

But it was found, in the case of incorporated companies, having numerous stockholders, a matter of great difficulty to procure all the individual affidavits, owing to the fact that the parties were often scattered in different parts of the country, or in some instances traveling abroad. The previous requirements were therefore again changed, and when the application was made by a person, or by an association of persons, not incorporated, claiming to be native-born or naturalized citizens of the

---

[1] Instructions Aug. 8th, 1870, Copp's U. S. Mining Decisions, 257.
[2] Ibid. Aug. 3d, 1871, Ibid. 267.
[3] Ibid. Sept. 7th, 1871, Ibid.

United States, there was required the affidavit of each person so claiming, that he was such a citizen; these affidavits to be made before a notary public, officer of a Court of Record, or the register or receiver of the land district, and in any case where it might be satisfactorily shown, under oath, that the affidavit of any claimant could not be readily obtained, by reason of his absence in a foreign country, or in consequence of his whereabouts or place of residence being unknown, the citizenship of such claimant might be established by the affidavit of another person, who must not only testify to the citizenship of such claimant, but also state the facts upon which his knowledge was based, such as when, where, and for how long he had known him; whether he had exercised the elective franchise in the United States, and any other points proper to be received as evidence of citizenship.[1]

It is the announced intention of the Department to so construe the acts as to enable applicants for patents who are in the actual and rightful possession of mining claims by virtue of compliance with the local laws and regulations and the Congressional enactments, to make the proof required before patents can issue, at the least expense and inconvenience possible.[2]

It is sufficient to allege citizenship or declaration of citizenship, though the Department may prescribe the form of the required affidavit.

The portion of a mining claim sold to an alien cannot be patented while such owner is an alien, but on his declaration to become a citizen, his right dates back to his purchase, and he may thereupon secure a U. S. patent for his claim.

Naturalization has a retroactive effect, and is a waiver of all liability to forfeiture, and a confirmation of the alien's former title.[3] Certified copies of certificates of naturalization are not

---

[1] Instructions March 26th, 1872, Copp's U. S. Mining Decisions, 268; Ibid. Sept. 7th, 1871.

[2] Decision of Commissioner, February 3d, 1873, Copp's U. S. Mining Decisions, 158. Instructions Feb. 1st, 1877, paragraphs 93, 94. See Instructions, Sept. 7th, 1871; Copp's Mining Decisions, 268, March 26th, 1872; Ibid. Decision of Commissioner, Sept. 11th, 1873; Ibid. 223.

[3] In re Kempton Mine, Decision Secretary of the Interior, January 2d, 1875, 1 Copp's Land-owner, 178; Decision of Commissioner, July 18th, 1876, 3 Copp's Land-owner, 69; Osterman v. Baldwin, 6 Wall. 116; Jackson v. Beach, Johnson's Cases, 401. See, also, Fairfax v. Hunter, 7 Cranch, 603; Orr v. Hodgson, 4

necessary, and the Land Office has no power to require such proof. But the affidavits must state whether the applicants are native or naturalized citizens, and when and where born. In case an applicant has declared his intention, or has been naturalized, his affidavit must also show the date, place, and the Court before which he declared his intention, or from which his certificate of citizenship is issued, and present residence.[1]

§ 49. Affidavits of citizenship—Requiring certificates of naturalization.—From Sec. 2321, Rev. Stats. U. S., it will be seen that the method to be pursued for the purpose of establishing the qualification of citizenship is explicitly set forth, and it is not within the jurisdiction of the Department to impose an additional condition, if such condition is at variance with the terms of the act. The Land Office had required applicants, who alleged that they were naturalized citizens, to furnish certified copies of their certificates of naturalization. But this additional condition required by the Office in the matter of the proof of citizenship, was not sustained, and it was thought could not be justified by a consistent interpretation of the law.

No discretion is allowed the Office, under the "Act to promote the development of the mining resources of the United States," as to what shall constitute sufficient proof of citizenship, as in the pre-emption and homestead laws. Applicants for lands under said laws are required to be citizens, or to have declared their intention to become such, and what shall constitute proof of citizenship by declaration is a matter for the consideration of the Office; but in the act under consideration, the manner of making satisfactory proof on this point is expressly prescribed.

Where citizenship, therefore, is properly alleged, copies of the certificates of naturalization are not required to be filed, applicants must file their affidavits, showing whether they are native or naturalized citizens, and when and where born. In case an applicant has declared his intention to become a citizen, or has

---

Wheat. 453; Craig v. Leslie, 9 Wheat. 563; Craig v. Radford, 3 Wheat. 594; Cross v. DeValle, 1 Wall. 1; Heirs v. Robertson, 11 Wheat. 332.

[1] Decision of Secretary of the Interior, July 29th, 1876, 3 Copp's Land-owner, 68; Instructions, February 1st, 1877, Subdivisions 78–81.

been naturalized, his affidavit must also show the date, place, and the Court before which he declared his intention, or from which his certificate of citizenship issued.[1]

§ 50. **Foreign corporation.**—A corporation created and existing under the laws of England is not a citizen of the United States, and not capable of asserting a claim to any portion of the public land of the United States, or of receiving from the Government a title therefor. A "fund," being neither a person nor an association, without legal existence, and powerless to "occupy and improve" a claim, or perform those acts of ownership or possession required of miners, as conditions essential to the holding of claims, cannot make locations under the United States mining laws.[2]

§ 51. **Restriction as to proof.**—The operative sections of the Act of 1872 bring to the executive cognizance the applicant and the adverse claimant, and to them applies its rule of proof of citizenship. Proof of citizenship of the applicant for the patent, is sufficient; proof of citizenship of the original locators, and intermediate owners, is not necessary. The rule applies to the *applicant* and no one else, unless it be the adverse claimant.[3]

---

[1] Application of Mooney, Decision of Acting Secretary, July 29th, 1876, reversing Decision of Commissioner, S. C. 3 Copp's Land-owner, 68.

Alien soldiers.—A party made affidavit that he was born in Germany; that he came to this country at the age of six years, and that he had an honorable discharge from the army. In case his parents became naturalized before he arrived at the age of twenty-one, it was held that proof should be made of this point, as he would then be regarded as a citizen. The 21st section of the Act of Congress, approved July 17th, 1872, (12 Stats. 597) provides that any alien who has an honorable discharge from the regular or volunteer army may become a citizen of the United States, upon his petition, without any previous declaration of intention. The applicant afterward made affidavit that his father was naturalized in Wisconsin before he, the son, arrived at the age of twenty-one, and, in the absence of an adverse claim, a patent was issued to him. Decision of Commissioner, August 13th, 1872, Copp's U. S. Mining Decisions, 134.

[2] In re Gunboat Lode; Decision of Commissioner, June 7th, 1871; Copp's U. S. Mining Decisions, 43.

[3] Decision of Commissioner, Dec. 14th, 1874; In re King of the West Lode; City Rock & Utah Claimants v. Pitts, 1 Copp's Land-owner, 146; In re Cash Lode Sept. 7th, 1871, 1 Copp's Land-owner, 98; Opinion of Assistant Attorney-General, New Idria Case, Land Office Report, 1871, 58, 59, 60; Ibid. pp. 81, Circular Instructions, Aug. 3d, 1871, Sept. 7th, 1871, March 26th, 1872, June 10th, 1872, paragraph 93.

CHAPTER IV.

DIMENSIONS OF CLAIMS AND LOCATIONS UPON VEINS OR LODES.

§ 52. Length and width of lode-claims.
§ 53. Veins or lodes of quartz or other rock in place.
§ 54. Location previous to the Mining Acts.
§ 55. Width of lode-claims—Rights granted by the patent.
§ 56. Survey must conform to the patent.
§ 57. Manner of locating prior to 1872.
§ 58. Several locations may be made.
§ 59. Local regulations.

§ 52. Length and width of lode-claims.—Section 2320 of the Revised Statutes of the United States reads as follows: "Mining claims upon veins or lodes of quartz or other rock in place bearing gold, silver, cinnabar, lead, tin, copper, or other valuable deposits, heretofore located, shall be governed as to length along the vein or lode by the customs, regulations, and laws in force at the date of their location. A mining claim located after the tenth day of May, one thousand eight hundred and seventy-two, whether located by one or more persons, may equal, but shall not exceed, one thousand five hundred feet in length along the vein or lode; but no location of a mining claim shall be made until the discovery of the vein or lode within the limits of the claim located; no claim shall extend more than three hundred feet on each side of the middle of the vein at the surface, nor shall any claim be limited by any mining regulation to less than twenty-five feet on each side of the middle of the vein at the surface, except where adverse rights existing on the tenth of May, eighteen hundred and seventy-two, render such limitation necessary. The end-lines of each claim shall be parallel to each other."[1]

---

[1] Rev. Stat. 2320, Sec. 2, Act of 1872; 17 U. S. Stats. 91. Sec. 4 of the Act of 1866, 14 U. S. Stats. 252, read: "That when such location and entry of a mine shall be upon unsurveyed lands, it shall and may be lawful, after the extension thereto of the public surveys, to adjust the surveys to the limits of the premises.

§ 53. Veins or lodes of quartz, or other rock in place.—Mineral-producing lands are divided by miners into two classes: 1st. Where the mineral matter is within "rock in place," or "*in situ*"; 2d. Placers, and all other forms of deposit. In geology, and among miners, veins or lodes imply generally an aggregation of mineral matter found in the fissures of the rocks which inclose it, but are of great variety—veins differing very much in their formation and appearance. "Lode" is a term in general use among the tin-miners of Cornwall, and was introduced on the Pacific Coast by emigrants from the Cornish mines, and signifies a fissure, filled either by metallic or earthy matter. In several of the mining districts, the terms "lead" and "ledge" are employed in the local regulations concerning mines. "Lead" is used to convey the same idea as "lode," while "ledge" would seem to indicate a layer or stratum of mineral, interposed between a course or ridge of rocks.

The terms were probably not employed in the acts in their strict geological signification. The plain object of the law is to dispose of the mineral lands of the United States for money value, and whatever form of deposit can be embraced in the general phrase "vein or lode of quartz, or other rock in place" must be sold at the rate of five dollars per acre. There is to be included in the first class, all lands wherein the mineral matter is contained in veins or ledges occupying the original habitat or location of the metal or mineral, whether in true or false veins, in zones, in pockets, or in the several other forms in which minerals are found in the original rock, whether the gangue or matrix is disintegrated at the surface or not.[1]

---

according to the location and possession and plat aforesaid, and the Surveyor-General may, in extending the surveys, vary the same from a rectangular form to suit the circumstances of the country and the local rules, laws, and customs of miners: *Provided*, that no location hereafter made shall exceed two hundred feet in length along the vein for each locator, with an additional claim for discovery to the discoverer of the lode, with the right to follow such vein to any depth, with all its dips, variations, and angles, together with a reasonable quantity of surface for the convenient working of the same as fixed by local rules: *and provided further*, that no person may make more than one location on the same lode, and not more than three thousand feet shall be taken in any one claim by any association of persons." See Secs. 2323, 2337 Rev. Stat.

[1] Decision of Commissioner, July 15th, 1873, Copp's U. S. Mining Decisions, 318. See Ante, Chap. 3.

§ 54. Locations previous to the mining acts of Congress—Limitation and size.—Lodes discovered and located previously to the passage of the mining acts are governed in regard to their size by the local laws, notwithstanding that the record thereof may not have been made till afterward.[1]

Prior to 1866, as a matter of general notoriety and history in all the mineral regions, miners held possessory rights by "locations" under local laws. These rights were always locally respected by the citizens and the Courts. Congress, by the Act of 1866, attempted to establish a general rule, by which these local rights should be recognized by the Government, so far as not in conflict with the laws of the United States. It for this purpose recognized those local laws, customs, and usages, and with reference to the quantity of any lode to be entered, applied two limitations, viz: A general one of all claimants to 3,000 feet; and with reference to a certain specified class of locations, those made *thereafter* a further limitation to 200 feet, and an additional claim for discovery. The intent was considered to be to recognize all locations in accordance with existing law, subject to these limitations.[2]

Where, therefore, a person, previously to the Act of 1866, located a claim of 1,400 feet, (a claim of that size being allowable under the local law) but did not record it till afterward, but the local law required that certain labor, necessarily requiring considerable time, should be performed prior to recording, and the location was not susceptible of being recorded at the date of the Act of 1866, yet was nevertheless good and valid so far as it had progressed, and was recorded in proper time, and as soon as the work was performed, the location was held complete previous to the passage of the act, and therefore governed by the local laws.[3]

In those cases where there was no general law in force at the date of location, regulating the size of the claim, and no written district laws, the matter was governed by the local customs or

---

[1] See Ante, Sec. 24.
[2] See Ante, Chap. 1.
[3] In re Silver Ore Lode, Decision of Commissioner Aug. 26th, 1874, 1 Copp's Land-owner. 83. See Ante. Sec. 24.

rules of miners in the district where the claim was situate.[1] If there were district laws, or local laws, the number of feet on a lode that might be located prior to the acts of Congress was governed by those laws.[2]

Since the Act of 1872, and the Revised Statutes, the size is limited by the provisions of those enactments.[3]

It was held, under the Act of 1866, that if ten men should locate 200 feet each on a ledge, one or two of them might buy out the rest, and apply for and secure a patent for all the ground, by showing the title by deeds, if the purchase was sanctioned by the local mining regulations, and the necessary expenditures had been made.[4]

§ 55. Width of lode-claims—Rights granted by the patent.—The uniform construction is, that no claim located after May 10th, 1872, can exceed 600 feet in width, under any circumstances; whether a location made after that date can equal 600 feet in width depends entirely upon the local regulations, or State or Territorial laws, in force in the several mining districts. But the latter cannot limit the surface width to less than fifty feet, unless adverse claims existing on the 10th of May, 1872, render such latter limitation necessary. In other words, the miners of the district, or the State or Territorial legislatures, are authorized by the act to regulate and control the width of a location, providing that the width shall not exceed 600 feet, nor be limited to less than fifty feet.

Where an application is made for a patent for a mine located prior to May 10th, 1872, the patent, when issued, conveys to the grantee the right to follow the particular lode named in the patent, to the number of feet expressed in the conveyance, although the lode should, in its course, leave the surface ground described in the patent, and enter the land adjoining. The patent not only grants him the right to follow the particular

---

[1] In re South Comstock G. & S. M. Co. Decision of Commissioner, Dec. 29th, 1875, 2 Copp's Land-owner, 146.
[2] In re San Augustin Mining Co. Decision of Commissioner, Sept. 22d, 1870, Copp's U. S. Mining Decisions, 32.
[3] Decision of Commissioner, Nov. 18th, 1873, Copp's U. S. Mining Decisions, 235.
[4] Decision of Comr. Nov. 6th, 1869, Copp's U. S. Mining Decisions, 23.

lode named, to the number of feet expressed in the patent, along the course thereof, but also grants him the right to follow said lode to any depth. It also conveys to the grantee the right to follow all other veins, lodes, or ledges, the tops or apexes of which lie within the exterior boundary lines of his survey, if the same were not adversely claimed on the 10th day of May, 1872, only to such an extent, however, along the course thereof, as may be embraced by such exterior boundaries, but to any depth.[1]

If a greater width of surface ground is embraced in the application than the local law allows, the size of the claim must be diminished to conform to that law before entry, and the Surveyor-General will be ordered to make new plat and field-notes.[2]

The Act of 1872 does not fix the width of a lode-claim, but simply provides that it shall not exceed 600 feet in width, nor be limited to less than fifty feet in width, except where adverse rights, existing on the 10th of May, 1872, shall render such limitation necessary. The width is regulated by local laws, customs, or rules of miners, not in conflict with the act.

No location of a lode made since May 10th, 1872, can exceed 1,500 feet in length. No survey of a lode-claim will be approved which exceeds, in length or width, the number of feet prescribed by law, at the date of the location.[3]

§ 56. Survey must conform to the patent.

—The survey should conform to the application for patent so far as the courses are concerned. Where the final survey covered but a small portion of the premises described in the application for patent, a re-survey was required, embracing only the number of feet to which the parties were entitled under the local law, and conforming to the patent in regard to courses.[4]

§ 57. Manner of locating claims on veins or lodes subsequently to May 10th, 1872.

—No lode-claim, located after

---

[1] Decision of Acting Commissioner, May 20th, 1873, Copp's U. S. Mining Decisions, 201.

[2] In re War Eagle Mine, Decision of Commissioner. May 1st, 1873, Copp's U. S. Mining Decisions, 195.

[3] Decision of Commissioner, Feb. 11th, 1875; 1 Copp's Land-owner, 179.

[4] In re Gus Belmont Lode, Decision of Commissioner. August 14th, 1873, Copp's U. S. Mining Decisions, 215.

May 10th, 1872, can exceed a parallelogram 1,500 feet in length by 600 feet in width, but whether surface ground of that width can be taken, depends upon the local regulations, or State or Territorial laws in force in the several mining districts; and no such local regulations, or State or Territorial laws, shall limit a vein or lode-claim to less than 1,500 feet along the course thereof, whether the location is made by one or more persons, nor can surface rights be limited to less than fifty feet in width, unless adverse claims, existing on the 10th day of May, 1872, render such lateral limitation necessary.[1]

The provision that no location can be made until after the discovery of the vein or lode, within the limits of the claim located, was evidently to prevent the incumbering of the district mining records with useless locations, before sufficient work has been done thereon to determine whether a vein or lode has really been discovered or not.[2]

The claimant should, therefore, prior to recording his claim, unless the vein can be traced upon the surface, sink a shaft or run a tunnel or drift to a sufficient depth therein to discover and develop a mineral-bearing vein, lode, or crevice; should determine, if possible, the general course of such vein in either direction from the point of discovery, by which direction he will be governed in making the boundaries of his claim on the surface, and should give the course and distance as nearly as practicable from the discovery shaft on the claim, to some permanent, well-known points or objects, such, for instance, as stone monuments, blazed trees, the confluence of streams, points of intersection of well-known gulches, ravines, or roads, prominent buttes, hills, etc., which may be in the immediate vicinity, and which will serve to perpetuate and fix the *locus* of the claim, and render it susceptible of identification from the description thereof given in the record of locations in the district.[3]

The claimant should also state the names of adjoining claims, or, if none adjoin, the relative positions of the nearest claims; should drive a post or erect a monument of stones at each cor-

---

[1] Instructions June 10th, 1872. Subdivision 11; Instructions February 1st, 1877, Subdivisions 9-19.
[2] Ibid. Subdivision 13; Ibid.
[3] Ibid. Subdivision 14; Ibid. Subdivisions 12-15.

ner of his surface ground, and at the point of discovery, or discovery shaft; should fix a post, stake, or board upon which should be designated the name of the lode, the name or names of the locators, the number of feet claimed, and in which direction from the point of discovery, it being essential that the location notice filed for record, in addition to the foregoing description, should state whether the entire claim of 1,500 feet is taken on one side of the point of discovery, or whether it is partly upon one and partly upon the other side thereof, and in the latter case, how many feet are claimed upon each side of such discovery point.[1]

Within a reasonable time (twenty days) after the location shall have been marked on the ground, notice thereof, accurately describing the claim in manner aforesaid, should be filed for record with the proper recorder of the district, who will thereupon issue the usual certificate of location.[2]

In order to hold the possessory right to a claim of 1,500 feet of a vein or lode located since May 10th, 1872, the act requires that until a patent shall have been issued therefor, not less than $100 worth of labor shall be performed, or improvements made thereon within one year from the date of such location, and annually thereafter, in default of which the claim will be subject to re-location by any other party having the necessary qualifications, unless the original locator, his heirs, assigns, or legal representatives, have resumed work thereon after such failure and before such re-location.[3]

The importance of attending to these details in the matter of location, labor, and expenditure is obvious, when it is understood that a failure to give the subject proper attention may invalidate the claim.[4]

§ 58. Several locations may be made.—There is no provision of law to prevent parties from locating other claims upon the same lode, outside of the first location made on the vein or

---

[1] Instructions June 10th, 1872, Subdivision 15; Instructions February 1st, 1877, Subdivisions 9–19.
[2] Ibid. Subdivision 16; Ibid.
[3] Ibid. Subdivision 17.
[4] Ibid. Subdivision 18; Instructions Feb. 1st, 1877, Subdivisions 17–19.

lode. If a lode or vein 3,000 feet in length is discovered, two locations may be made, each of 1,500 feet thereon.[1]

§ 59. Local regulations.

—The mining regulations of the different mining districts remain intact and in full force with regard to the size of locations, where they do not permit locations in excess of the limits fixed by Congress. Where such regulations permit locations in excess of the maximum fixed by Congress, they are restricted accordingly. A local regulation is valid, providing that a placer claim, for instance, shall not exceed 100 feet square.[2]

The local rules and laws must not be inconsistent with the laws of the United States. Where the local law conflicts with the acts of Congress, the local law must give way, whether it be a State or Territorial statute, a regulation, or a custom. The act will control as to the number of feet claimed. The claim must conform to the laws in force at the date of its location.[3]

In the absence of any State or Territorial enactment regulating the occupancy and possession of mining claims, miners may alter or amend the laws of the district; but this action will not affect claims already located. Should the miners deem it advisable to amend their district laws, they may re-locate their claims under and conformably to such enacted laws, and upon complying with the acts of Congress and the instructions of the Office, may enter and receive patents for the same.[4]

---

[1] Decision of Acting Commissioner, June 17th, 1873, Copp's U. S. Mining Decisions, 207.

[2] Ibid. March 19th, 1873, Ibid. 164.

[3] Decision of Commissioner, August 4th, 1871, Copp's U. S. Mining Decisions, 57; Ibid. August 25th, 1871; Ibid. 59.

[4] Decision of Commissioner, August 25th, 1871, Copp's U. S. Mining Decisions, 59.

CHAPTER V.

LOCATORS' RIGHT OF POSSESSION AND ENJOYMENT OF SURFACE GROUND, AND OF THE LODE.

§ 60. Locators' rights of possession and enjoyment.
§ 61. Status of lode-claims previously located.
§ 62. Patents for veins or lodes previously issued.
§ 63. Priority of location, importance of.

§ 60. Locators' rights of possession and enjoyment.—Section 2322 of the Revised Statutes reads as follows: "The locators of all mining locations heretofore made or which shall hereafter be made, on any mineral vein, lode, or ledge, situated on the public domain, their heirs and assigns, where no adverse claim exists on the tenth day of May one thousand eight hundred and seventy two, so long as they comply with the laws of the United States, and with State, territorial, and local regulations not in conflict with the laws of the United States governing their possessory title, shall have the exclusive right of possession and enjoyment of all the surface included within the lines of their locations, and of all veins, lodes, and ledges throughout their entire depth, the top or apex of which lies inside of such surface lines extended downward vertically, although such veins, lodes, or ledges may so far depart from a perpendicular in their course downward as to extend outside of the vertical side-lines of such surface locations. But their right of possession to such outside parts of such veins or ledges shall be confined to such portions thereof as lie between vertical planes drawn downward as above described, through the end-lines of their locations, so continued in their own direction that such planes will intersect such exterior parts of such veins or ledges. And nothing in this section shall authorize the locator or possessor of a vein or lode which extends in its downward course beyond the vertical lines of his claim to enter upon the surface of a claim owned or possessed by another." [1]

---

[1] Rev. Stats. 2322, Sec. 3, Act 1872, 17 Stats. 91. See Secs. 2320, 2324, Rev. Stats. U. S.

§ 61. Status of lode-claims previously located.—The status of lode-claims located previously to the Act of 1872 was not changed by that act with regard to their extent along the lode or width of surface, such claims being restricted and governed, both as to their lateral and linear extent, by the State, Territorial, or local laws, customs, or regulations which were in force in the respective districts at the date of such location, in so far as the same did not conflict with the limitations fixed by the Act of 1866. (14 U. S. Stats. 251.[1])

Mining rights, acquired under such previous locations, were, however, enlarged by the Act of May 10th, 1872, in the following respect, viz: The locators of all such previously taken veins or lodes, their heirs and assigns, so long as they comply with the laws of Congress, and with State, Territorial, or local regulations, not in conflict therewith, governing mining claims, are invested by the act with the exclusive possessory right of all the surface inclosed within the lines of their locations, etc., as fully provided in Sec. 3, Act 1872.[2]

The law limits the possessory right to veins, lodes, or ledges, other than the one named in the original location, to such as were not adversely claimed at the date of the Act of May 10th, 1872; and where such other vein or ledge was so adversely claimed at that date, the right of the party so adversely claiming is in no way impaired by said act,[3] or by the Revised Statutes.

§ 62. Patents for veins or lodes previously issued.—Rights under patents for veins or lodes, granted under previous legislation of Congress, were enlarged by the Act of 1872, so as to invest the patentee, his heirs or assigns, with title to all veins, lodes, or ledges, as they are fully described in Section 3, Act of 1872, (Revised Statutes, Sec. 2322) providing for a locator's rights of possession and enjoyment; but all veins, lodes, or ledges, the top or apex of which lies inside such surface

---

[1] Instructions June 10th, 1872, Subdivision 2; Instructions February 1st, 1877, Subdivisions 2–6.

[2] Rev. Stats. 2322; Instructions June 10th, 1872, Subdivision 3; Instructions February 1st, 1877, Subdivisions 2–6.

[3] Instructions June 10th, 1872, Subdivision 4; Instructions February 1st, 1877, Subdivisions 2–6.

locations, other than the one named in the patent, which were adversely claimed at the date of the act, are excluded from such conveyance by patent.[1]

Applications for patents for mining claims, pending at the date of the Act of May 10th, 1872, may be prosecuted to final decision in the General Land Office, and where no adverse rights are affected thereby, patents will be issued, in pursuance of the provisions of the law.[2]

§ 63. **Priority of location** is of great importance in the title of mining property; the older the better. The Act of 1872 protects mining claims located previous thereto, and gives the owners all lodes within their surface ground, not adversely claimed at the date of the act. After that date, no person has the right to prospect for veins on another party's surface ground; and where the old mine is held in accordance with local and Congressional law, another claim cannot lawfully be extended so as to embrace any part of the surface ground or veins owned under the old location.[3]

In all cases where a party claims a lode which has been re-located, he should furnish proof that the re-location was made in accordance with the local law, and that he was entitled to re-locate it.[4]

---

[1] Instructions June 10th, 1872, Subdivisions 7. 8; Instructions February 1st, 1877, Subdivisions 2-6.
[2] Ibid. Subdivision 8; Ibid.
[3] Copp's Land-owner, 31.
[4] Decision of Commissioner, Sept. 25th 1873; Copp's U. S. Mining Decisions, 225. See, as to locations under mining customs and regulations, Blanchard & Weeks' Leading Cases on Mines and Mining Water Rights, Chaps. 7-9.

CHAPTER VI.

TUNNEL RIGHTS.

§ 64. Owners of tunnel rights.
§ 65. Patenting tunnel rights.
§ 66. Expenditures upon tunnel.

§ 64. Owners of tunnels, rights of.—Sec. 2323, Rev. Stats. U. S., reads: " Where a tunnel is run for the development of a vein or lode, or for the discovery of mines, the owners of such tunnel shall have the right of possession of all veins or lodes within three thousand feet from the face of such tunnel on the line thereof, not previously known to exist, discovered in such tunnel, to the same extent as if discovered from the surface; and locations on the line of such tunnel of veins or lodes not appearing on the surface, made by other parties after the commencement of the tunnel, and while the same is being prosecuted with reasonable diligence, shall be invalid; but failure to prosecute the work on the tunnel for six months shall be considered as an abandonment of the right to all undiscovered veins on the line of such tunnel." [1]

An Act of Feb. 11th, 1875, provided as follows: " That Sec. 2324 of the Revised Statutes be, and the same is hereby, amended so that where a person or company has or may run a tunnel for the purpose of developing a lode or lodes, owned by said person or company, the money so expended in said tunnel shall be taken and considered as expended on said lode or lodes, whether located prior to or since the passage of said Act; and such person or company shall not be required to perform work on the surface of said lode or lodes in order to hold the same as required by said act."

The effect of Sec. 2323, R. S., is simply to give the proprietors of a mining tunnel, run in good faith, the possessory right to 1,500 feet of any blind lodes cut, discovered, or intersected

---

[1] Rev. Stats. 2323, Sec. 4. Act 1872, 17 U. S. Stats. 92. See Sec. 2320.

by such tunnel, which were not previously known to exist, within 3,000 feet from the face or point of commencement of such tunnel, and to prohibit other parties, after the commencement of the tunnel, from prospecting for and making location of lodes on the line thereof, and within the distance of 3,000 feet, unless such lodes appear upon the surface, or were previously known to exist.[1]

The term "face," as used in the section, is construed and held to mean the first working face formed in the tunnel, and to signify the point at which the tunnel actually enters cover, it being from this point that the 3,000 feet are to be counted, upon which prospecting is prohibited.[2]

To avail themselves of the benefits of the act, the proprietors of a mining tunnel are required, at the time they enter cover as mentioned, to give proper notice of their tunnel location, by erecting a substantial post, board, or monument at the face or point of commencement of the tunnel, upon which there should be posted a good and sufficient notice, giving the names of the parties or company claiming the tunnel right, the actual or proposed course or direction of the tunnel, the height and width thereof, and the course and distance from such face or point of commencement to some permanent well-known objects in the vicinity, by which to fix and determine the *locus* in the manner shown to be applicable to locations of veins or lodes; and at the time of posting such notice they shall, in order that miners or prospectors may be enabled to determine whether or not they are within the lines of the tunnel, establish the boundary lines thereof by stakes or monuments placed along such lines at proper intervals, to the terminus of the 3,000 feet from the face or point of commencement of the tunnel; and the lines so marked will define and govern as to the specific boundaries, within which prospecting for lodes not previously known to exist is prohibited while work on the tunnel is being prosecuted with reasonable diligence.[3]

At the time of posting notice and marking out the lines of

---

[1] Instructions June 10th, 1872, Subdivision 20; Instructions February 1st, 1877, Subdivisions 20-26.
[2] Ibid. Subdivision 21; Ibid.
[3] Ibid. Subdivision 22; Ibid.

the tunnel, a full and correct copy of such notice of location, defining the tunnel claim, must be filed for record with the mining recorder of the district, to which notice must be attached the sworn statement or declaration of the owners, claimants, or projectors of such tunnel, setting forth the facts in the case; stating the amount expended by themselves and their predecessors in interest in prosecuting work thereon; the extent of the work performed; and that it is bona fide their intention to prosecute work on the tunnel so located and described with reasonable diligence, for the development of a vein or lode, or for the discovery of mines, or both, as the case may be.[1]

The notice of location must be duly recorded, and, with the sworn statement attached, kept on the recorder's files for future reference.[2]

The Land Office takes particular care that no improper advantage is taken by parties making or professing to make tunnel locations, ostensibly for the purposes named in the statute, but really for the purpose of monopolizing the lands lying in front of their tunnels, to the detriment of the mining interests, and to the exclusion of bona fide prospectors or miners; but will hold such tunnel claimants to a strict compliance with the terms of the act; and, as reasonable diligence on their part in prosecuting the work is one of the essential conditions of their implied contract, negligence or want of due diligence will be construed as working a forfeiture of their right to all undiscovered veins on the line of such tunnel.[3]

§ 65. **Patenting tunnel locations.**—There is no provision of law for patenting tunnel locations, but such lodes as are discovered in running a tunnel may be patented upon a full compliance with the law.[4]

The uniform construction given by the Land Office to these provisions in regard to tunnel locations is as follows: The line of the tunnel is held to be the width thereof and no more, and

---

[1] Instructions June 10th, 1872, Subdivision 23; Instructions February 1st, 1877, Subdivisions 20–26.

[2] Ibid. Subdivision 24.

[3] Ibid. Subdivision 26; Instructions February 1st, 1877, Subdivisions 20–26

[4] Decision of Commissioner, April 15th, 1873; Copp's U. S. Mining Decisions, 193.

upon this line only is prospecting for blind lodes prohibited while the tunnel is in progress, and the right is granted to the tunnel-owners to 1,500 feet of each blind lode not previously known to exist, which may be discovered in such tunnel; but that other parties are in no way debarred from prospecting for blind lodes, or running tunnels, so long as they keep without the line of the tunnel as above defined, the said line being required by the regulations to be marked on the surface by stakes or monuments placed along the same from the face or point of commencement to the terminus of the tunnel line. When a lode is struck or discovered for the first time by running a tunnel, the tunnel-owners have the option of recording their claim of 1,500 feet all on one side of the point of discovery or intersection, or partly upon one and partly upon the other side thereof, but in no case can they record a claim so as to absorb the actual or constructive possession of other parties on a lode which had been discovered and claimed outside the line of the tunnel before the discovery thereof in the tunnel.[1]

Where, therefore, a location described a tract of land 3,000 by 1,500 feet, and embraced more than 100 acres of land, it was held that there was no authority for locations of this size.

The law gives to the tunnel-owner only such lodes as may be *discovered* in such tunnel, and only prevents the location by other parties of lodes *upon the line of such tunnel,* " not appearing on the surface."[2]

§ 66. Expenditures upon a tunnel regarded as expenditures upon a lode.

—The Act of 1866 did not fix any amount of work or expenditure as necessary to hold a claim, but left that to be regulated by the miners themselves. It did, however, prescribe that an amount of not less than $1,000 should be expended on the claim, as one of the conditions precedent to obtaining a patent. The Act of 1872 repealed the Act of 1866, in part, and after its passage, permitted 1,500 linear feet to be

---

[1] Corning Tunnel Mining and Reduction Co. *v.* Bell; In re Slide Lode, Decision of Commissioner Nov. 3d, 1876; 3 Copp's Land-owner, 130, 131, 195. Land Office Report, 1872, 60, 61; Decision of Commissioner, Sept. 20th, 1872; Copp's U. S. Mining Decisions, 144.

[2] Decision of Commissioner, Sept. 20th, 1872; Copp's U. S. Mining Decisions, 144.

located as one claim on a lode, which location may be made by an individual, or by an association of persons jointly; but no lode-claim, located after the passage of the Act of May 10th, 1872, can exceed 1,500 feet, whether located by one or more persons. The interpretation given to the provisions in regard to expenditure, is this: That a claim on a lode, located subsequently to May 10th, 1872, may be 1,500 feet, and no more, whether located by one or more persons, and that to hold such claim of 1,500 feet requires an annual expenditure of $100 thereon, and that, on all lodes located prior to May 10th, 1872, there must be an annual expenditure of not less than ten dollars, in labor or improvements, for each 100 feet so claimed, along the lode; and that where a number of such claims, of 100 or 200 feet each, as the case may be, upon the same lode, are held in common by one or more persons, the aggregate amount necessary to hold all the claims so held in common, on a lode, at the rate of ten dollars per 100 feet, may be expended upon any one claim thereon, or in other words, at any one point on the lode, so held in common; the words "where such claims are held in common, such expenditure may be made on any one claim," being construed to mean that where several of these individual locations, made previous to May 10th, 1872, upon the same lode, are held in common by one or more persons, the entire expenditure necessary to hold all the claims so held in common on such lode, may be made upon any one claim thereon, but that expenditures made upon any one lode or claim, however great, can in no way be made to apply to other lodes claimed by the same parties.[1]

This interpretation of the law remained in force, with the following modification: Under this view, it was at first held that work done and expenditures made in constructing a tunnel intended for the development and improvement of lodes, would not satisfy the legal requirement as to expenditure, but such expenditure or labor was required to be made in good faith upon each lode claimed, otherwise the same would be subject to re-location by other parties, as provided by law; and where a company were the claimants of nine separate lodes, all of which it was their purpose to develop and improve by a mining tunnel,

[1] Instructions February 1st, 1877, Subdivision 5.

run in order to intersect the lodes below the surface, this was held not a sufficient compliance with the act.[1]

But this last ruling has been superseded by the amendment providing that "where a person or company has or may run a tunnel for the purposes of developing a lode or lodes, owned by said person or company, the money so expended in said tunnel shall be taken and considered as expended on said lode or lodes, whether located prior to or since the passage of said act; and such person or company shall not be required to perform work on the surface of said lode or lodes, in order to hold the same, as required by said act."

By this legislation, the requirements of Section 2324, Revised Statutes, in regard to the expenditure upon mining claims, are so modified that money which has been or may be expended in running a tunnel for the purpose of developing one or more lodes, owned by such person or company, shall be considered as expended upon such lodes. The expenditures required upon mining-claims may be made from the surface or in running a tunnel for the development of such claims.[2]

Tunnel rights, diligence, expenditure.—Locators of tunnels, under the Act of May 10th, 1872, are required to use reasonable diligence in working and advancing their tunnels; otherwise, such tunnel locations will be treated as abandoned. There is no specified amount to be expended to retain the ownership of a tunnel location. The Act, approved March 1st, 1873, amending the Act of May 10th, 1872, only refers to lode-claims located prior to the passage of the Act of May 10th, 1872.[3]

[1] Rev. Stats. U. S. Secs. 2324, 2323, Act May 10th, 1872; 17 U. S. Stats. 92, Act March 1st, 1873, Sec. 5; 17 U. S. Stats. 92, Act June 6th, 1874; Act July 26th, 1866, Sec. 4, 14 U. S. Stats. 252; In re Helmick Silver Mining Company, Decisions of Commissioner and Acting Secretary, Aug. 27th and Sept. 4th, 1872; Copp's U. S. Mining Decisions, 136.

[2] Rev. Stats. Sec. 2324, as amended by Act of February 11th, 1875; Decision of Commissioner, March 11th, 1875; Skidmore's Mining Stats. 47; Instructions February 1st, 1877, Subdivision 18.

[3] Decision of Commissioner, August 1st, 1873, Copp's U. S. Mining Decisions, 215.

CHAPTER VII.

REGULATIONS AND CUSTOMS—EXPENDITURES AND IMPROVEMENTS—SURVEYS AND BOUNDARIES.

§ 67. Regulations and customs.
§ 68. Definition of "claim."
§ 69. Annual expenditures on placer-claims.
§ 70. Annual expenditures on lode-claims.
§ 71. Neglect of co-claimants to contribute.
§ 72. Re-located mines—expenditure.
§ 73. Amount of expenditure shown upon plat and field-notes.
§ 74. Location and survey—Boundaries.
§ 75. Certificate as to improvements.
§ 76. Fixed monuments—Courses—Distances.

§ 67. Regulations and customs, expenditures and improvements.—Section 2324 of the Revised Statutes reads as follows:[1] "The miners of each mining district may make regulations not in conflict with the laws of the United States, or with the laws of the State or Territory in which the district is situated, governing the location, manner of recording, amount of work necessary to hold possession of a mining claim, subject to the following requirements: The location must be distinctly marked on the ground so that its boundaries can be readily traced. All records of mining claims hereafter made shall contain the name or names of the locators, the date of the location, and such a description of the claim or claims located by reference to some natural object or permanent monument as will identify the claim. On each claim located after the tenth day of May, eighteen hundred and seventy-two, and until a patent has been issued therefor, not less than one hundred dollars worth of labor shall be performed or improvements made during each year. On all claims located prior to the tenth day of May, eighteen hundred and seventy-two, ten dollars worth of labor shall be performed or improvements made by the tenth

---

[1] Rev. Stat. 2324; See Sec. 5, Act of 1872; 17 U. S. Stats. 92; See Secs. 2331, 2332, Rev. Stats.

day of June, eighteen hundred and seventy-four, and each year thereafter, for each 100 feet in length along the vein, until a patent has been issued therefor; but where such claims are held in common, such expenditure may be made upon any one claim; and upon a failure to comply with these conditions, the claim or mine upon which such failure occurred shall be open to relocation in the same manner as if no location of the same had ever been made, provided, that the original locators, their heirs, assigns or legal representatives, have not resumed work upon the claim after failure and before such location. Upon the failure of any one of several co-owners to contribute his proportion of the expenditures required hereby, the co-owners who have performed the labor or made the improvements may, at the expiration of the year, give such delinquent co-owner personal notice in writing or notice by publication in the newspaper published nearest the claim, for at least once a week for ninety days, and if at the expiration of ninety days after such notice in writing or by publication such delinquent should fail or refuse to contribute his proportion of the expenditure required by this section, his interest in the claim shall become the property of his co-owners who have made the required expenditures."[1]

The time for the first annual expenditure on claims located prior to the passage of the act, was extended to the 10th day of June, 1874, by the Act of March 1st, 1873, amending Sec. 5 of the Act of 1872, (see 17 U. S. Stat. 92) and again extended to January 1st, 1875, by the Act of June 6th, 1874.

By the Act of Feb. 11th, 1875, amending the above section, R. S. 2324, the latter was amended so that where a person or company has or may run a tunnel for the purposes of developing a lode or lodes, owned by said person or company, the money so expended in said tunnel shall be taken and considered as expended on said lode or lodes, whether located prior to or since the passage of said act; and such person or company shall not be required to perform work on the surface of said lode or lodes in order to hold the same as required by said act.

---

[1] *Note.*—Sec. 5 of the Act of 1872, 17 U. S. Stat. 92, said: "Each year for each hundred feet," instead of "by the 10th day of June, 1874, and each year thereafter." In other respects the sections are identical. Instructions Feb. 1, 1877, Subd. 6.

§ 68. **Definition of "claim."**—To obtain a patent under the Act of 1866, one of the conditions precedent was that not less than $1,000 should be expended or improvements made upon the "claim." The term "claim," as used in the condition, being held to mean that portion of the vein or lode, and adjoining surface, to which the claimant had the right of possession by virtue of a compliance with the laws of the United States, and the local customs or rules of miners not in conflict therewith.[1]

From and after the date of the Act of 1872, in order to hold the possessory title to a mining claim previously located, and for which a patent has not been issued, the law requires that ten dollars shall be expended annually in labor or improvements, on each claim of 100 feet on the course of the vein or lode, until a patent shall have been issued therefor; but where a number of such claims are held in common upon the same vein or lode, the aggregate expenditure that would be necessary to hold all the claims at the rate of ten dollars per hundred feet, may be made upon any one claim.[2] A failure to comply with this requirement in any one year subjects the claim upon which such failure occurred to re-location by other parties, the same as if no previous location thereof had ever been made, unless the claimants under the original location shall have resumed work thereon after such failure and before such re-location. The first annual expenditure upon claims of this class should have been performed subsequent to May 10th, 1872, and prior to January 1st, 1875. From and after January 1st, 1875, the required amount must be expended *annually* until patent issues.

§ 69. **Annual expenditure not required on placer claims.**—The Act of 1866 only applied to veins or lodes of quartz, or other rock in place, bearing gold, silver, cinnabar, or copper; but by the Act of 1870, provision was made for the disposal of "claims usually called placers," "including all forms of deposit, excepting veins of quartz or other rock in place." Neither of these acts prescribed the amount of work or expenditure which should be made annually upon mining

---

[1] Decision of Commissioner, Sept 9th, 1872, Copp's U. S. Mining Decisions, 136, 142.

[2] Instructions June 10th, 1872, Subdivision 5; Ibid. Feb. 1st, 1877, Subd. 5.

claims, to enable parties to hold the same, but left this matter to be determined by the local laws, rules, regulations, and customs. The Act of 1872 repealed certain portions of the mining acts then in force, and among other things prescribed a new mode of procedure for obtaining patents to mining claims.

The only reference made to the subject of annual expenditures is found in the fifth section. The tenth section provided that the Act of 1870, " shall be and remain in full force, except as to the proceedings to obtain a patent which shall be similar to the proceedings prescribed by sections six and seven of this act for obtaining patents to vein or lode claims." It was therefore the intention of Congress to require annual expenditures only upon vein or lode claims, leaving placer claims as they had been previous to the passage of the Act of 1872, subject to local laws, rules, regulations, and customs.[1]

Extensions of time—Re-location.—Mines located prior to May 10th, 1872, upon which the required amount has been expended, and improvements made at any time since May 10th, 1872, and prior to January 1st, 1875, are not subject to re-location at the latter date. A claim located prior to May 10th, 1872, upon which the required amount was expended in actual labor and improvements, at any time since the 10th of May, 1872, was not subject to re-location on the 1st of January, 1875; providing the claimants thereof had in all respects complied with the local laws. But a claimant must make the annual expenditure upon his claim, which was required by the act, each and every year after January 1st, 1875, until the patent shall have been issued therefor, to entitle him to the possession of the location.

Claims located since the 10th of May, 1872, become liable to re-location in case the required amount of labor and improvements has not been expended thereon, within one year from the date of such location, and thereafter yearly.[2]

After January 1st, 1875, upon locations prior to May 10th,

---

[1] Rev. Stat. 2324, 2331, Decision of Acting Commissioner, April 25th, 1874, 1 Copp's Land-owner, 18.

[2] Decision of Commissioner, December 2d, 1874, 1 Copp's Land-owner, 138, 184; 2 Ibid. 31.

1872, the like amount of labor is required each calendar year. On claims located since May 10th, 1872, although it seems not generally to have been so understood, annual labor has been required upon all locations made since May 10th, 1872, without any postponement. The acts extending time to June 10th, 1874, and afterward to January 1st, 1875, only apply to locations prior to the Act of 1872.

Persons who own claims located prior to May 10th, 1872, had until Jan. 1st, 1875, to perform the assessment work thereon, as required under the provisions of the Congressional mining law. On all claims located subsequent to May 10th, 1872, (the date of the passage of the law) the yearly expenditure of $10 worth of work for each 100 feet was required to be made previous to January 1st, 1875. In other words, all claims located prior to the passage of the act went over until Jan. 1st, 1875; while those located since the passage were required to comply with its provisions prior to January 1st, 1875.[1]

The requirements in regard to expenditures upon claims located since May 10th, 1872, are in no way changed by the amendatory acts.[2]

§ 70. **Annual expenditure on lode-claims, etc.**—The construction given by the office is, that upon all claims located after May 10th, 1872, not less than $100 shall be expended in labor or improvements during each year, and that the year shall commence from the date of the location of the claim. The annual expenditure under the Act of 1872 is required until a patent shall have been issued.[3]

The required amount might have been expended at any time prior to the date of the amendatory extending acts, but after that the law required annual expenditure.[4]

The office has no power to go outside of the law, and rule that Sec. 5 only had reference to such claims as have not been improved to the amount of $500, the amount required by Sec.

---

[1] Decision of Commissioner, December 2d, 1874, 1 Copp's Land-owner, 138, 184; 2 Ibid. 31.

[2] Instructions of Commissioner, June 9th, 1874; March 11th, 1875.

[3] Decision of Commissioner, September 14th, 1872, Copp's U. S. Mining Decisions, 112; 1 Copp's Land-owner, 31.

[4] Decision Acting Commissioner, April 20th, 1874, 1 Copp's Land-owner, 18.

6 to be expended upon a claim before patent can issue. Sec. 5 applies to all claims which have not been patented.[1]

Work done on a tunnel.—Expenditures in running a mining tunnel, before a lode is struck therein, were held not tantamount to expenditures on the lode. But they were made so by the Act of February 11th, 1875.[2]

Work done on a tunnel, run to develop a particular lode, is now considered as done on the lode. The required expenditures may be made from the surface or in running a tunnel for the purpose of developing the lode or claims.[3]

§ 71. **Neglect or refusal of co-claimants to contribute.**—As the first annual expenditure on mines located prior to May 10th, 1872, should have been made prior to January 1st, 1875, notice to delinquent co-owners of such mines could not be legally given until after the latter date.

Clear, full, and explicit proof should be presented of the neglect or refusal of co-claimants to contribute their proportion of the annual expenditures required by law, and notice should be given to the co-claimants in the manner prescribed.[4]

§ 72. **Re-located mines—Expenditure.**—If a party applies for a patent for a re-located mine, it will be necessary for him to offer satisfactory proof that a sum of not less than $500 has been expended upon the mine *by the applicant or his grantors*. The fact that $500 had been expended upon the claim by a person or persons who subsequently abandoned it, will not relieve the applicant from the necessity of showing *that he or his grantors* have expended thereon the amount required by law.[5]

§ 73. **Amount of expenditure shown upon plat and field-notes of survey.**—The $500 expenditure must be shown upon the plat and field-notes of the classes of claims mentioned in the Mining Statutes, but where a mill site is applied

---

[1] Decision Commissioner, August 17th, 1872, Copp's U. S. Mining Decisions, 135.
[2] Ibid. August 27th, 1872, Ibid. 136.
[3] Rev. Stat. 2324, as amended; Letter of Commissioner, March 11th, 1875: June 9th, 1874; 1 Copp's Land-owner, 34.
[4] Minnie Tunnel & M. Co. In re Little Fred Mine; Decision of Commissioner, July 19th, 1876, 3 Copp's Land-owner, 66.
[5] Decision of Commissioner, January 30th, 1875, 1 Copp's Land-owner, 179.

for, together with a lode-claim, the $500 expenditure is only required to be upon the lode-claim, not upon the mill site also.[1]

§ 74. Location and survey — Boundaries. — The provision of Sec. 2324 Rev. Stats. Sec. 5, Act 1872, requiring that the location must be so distinctly marked upon the ground that its boundaries may be readily traced, is an important one, and locators cannot exercise too much care in defining their locations at the outset, inasmuch as the law requires that all records of mining locations made subsequent to its passage, shall contain the name or names of the locators, the date of the location, and such a description of the claim or claims located, by reference to some natural object or permanent monument, as will identify the claim.[2] Deputies, in surveying mining claims, were frequently in the habit of following the direction of the parties in interest, instead of adhering to the lines established in the original location of such claims, and thus, in effect, making a private, instead of an official survey. Under all laws and regulations, whether local or general, the location of a claim in such a manner as to give notice to all the world of the nature and extent of the same, is not only indispensable, but in most cases mining claims are initiated thereby, and all subsequent proceedings are based upon and must conform to such location. A failure to make and record the location in accordance with the law and regulations in force at the date of the location, will defeat the claim, and if it is not made with such definiteness as to operate as notice to all persons seeking to acquire rights to mining lands, it will be void for uncertainty. In making surveys of mining claims it becomes, therefore, essentially necessary to ascertain the boundaries thereof, as established by the original location, for the rights of the claimant are limited and defined by such boundaries. To make a survey in accordance with other lines or boundaries is tantamount to making a new location of the claim, and the rights of adjoining locators, who have complied with the requirements of the law, may be interfered with, and defeated thereby. The practice of making surveys according to the dictation of parties in interest,

---

[1] 1 Copp's Land-owner, 2.
[2] Instructions June 10th, 1872, and Feb. 1st, 1877, Subdivision 12.

instead of in accordance with the original location, had been productive of great confusion and injury to bona fide claimants.[1]

The applicant for a survey is, therefore, now required to furnish a copy of the original record of location, properly certified to by the recorder having charge of the records of the mining locations in the district where the claim is situate, and cause all official surveys of mining claims to be made in strict conformity to the lines established by the original location as recorded; and if the records of locations made prior to the passage of the Mining Act of May 10th, 1872, are not sufficiently definite and certain to enable the deputy to make a correct survey therefrom, he should, after reasonable notice in writing, to be served personally, or through the United States mail, on the applicant for survey, and adjoining claimants, whose residence or post-office address he may know or can ascertain by the exercise of reasonable diligence, take the testimony of neighboring claimants, and other persons who are familiar with the boundaries thereof, as originally located, and asserted by the locators of the claim, and after having ascertained, by such testimony, the boundaries as originally established, he should make a survey in accordance therewith, and transmit full and correct returns of survey, accompanied by the copy of the record of location, the testimony, and a copy of the notice served on the claimant and adjoining proprietors, certifying thereon, when, in what manner, and on whom service was made.[2]

The provisions of Rev. Stat. 2324, (Act of 1872, fifth section) must be strictly complied with in each case to entitle the claimant to a survey and patent, and should a claimant, under a location made subsequent to the Act of 1872, who has not complied with said requirements in regard to marking the location upon the ground, and recording the same, apply for a survey, the local officers are instructed to decline to make it. The only relief for a party, under such circumstances, will be to make a new location in conformity to law and regulation, as no case will be approved and patented by the office unless these and all other provisions of law are substantially complied

---

[1] Instructions Nov. 20th, 1873; Copp's U. S. Mining Decisions, 319.
[2] Ibid.

with. If the law has been complied with in the matter of marking the location on the ground, and recording the same, and any question arises in the execution of the survey, as to the identity of monuments, marks, or boundaries, which cannot be determined by a reference to the record, the deputy should take testimony in the manner prescribed for surveys of claims located prior to May 10th, 1872, and having thus ascertained the true and correct boundaries originally established, marked, and recorded, make the survey accordingly.[1]

§ 75. Improvements, certificates as to.—In many instances, deputy-surveyors certified to the value of improvements without ascertaining whether such improvements were made by the claimant or his grantors, or not. No improvements should be included in the estimate unless they have been made by the applicant for survey or by those from whom he derives title.[2] The value of improvements made upon other locations, or by other claimants, should not be taken into consideration, but excluded by deputies in their estimate of improvements upon the claim. Deputies are required to certify in each instance that the improvements and expenditures considered by them in their estimate, and which they must describe in their report, were made by the applicant, or by the persons from whom he derives title.[3]

---

[1] Instructions Nov. 20th, 1873; Copp's U. S. Mining Decisions, 319.
[2] Ibid.
[3] Ibid.

The following certificate will be attached to the field-notes of survey by the Surveyor-General: "I certify that the foregoing transcript of the field-notes of the survey of the —— mining claim, situate in —— mining district, county of —— and —— of ——, has been correctly copied from the original notes of said survey on file in this office; that said field-notes furnish such an accurate description of said mining claim as will, if incorporated into a patent, serve fully to identify the premises; and that such reference is made therein to natural objects and permanent monuments as will perpetuate and fix the *locus* thereof.

"I further certify that the value of the labor and improvements upon the said mining claim, placed thereon by the claimant and his grantors, is not less than five hundred dollars, and that said improvements consist of —— [here describe the improvements made by the applicant and his grantors upon the claim]. I further certify that the plat thereof filed in the U. S. Land Office at ——, is correct and in conformity with the foregoing field-notes. U. S. Surveyor-General, for —— U. S. Surveyor-General's Office." (Date.)

The following certificate will be indorsed upon each plat by the Surveyor-General: "The original field-notes of the survey of the ——, from which this

§ 76. Fixed monuments, courses, and distances.— Courses and distances must give way, when in conflict with fixed objects. Where an application called for two well-defined points, the mouth of a tunnel and the discovery, and the course and distance given would not lead to the discovery, they were rejected, as the call required a straight line from the mouth of the tunnel to the discovery. Where there was, with this exception, no discrepancy between the application and final survey, the application was allowed.[1]

Where a party establishes monuments at the corners of his claim at the time of location, and makes record of such location, the boundaries of his claim will be established by such monuments.[2]

---

plat has been made, have been examined and approved and are on file in this office, and I hereby certify that they furnish such description of said —— mining claim as will, if incorporated into a patent, serve fully to identify the premises; and that such reference is made therein to natural objects and permanent monuments as will perpetuate and fix the *locus* thereof.

"I further certify that the value of the labor and improvements upon the said mining claim, placed thereon by the applicant and his grantors, is not less than five hundred dollars, and that said improvements consist of —— [here describe the improvements made by the applicant or his grantors upon the claim]. And I further certify that this is a correct plat of said —— mining claim or premises, made in conformity with said original field-notes of survey thereof. U. S. Surveyor-General, for —— U. S. Surveyor-General's Office." (Date.)

[1] In re Mammoth Lode; Decisions of Secretary of Interior and Assistant Attorney-General, July 15th, 1873, reversing S. C. Decision Commissioner; Copp's U. S. Mining Decisions, 211, 212.

[2] Decision of Commissioner, June 13th, 1876, 3 Copp's Land-owner, 50.

CHAPTER VIII.

PATENTS TO MINERAL LANDS—MODE OF PROCURING GOVERNMENT TITLE.

§ 77. Patents for vein or lode-claims, how obtained.
§ 78. Details of procedure.
§ 79. Duties of registers and receivers.
§ 80. Nature of the patent.
§ 81. Impeachment of patent.
§ 82. Adverse possession as against a patent.
§ 83. What is granted.
§ 84. Who may apply.
§ 85. Evidence of ownership — Deraigning title — Identity of applicant — Transfers.
§ 86. Claim through an executor—Where an alien is grantee of a claim.
§ 87. United applications—Unincorporated associations.
§ 88. Several claims cannot be embraced in one application.
§ 89. Grantee of several locators may obtain patent for the whole tract.
§ 90. Conflicting patents.
§ 91. Errors in description in patent—Relinquishment—Calls for the relinquishment of land inadvertently patented.
§ 92. Second patent—Entries of mineral lands by settlers and corporations.
§ 93. Minerals discovered after agricultural patent.
§ 94. Setting aside patent.
§ 95. Number of patents.
§ 96. Protests against issuance of patents—Status of protestants.
§ 97. An illegal location invalidates subsequent proceedings.
§ 98. Location by a minor.
§ 99. Application for several lodes and a mill site—Claim partly in one district and partly in another.
§ 100. Delaying action at request of Congressional Committees.
§ 101. The affidavit—Proper party to make it.
§ 102. Verification of affidavits.
§ 103. The location notice.
§ 104. Parol evidence to aid the notice.
§ 105. Plat must show the boundaries of the claim.
§ 106. Surveys to show exterior boundaries.
§ 107. Specific surface ground.
§ 108. Posting on claim, and proof thereof.
§ 109. Publication of the notice.
§ 110. Time of publication.
§ 111. Counting the sixty days.
§ 112. Proof of publication.
§ 113. The newspaper in which the notice is to be published.
§ 114. Defects in the published notice.

§ 115. Discrepancies between final survey and patent, and the application and published notice.
§ 116. Discrepancies between the published notice and the notice and diagram filed.
§ 117. Discrepancies between the published notice and the diagram and posted notice.
§ 118. Discrepancies between the final survey and patent and the application.
§ 119. New survey, pending another application.
§ 120. Discrepancies between survey and diagram.
§ 121. Discrepancies between survey and notice, matter of description.
§ 122. Errors in survey.
§ 123. When application will be rejected.
§ 124. Sworn statement.
§ 125. Approval of survey—Jurisdiction of Surveyor-General.
§ 126. Proof of citizenship.
§ 127. Miscellaneous.

§ 77. Patents for vein or lode-claims, how obtained.—

Sec. 2325 of the Revised Statutes provides as follows:[1] "A patent for any land claimed and located for valuable deposits may be obtained in the following manner: Any person, association, or corporation authorized to locate a claim under this chapter, having claimed and located a piece of land for such purposes, who has, or have, complied with the terms of this chapter, may file in the proper land office an application for a patent, under oath, showing such compliance, together with a plat and field-notes of the claim or claims in common, made by or under the direction of the United States Surveyor-General, showing accurately the boundaries of the claim or claims, which shall be distinctly marked by monuments on the ground, and shall post a copy of such plat, together with a notice of such application for a patent, in a conspicuous place on the land embraced in such plat previous to the filing of the application for a patent, and shall file an affidavit of at least two persons that such notice has been duly posted, and shall file a copy of the notice in such land office, and shall thereupon be entitled to a patent for the land, in the manner following: The register of the land office, upon filing of such application, plat, field-notes, notices, and affidavits, shall publish a notice that such application has been made, for the period of sixty days in a newspaper to be by him designated as published nearest to such claim; and he shall also post such notice in his office for the same

---

[1] Rev. Stat. 2325, Sec. 6, Act 1872, 17 U. S. Stat. 92. See Secs. 2, 3, Act 1866, 14 U. S. Stat. 251. See, also, Secs. 2325, 2327, 2328, 2333, U. S. Rev. Stats.

period. The claimant, at the time of filing this application, or at any time thereafter, within the sixty days of publication, shall file with the register a certificate of the United States Surveyor-General that five hundred dollars worth of labor has been expended or improvements made upon the claim by himself or grantors; that the plat is correct, with such further description by such reference to natural objects or permanent monuments as shall indentify the claim, and furnish an accurate description, to be incorporated in the patent. At the expiration of the sixty days of publication the claimant shall file his affidavit, showing that the plat and notice have been posted in a conspicuous place on the claim during such period of publication. If no adverse claim shall have been filed with the register and the receiver of the proper land office at the expiration of the sixty days of publication, it shall be assumed that the applicant is entitled to a patent, upon the payment to the proper officer, of five dollars per acre, and that no adverse claim exists; and thereafter no objection from third parties to the issuance of a patent shall be heard, except it be shown that the applicant has failed to comply with the terms of this chapter." [1]

§ 78. **Details of procedure.**—The claimant is required, in the first place, to have a correct survey of his claim made under authority of the Surveyor-General of the State or Territory in which the claim lies; such survey to show with accuracy the exterior surface boundaries of the claim, which boundaries are required to be distinctly marked by monuments on the ground.[2] Four plats and one copy of the original field-notes, in each case, will be prepared by the Surveyor-General: one plat and the original field-notes to be retained in the office of the Surveyor-General; one copy of the plat to be given the claimant for posting upon the claim; one plat and a copy of the field-notes to be given the claimant for filing with the proper register, to be finally transmitted by that officer, with the other papers in the case, to the General Land Office; and one plat to be sent by

---

[1] Section 6 of the Act of May 20th, 1872, 17 U. S. Stats. 92, was identical with this section. For the corresponding repealed sections of the Act of 1866, see Ante, Secs. 4, 11.
[2] Instructions June 10th, 1872, Subdivision 28; Report of Commissioner of the General Land Office, 1872, p. 44; Instructions February 1st, 1877, Subdivision 28.

the Surveyor-General to the register of the proper land district, to be retained on his files for future reference. The claimant is then required to post a copy of the plat of such survey in a conspicuous place upon the claim, together with notice of his intention to apply for a patent therefor, which notice will give the date of posting, the name of the claimant, the name of the claim, mine, or lode; the mining district and county; whether the location is of record, and if so, where the record may be found; the number of feet claimed along the vein, and the presumed direction thereof; the number of feet claimed on the lode in each direction from the point of discovery, or other well-defined place on the claim; the name or names of adjoining claimants on the same or other lodes; or if none adjoin, the names of the nearest claims.[1]

After posting the said plat and notice upon the premises, the claimant will file, with the proper register and receiver, a copy of such plat, and the field-notes of survey of the claim, accompanied by the affidavit of at least two credible witnesses, that such plat and notice are posted conspicuously upon the claim, giving the date and place of such posting, a copy of the notice so posted to be attached to, and form a part of, said affidavit.[2] Attached to the field-notes so filed, must be the sworn statement of the claimant, that he has the possessory right to the premises therein described, in virtue of a compliance by himself, (and by his grantors, if he claims by purchase) with the mining rules, regulations, and customs of the mining district, State, or Territory in which the claim lies, and with the mining laws of Congress; such sworn statement to narrate briefly, but as clearly as possible, the facts constituting such compliance, the origin of his possession, and the basis of his claim to a patent.[3]

The affidavit should be supported by appropriate evidence, from the mining recorder's office, as to his possessory right as follows, viz: Where he claims to be a locator, a full, true, and correct copy of such location should be furnished,

---

[1] Instructions June 10th, 1872, Subdivision 29; Report of Commissioner of the General Land Office, 1872, p. 44; Instructions February 1st, 1877, Subdivision 29.
[2] Ibid. Subdivision 30; Ibid. Subdivision 30.
[3] Ibid. Subdivision 31; Ibid. Subdivision 31.

as the same appears upon the mining records; such copy to be attested by the seal of the recorder, or if he has no seal, then he should make oath to the same being correct, as shown by his records; where the applicant claims as a locator in company with others, who have since conveyed their interests in the lode to him, a copy of the original record of location should be filed, together with an abstract of title from the proper recorder, under seal or oath, tracing the co-locator's possessory rights in the claim to such applicant for patent; where the applicant claims only as a purchaser for valuable consideration, a copy of the location record must be filed, under seal or upon oath, with an abstract of title certified as above by the proper recorder, tracing the right of possession by a continuous chain of conveyances from the original locators to the applicant.[1]

In the event of the mining records in any case having been destroyed by fire, or otherwise lost, affidavit of the fact should be made, and secondary evidence of possessory title will be received, which may consist of the affidavit of the claimant, supported by those of any other parties cognizant of the facts relative to his location, occupancy, possession, improvements, etc., and in such case of lost records, any deeds, certificates of location or purchase, or other evidence which may be in the claimant's possession and tend to establish his claim, should be filed.[2]

Upon the receipt of these papers, the register will, at the expense of the claimant, publish a notice of such application for the period of sixty days, in a newspaper published nearest to the claim, and will post a copy of such notice in his office for the same period. In all cases, sixty days must intervene between the first and the last insertion of the notice in such newspaper. The notices so published and posted must be as full and complete as possible, and embrace all the data given in the notice posted upon the claim. Great care should be exercised in the preparation of these notices, inasmuch as upon

---

[1] Instructions June 10th, 1872, Subdivision 32; Report of Commissioner of the General Land Office, 1872, p. 41; Instructions February 1st, 1877, Subdivision 32.
[2] Ibid. Subdivision 33; Ibid. Subdivisions 32-36.

their accuracy and completeness will depend, in a great measure, the regularity and validity of the whole proceedings.[1]

The claimant, either at the time of filing these papers with the register, or at any time during the sixty days' publication, is required to file a certificate of the Surveyor-General, that not less than $500 worth of labor has been expended or improvements made upon the claim by the applicant or his grantors; that the plat filed by the claimant is correct; that the field-notes of the survey, as filed, furnish such an accurate description of the claim as will, if incorporated into a patent, serve to fully identify the premises, and that such reference is made therein to natural objects or permanent monuments as will perpetuate and fix the *locus* thereof.[2]

It is the more convenient way to have this certificate indorsed by the Surveyor-General, both upon the plat and field-notes of survey filed by the claimant.[3]

After the sixty-days period of newspaper publication has expired, the claimant must file his affidavit, showing that the plat and notice remained conspicuously posted upon the claim sought to be patented during said sixty days of publication. Upon the filing of this affidavit, the register will, if no adverse claim was filed in his office during the period of publication, permit the claimant to pay for the land according to the area given in the plat and field-notes of survey, at the rate of five dollars for each acre and five dollars for each fractional part of an acre, the receiver issuing the usual duplicate receipt therefor, after which the whole matter will be forwarded to the Commissioner of the General Land Office, and a patent issued thereon if found regular.[4]

When a duplicate receipt for a mineral entry has been lost, a patent may be transmitted upon affidavit of a secretary under the corporate seal of the company, that he is the duly elected secretary of the company, and authorized to receive a patent; that the party had the receipt in possession; that it has been

---

[1] Instructions June 10th, 1872, Subdivisions 34, 35, 36; Report of Commissioner of the General Land Office, 1872, p. 44; Instructions February 1st, Subdivisions 32-36.

[2] Ibid. Subdivision 37; Instructions February 1st, 1877, Subdivisions 35-43.

[3] Ibid. Subdivision 38; Ibid.

[4] Ibid. Subdivisions 39-40; Ibid.

lost, and that up to the time of making the affidavit, after careful and diligent search, he had been unable to find it. This affidavit is to be taken before an officer duly authorized to administer oaths, and attested by his seal, and the same filed.[1]

In sending up the papers in the case, the register must not omit certifying to the fact that the notice was posted in his office for the full period of sixty days, the certificate to state distinctly when such posting was done and how long continued. The consecutive series of numbers of mineral entries must be continued, whether the same are of lode or placer claims. The Surveyor-General must continue to designate all surveyed mineral claims as heretofore, by a progressive series of numbers, beginning with lot No. 37 in each township, the claim to be so designated at date of filing the plat, field-notes, etc., in addition to the local designation of the claim; it being required in all cases that the plat and field-notes of the survey of a claim must, in addition to the reference to permanent objects in the neighborhood, describe the locus of the claim with reference to the lines of public surveys by a line connecting a corner of the claim with the nearest public corner of the United States surveys, unless such claim be on unsurveyed lands at a remote distance from such public corner; in which latter case the reference by course and distance to permanent objects in the neighborhood will be a sufficient designation by which to fix the *locus* until the public surveys shall have been closed upon its boundaries.[2]

§ 79. Duties of registers and receivers.—On receiving applications for mining patents, registers and receivers are required to ascertain from the claimants whether they claim the right of possession under the local customs or rules of miners, as the same existed in the district prior to the adoption of local legislative regulations, and if so, require satisfactory proof that the claim is occupied in accordance with such customs or rules,

---

[1] In re Cascade Silver Mining Co. Decision of Commissioner, April 18th, 1870, Copp's U. S. Mining Decisions, 30.

[2] Instructions June 10th, 1872, Subdivisions 41, 42, 43; Report of Commissioner of General Land Office, 1872, p. 44; Instructions February 1st, 1877, Subdivisions 35-43.

certified copies of the regulations in force at the date of location to be transmitted with the case.

To ascertain whether the application is for a claim located in pursuance of local legislative regulations, and if so, require satisfactory proof that the claimants have, in making their locations, complied with such regulations. If the claimants desire patent for a claim located in accordance with the Act of Congress, approved July 26th, 1866, the officers are to observe that the location does not exceed 200 feet on the course of the vein or lode for each person who is a party to such location, with 200 additional feet for the discoverer, or 3,000 feet for any association of persons, which 3,000 feet can only be taken at the rate of 200 feet to each individual comprising such association, 200 additional feet being allowed the discoverer. By which it will be perceived that to locate 3,000 feet on any vein or lode under the Act of 1866 required not less than fourteen persons where one was the discoverer, or fifteen persons if taken without reference to the discovery claim.

If the application be for a mine located since May 10th, 1872, the maximum along a vein or lode that can be located by one person or several persons is 1,500 feet, and 300 feet on each side of the center of the vein at the surface is the greatest width of surface ground permitted under the Mining Act of May 10th, 1872.[1]

§ 80. Nature of the patent.—The patent is evidence of the series of proceedings recited in it; it is the deed of the United States, and is a solemn record of the Government, of its action and judgment with respect to the title of the claimants. As such, it imports absolute verity. Both the officers of the Government and the grantee, as well as those in privity with him, are bound by the recital of facts contained in it. But it has been held, in California, that neither the President nor any officer has other power to dispose of the public domain, or to sign or cause the seal of the Land Office to be affixed to patents, than such as is conferred by statutes of the United States. That while the recitals of fact are binding on all concerned, an

---

[1] In re San Xavier Mine, Decision Commissioner, July 10th, 1873, Copp's U. S. Mining Decisions, 209.

opinion of the executive officers in respect to matters of law, as indicated either by the ultimate act of issuing the patent or by recitals inserted in the instrument, is not conclusive, and a patent issued without legal authority is void.[1]

§ 81. **Impeachment of a patent.**—A United States patent is conclusive evidence in an action at law of the nature of the land conveyed. Where it is the duty of executive officers (as, for instance, the Secretary of the Interior) to identify lands and make lists, and issue patents for them, a patent so issued cannot be impeached at law, by extrinsic proof showing that the land which it conveys is not in fact what the patent states it to be. The subject is important, and a late enunciation of the doctrine maintained by the highest judicial tribunal of the country upon a subject peculiarly within its province, is to be found in the case of French v. Fyan, in the Supreme Court of the United States.[2]

The subject-matter was a swamp-land grant, but it is presumed that the principles set forth are of general application, and that they will not be materially changed.

The action was ejectment, and the single question raised was on the refusal of the lower Court to receive oral testimony to impeach the validity of a patent issued by the United States to the State of Missouri for the land in question, under the Act of 1850, known as the swamp-land grant, the purpose being to show by such testimony that it was not, in point of fact, swamp-land within the meaning of that act.

The bill of exceptions showed that the land was certified, in March, 1854, to the Missouri Pacific Railroad Company, as part of the land granted to aid in the construction of said road by the Act of June 10th, 1852, and the plaintiff, by purchase made in 1872, became vested with such title as this certificate gave.

To overcome this prima facie case, defendant gave in evidence the patent issued to Missouri, in 1857, under the Swamp-land Act, and it was admitted that defendant had a regular chain of title under this patent.

---

[1] McGarrahan v. New Idria M. Co. 49 Cal. 335; Teschemacher v. Thompson, 18 Cal. 11; Parker v. Duff, 47 Cal. 554; Foscalina v. Doyle, 47 Cal. 437.

[2] No. 42, Oct. T. 1876, 3 C. L. O. 134, to be reported in about the 93d or 94th U. S. R. 3d or 4th Otto.

It was at this stage of the proceedings that "the plaintiff offered to prove, in rebuttal, by witnesses who had known the character of the land in dispute since 1849 till the time of trial, that the land in dispute was not swamp and overflowed land, made unfit thereby for cultivation, and that the greater part thereof is not and never has been, since 1849, wet and unfit for cultivation.

"But the Court ruled that since the defendant had introduced a patent from the United States to the State for the said land under the Act of September 28th, 1850, as swamp land, this concluded the question, and the Court, therefore, rejected said parol testimony; to which ruling of the Court the plaintiff then and there excepted."

Mr. Justice Miller, delivering the opinion of the Court, said: "This Court has decided more than once that the Swamp-land Act was a grant *in præsenti*, by which the title to those lands passed at once to the State in which they lay, except as to States admitted to the Union after its passage. The patent, therefore, which is the evidence that the lands contained in it had been identified as swamp-lands under that act, relates back and gives certainty to the title as of the date of the grant. As that act was passed two years prior to the act granting lands to the State of Missouri for the benefit of the railroad, the defendant had the better title on the face of the papers, notwithstanding the certificate to the railroad company for the same land was issued three years before the patent to the State under the Act of 1850. For while the title under the Swamp-land Act, being a present grant, takes effect as of the date of that act, or of the admission of the State into the Union, when this occurred afterward, there can be no claim of an earlier date than that of the Act of 1852, two years later, for the inception of title of the railroad company.

"The only question that remains to be considered is whether, in an action at law in which these evidences of title come in conflict, parol testimony can be received to show that the land in controversy was never swamp-land, and, therefore, the patent issued to the State under that act is void.

"The second section of the Swamp-land Act declares, 'that it shall be the duty of the Secretary of the Interior, as soon as practicable after the passage of this act, to make out an accurate list and plats of the land described as aforesaid, and transmit the same to the governor of the State, and, at the request of the governor, cause a patent to be issued to the State therefor, and on that patent the fee-simple to said lands shall vest in said State, subject to the disposal of the legislature thereof.' It was under the power conferred by this section that the patent was issued under which defendant holds the land. We are of opinion that this section devolved upon the Secretary, as the head of the department which administered the affairs of the public lands, the duty, and conferred on him the power of determining what lands were of the description granted by that act, and made his office the tribunal whose decision on that subject was to be controlling.

"We have so often commented in this Court on the conclusive nature and effect of such a decision when made and evidenced by the issuance of a patent, than we can do no better than to repeat what was said in the case of Johnson v. Towsley, 13 Wall. 72, where the whole question was reviewed both on principle and authority. In that case it had been strongly argued that the specific language of one of the statutes concerning pre-emption on the public lands, made the decision of the Commissioner of the General Land Office conclusive everywhere and under all circumstances. The Court responded to this argument in this language : ' But while we find no support to the proposition of the counsel for plaintiffs in error in the special provisions of the statute relied on, it is not to be denied that the argument is much stronger when founded on the general doctrine that when the law has confided to a special tribunal the authority to hear and determine certain matters arising in the course of its duties, the decision of that tribunal, within the scope of its authority, is conclusive upon all others. That the action of the Land Office in issuing a patent for any of the public land, subject to sale by pre-emption or otherwise, is conclusive of the legal title, must be admitted under the principle above stated ; and in all Courts, and in all forms of judicial proceedings where this title must control, either by reason of the limited powers of the Court or the essential character of the proceedings, no inquiry can be permitted into the circumstances under which it was obtained. On the other hand, there has always existed in the Courts of Equity the power in certain classes of cases to inquire into and correct mistakes, injustice, and wrong in both judicial and executive action, however solemn the form which the result of that action may assume, when it invades private rights; and by virtue of this power the final judgments of Courts of Law have been annulled or modified, and patents, and other important instruments issuing from the crown or other executive branch of the government, have been corrected, or declared void, or other relief granted.'

"We see nothing in the case before us to take it out of the operation of that rule, and we are of opinion that, in this action at law, it would be a departure from sound principle, and contrary to well-considered judgments in this Court and in others of high authority, to permit the validity of the patent to the State to be subjected to the test of the verdict of a jury on such oral testimony as might be brought before it. It would be substituting the jury, or the Court sitting as a jury, for the tribunal which Congress had provided to determine the question, and would be making a patent of the United States a cheap and unstable reliance as a title for lands which it purported to convey.

"The learned judge of this Court who presides in the California Circuit, has called our attention to a series of decisions of the Supreme Court of that State in regard to this swamp-land grant, commencing with 27 Cal. 87,[1] in which a different doctrine is announced. But with all the respect we have for that learned Court, we are unable to concur in the views therein expressed. The principle we have laid down is in harmony with the system which governs the relations of the Courts to the officers of the executive departments; especially those having charge of the public lands, as we have repeatedly decided, and we must abide by them.

"We do not mean to affirm that there is anything in the case before us, as it is here presented, which would justify a resort to a Court of Chancery; we merely mean to express our conviction that the only mode by which the conclusive effect of the

[1] Kernan v. Griffith.

patent in this case can be avoided, if it can be done at all, is by a resort to the equitable jurisdiction of the Courts.

"The case of the Railroad Company *v.* Smith, 9 Wall. 45, is relied on as justifying the offer of parol testimony in the one before us. In that case it was held that parol evidence was competent to prove that a particular piece of land was swamp land, within the meaning of the act of Congress.

"But a careful examination will show that it was done with hesitation, and with some dissent in the Court. The admission was placed expressly on the ground that the Secretary of the Interior had neglected or refused to do his duty; had made no selection or lists whatever, and would issue no patents, although many years had elapsed since the passage of the act. There was no means, as this Court has decided, to compel him to do so, and if the party claiming under the State in that case could not be permitted to prove that the land which the State had conveyed to him as swamp lands was in fact such, a total failure of justice would occur, and the entire grant to the State might be defeated by this neglect or refusal of the Secretary to perform his duty. (Gaines *v.* Thompson, 7 Wall. 347; Secretary *v.* McGarrahan, 9 Wall. 298; Litchfield *v.* The Register and Receiver, 9 Wall. 575.) The Court said in that case: "The matter to be shown is one of observation and examination; and whether arising before the Secretary, whose duty it was primarily to decide it, or before *the Court whose duty it became, because the Secretary had failed to do it*, this was clearly the best evidence to be had, and was sufficient for the purpose."

"There is in this no conflict with what we decide in the present case, but, on the contrary, the strongest implication that if, in that case, the Secretary had made any decision, the evidence would have been excluded." The judgment of the Circuit Court was affirmed.[1]

§ 82. **Adverse occupation as against a patent.**—A patent is the instrument which, under the laws of Congress, passes the title of the United States. It is the Government convey-

---

[1] See, also, Gaines *v.* Thompson, 7 Wall. 352; Kendall *v.* U.S. 12 Pet. 618; Comr. *v.* Whiteley, 4 Wall. 522; Reeside *v.* Walker, 11 How. 272; U. S. *v.* Guthrie, 17 Wall. 284; Decatur *v.* Paulding, 14 Pet. 479; Brashear *v.* Mason, 6 How. 92; U. S. *v.* Comr. 5 Wall. 563; U. S. *v.* Seaman, 17 How. 230.

ance. If other parties possess equities superior to those of the patentee, upon which the patent issued, a Court of Equity will, upon proper proceedings, enforce such equities, by compelling a transfer of the legal title, or enjoining its enforcement, or cancelling the patent. But, in an action of ejectment, the legal title must prevail in all the Federal Courts; and a patent, when regular on its face, is conclusive evidence of that title. So, also, in the action of ejectment in the State Courts, when the question presented is whether the plaintiff or the defendant has the superior legal title from the United States, the patent must prevail. Neither in a separate suit in a Federal Court, nor in an answer to an action of ejectment in the State Courts, can mere occupation of the demanded premises by either party for the period prescribed by the Statute of Limitations of the State, be held to constitute a sufficient equity in their favor to control the legal title subsequently conveyed to others by the patent of the United States. The power of the United States to dispose of its public lands cannot be defeated nor obstructed by any occupation of the premises before the issue of the patent, under State legislation, in whatever form of tribunal such occupation be asserted.[1]

§ 83. The patent, what is granted.

What is granted by the United States to the patentee of a vein or lode-claim may be thus stated: A patent granted for a mining claim under the Act of 1866, by the express provision of the act, conveyed to the grantee thererein named the *surface ground* embraced within the exterior boundaries of the survey, and the particular lode named in the patent for the number of feet patented along the course thereof, with all its dips, angles, and variations, although it might depart from the surface ground described in the survey and enter the land adjoining.

Where the application for patent was pending under the Act of 1866, on the 10th day of May, 1872, none of the rights which

---

[1] Gibson v. Chouteau, 13 Wall. U. S. 92, reversing S. C. 39 Mo. 588; Wilcox v. Jackson, 13 Pet. 516; Irvine v. Marshall, 20 How. U. S. 558; Fen v. Holme, 21 How. U. S. 481; Lindsey v. Miller, 6 Pet. 672; Stephenson v. Smith, 7 Mo. 610; Barry v. Gamble, 8 Id. 881; Cunningham v. Ashley, 14 How. 377; Lindsey v. Hawes, 2 Black. 554; Stark v. Starrs, 6 Wall. 402; Johnson v. Towsley, 13 Wall. 72; Bagnell v. Broderick, 13 Pet. 450.

the applicant had acquired by virtue of compliance with the Act of 1866 were in any way affected or impaired, and patents issued upon such applications conveyed the same rights which were conveyed under the Act of 1866, together with all other veins or lodes, the tops or apexes of which lie inside the exterior boundaries of the surface ground patented, to the extent and in the manner provided by the third section of the Act of 1872.[1]

§ 84. Who may apply.

—The real owners of the mine, having also the possessory title to the lode, are the persons to whom it is proper to deliver the patent, notwithstanding that they may not be parties named in it as it was originally made out.[2]

Joint owners must jointly apply. Where several parties own undivided interests in a mining claim, all the owners must join in an application; and where several parties own separate and distinct portions of a claim, application for a patent may be made by either for the portion he desires.[3]

A patent may issue to an assignee of the applicant. In such case it is necessary for the party to file in the commissioner's office the duplicate receiver's receipt, with an indorsement thereon of the applicants, of all right, title, and interest in the lodes.[4]

Patented ground is subject to entry by adjoining proprietors where the patent provides that the premises conveyed are subject to be entered by any adjoining proprietors of a vein or lode of gold, silver, cinnabar, or copper, in exploring or operating such vein or lode.[5]

§ 85. Evidence of ownership—Deraigning title—Identity of applicant—Transfers.

—Patents for mining claims are

---

[1] In re Hercules Lode and Seven-Thirty Lode, Decision Commissioner, August 17th, 1874, I Copp's Land-owner, 83; Decision Commissioner, December 26th, 1872, Copp's U. S. Mining Decisions, 154; Rev. Sts. U. S. 2322.

[2] In re Chicago and Clear Creek G. & S. M. Co. Decision Acting Commissioner, April 4th, 1872, Copp's U. S. Mining Decisions, 85.

[3] Decision of Commissioner, February 18th, 1873, Copp's U. S. Mining Decisions, 159.

[4] Decision of Commissioner, Oct. 2d, 1872; In re Vespasian Lode, Copp's U. S. Mining Decisions, 116.

[5] In re Idaho Lode, Decision of Commissioner, July 22d, 1869; Copp's U. S. Mining Decisions, 21.

issued to the parties named in the register's certificate of entry. If any conveyance has taken place after the original applicants have commenced proceedings for a patent, but before the entry is made at the local office, the register's certificate and the receiver's receipt must be made out in the name of the grantee. Upon filing a deed in the General Land Office, the register and receiver will be instructed to so make out the certificates and receipts.

If, however, the transfer takes place after the date of entry, an indorsement should be made upon the duplicate receipts by the applicant for a patent, assigning all right and title in and to the premises therein described. The patent will then issue in the name of the grantee.[1]

Transfer of interest from the original locators to the applicant for patent must be shown. The identity of the parties must also be satisfactorily established. Where this is done, a difference in the name of the same party, as used in the deeds, or the abstract of title, is not fatal to the application.

It is also well to have on file full and complete copies of the respective conveyances showing title in the applicants, but abstracts of title are held sufficient.[2] They must, however, be complete and not partial.[3]

In the case of the Kempton mine, it was objected that it did not appear that one *B. F. Buck*, one of the original locators, ever transferred his interest. The Secretary, in deciding the case, said: "The original application for the Kempton patent, which is sworn to by five different persons, alleges that 'Samuel Buck, under the name of B. F. Buck,' was one of the original locators, and that the said Samuel had transferred his interest in the mine to John Segus, who was one of the applicants for patent. There is, in the abstract of title furnished, a certificate of the recorder of the conveyance from Samuel Buck to the said Segus. I think this is sufficient. Names are arbitrary. Identity is the important matter, and the identity of Samuel

---

[1] Decision of Commissioner, March 8th, 1873, Copp's U. S. Mining Decisions, 162.

[2] In re Kempton Mine, Decision of Secretary of Interior, Jan. 2d. 1875, 1 Copp's Land-owner, 178.

[3] Decision of Comr. Jan. 6th, 1874; Copp's M. D. 340.

Buck with the B. F. Buck of the location is satisfactorily shown."[1]

Where the only record evidence of title was a bill of sale of one-third interest in the claim, and deeds from two of the applicants for patent to the party who made the entry, of all their interest in the mine, and the bill of sale stated that the claims were located as quartz claims and recorded as such, the applicants were required to furnish a copy of the location notice, and an abstract of all transfers of the claims, tracing the title from the original locators to the applicants for patent. A survey was also required to be made of the premises, and embracing only such surface ground as was originally located in conformity with local laws.[2]

§ 86. Claim through an executor.—Where an applicant claimed through a deed made by one of two executors of the estate of a deceased locator, and it appeared that two were appointed executors and that the letters had been revoked, the applicant was required to file a certified copy of the letters testamentary, with copy of will attached, a certificate of the clerk showing the date of the revocation, and evidence that *one* of the executors could legally pass title by deed.[3]

Where an alien is grantee of a claim—Holding until office found.—In an affidavit a party alleged that "he is informed by said John Henry, and the deponent verily believes, that said John Henry is an alien, and a subject of Great Britain; that deponent has frequently requested said Henry to make declarations of his intention to become a citizen of the United States, in order that said application for patent might proceed; but the said Henry has constantly, and does now positively refuse to make any such declaration, but still continues an alien, and declines and refuses to take any step toward becoming a citizen of the United States."

"Deponent further says that, by reason of the facts afore-

---

[1] In re Kempton Mine, Decision of Secretary of Interior, Jan. 2d, 1875, 1 Copp's Land-owner, 178.

[2] In re Live Oak Quartz Mine, Decision of Acting Commissioner, April 24th, 1876; 3 Copp's Land-owner, 18.

[3] In re N. E. Extension of Yosemite Mine, Decision of Acting Commissioner, April 29th, 1876, 3 Copp's Land-owner, 18.

said, the applicants are unable to present any abstract of title showing a right in them to all of the mining claim aforesaid, and that the undivided fifty feet thereof stands in the name of the said John Henry."

It was urged, for the applicants for patent, that an alien is incapable of acquiring a patentable interest in a mining location, and that the "attempt of a party to convey to Henry what the law prohibited the latter from holding, did not in any way affect the rights of the applicants, the act being void."

It was therefore asked that the patent issue to the applicants. The Commissioner said: "No patent can issue upon the application as it now stands, as they have not title to the entire premises for which patent is sought. It is true that John Henry, being an alien, has no patentable interest in said mine at the present time, but should he become naturalized, his right to a patent, upon compliance with the law, would be perfect, for naturalization has a retroactive effect, so as to be deemed a waiver of all liability to forfeiture, and a confirmation of his former title."[1]

It has been held by the Supreme Court of the United States, in numerous cases, that an alien can take by deed and hold until office found.[2]

The application for patent was, therefore, ordered to remain suspended until the applicants should show that they were in a condition to receive patent.[3]

§ 87. United applications—Unincorporated association.

—An application for patent may be filed by an association of two or more persons owning divided or undivided interests in the premises for which patent is sought, and where the required improvements have been made upon the premises described in the application, jointly by the several owners, the said association of

---

[1] Vide Osterman v. Baldwin, 6 Wall. 116; Jackson v. Beach, Johnson's Cases, 401.

[2] Vide Fairfax, Devisee, v. Hunter, 7 Cranch, 603; Orr v. Hodgson, 4 Wheat. 453; Craig v. Leslie et. al. 3 Wheat. 563; Craig v. Radford, 3 Wheat. 594; Cross v. De Vallie, 1 Wall. 1; Osterman v. Baldwin, 6 Wall. 116; Governeur's Heirs v. Robertson, 11 Wheat. 332.

[3] In re Lady Allen Lode. Decision of Commissioner, July 18th, 1876, 3 Copp's Land-owner, 69.

persons may receive patent therefor, upon full compliance with the law and instructions.

Where it appeared from the papers in the case that the several applicants owned separate and distinct interests; that the said applicants were an association of persons unincorporated; that the required amount had been expended upon the claim at the joint expense of the several members of the association, the applicants were required to furnish the following additional evidence:

1st. The affidavit of the applicants as to whether or not any known veins or lodes of gold, silver, cinnabar, lead, tin, copper, or other valuable deposits existed within the exterior boundaries of said premises. If any were known to exist, their names were required to be given, and the affidavits to show that no other known veins existed within the said premises other than those named.

2d. The applicants to file an abstract of conveyances from the original locators to the present applicants, properly certified to by the recorder. Also, copies of the several locations.

3d. Evidence to be filed to show that the person before whom some of the proofs submitted were verified, was a justice of the peace.[1]

§ 88. Several claims cannot be embraced in one application.

—Several claims, separate in their inception, should not be embraced in one application for patent. The slight saving in expense does not compensate for the delays in furnishing satisfactory proofs in the several claims sought to be patented.

No application for patent that shall embrace more than one vein or lode will now be received. This, however, does not apply to placers which embrace several lodes within the boundaries sought to be patented,[2] or to consolidate claims on the same vein or lode.[3]

§ 89. Grantee of several locators may obtain patent for the whole tract.

—A number of bona fide locators, having

---

[1] Decision of Commissioner, October 28th, 1875, 2 Copp's Land-owner, 114.
[2] Sec. 11, Act of 1872, Rev. Stats. 2333.
[3] Decision Acting Commissioner, August 17th, 1875, 2 Copp's Land-owner, 82; reversing decision, March 26th, 1874, 1 Copp's Land-owner, 2; See Ibid. 174.

complied with the laws and the local rules and regulations, may convey all their right, title, and interest in such locations to one person, and the latter may apply for a patent for the whole tract thus located. In this event, it is necessary for the applicant to file with the register and receiver copies of the original notices of location, and an abstract of title from the office of the proper recorder, showing the record-title to the premises claimed to be in the name of the applicant.[1]

§ **90. Conflicting patents.**—In cases where two applications for patent conflict with each other, and the applicants may desire to compromise or amicably settle their disputes by each party releasing to the other a portion of the premises embraced in the respective applications, a survey will be required of that portion of each claim which may be necessary to show the compromise line agreed upon between the parties, and the exterior boundaries of each claim to be patented. It is deemed unnecessary in such cases for the Land Office to direct the Surveyor-General to make such surveys in cases of this kind, as he will do so upon application of the parties in interest.[2]

A case arose where a lode did not follow the surface ground patented throughout its entire length, but left the surface near the southwesterly end of the survey of the surface ground and underlying a portion of another survey. It was found, upon inspection of the official plat of survey furnished and filed by the applicants for the patent, as well as the diagram posted with the notice on the claim, that the claim applied for did so embrace a portion of the surface ground embraced in another survey, and that it covered a part of another lode. The applicants in effect asked the United States to sell and convey to them, as a portion of the public domain, a tract of land and certain premises already sold and conveyed. This the office declined to do, and held that in such cases it is the duty of the office to protect a prior patentee by inserting in the subsequent patent such apt words as shall clearly except every right already conveyed. It has been the uniform practice and custom of the office in the

---

[1] Decision of Commissioner, January 22d, 1873, Copp's U. S. Mining Decisions, 157.

[2] Ibid. August 18th, 1874, 1 Copp's Land-owner, 83.

recitations of its mineral patents, to expressly convey the lode or vein named in the patent, to the number of feet named, as well as the surface ground described in the patent; and it is also held that in forming an exception it should be made equally broad.

The form of exception was ordered to be in the following words: "Excepting from this conveyance the surface ground and lode conveyed to the said I. M. & E. Company, by said patent, dated September 3d, A. D. 1872."[1]

§ 91. Errors in description in patent—Relinquishment.

—Where it appears that a claim is erroneously described in a patent, the applicant will be informed that a new patent will issue to him for his claim upon the receipt at the Land Office of the patent already issued, with a relinquishment indorsed thereon to the United States of the premises therein described, together with a certificate of the recorder that the relinquishment has been duly recorded in the records of his office.

The relinquishment should state that the same is made for the reason that the premises are erroneously described in said patent, and release all right, claim, title, or interest to the premises described. The recorder's certificate should also state as to whether or not his records show any conveyances of said premises.

If the applicant has conveyed the premises to any other person, it will be necessary for him to cause an abstract of such conveyances to be made, certified to by the recorder, and accompanied with a relinquishment from the parties named in the conveyances, and to forward the same with the inclosure (the patent) to the General Land Office.[2]

The relinquishment may be attested under seal by the clerk of any Court within the land district where the claim is situated.[3]

---

[1] In re Hercules & Seven-Thirty Lodes; Decision of Commissioner, Aug. 4th, 1871, 1 Copp's Land-owner, 82. See, also, In re Wandering Boy Lode; In re Prince of Wales Lode and Antelope Lode; Decision of Commissioner, May 6th, 1873, Copp's U. S. Mining Decisions, 197.

[2] Decision of Commissioner, June 22d, 1875, 2 Copp's Land-owner, 98; In re Empire Mining Co. Decision of Commissioner, April 11th, 1871, Copp's U. S. Mining Decisions, 41.

[3] Decision of Commissioner, June, 22d, 1785, 2 Copp's Land Owner, 98; In re

The relinquishment of a claim to a patent to a portion of a lode, closes and terminates the proceedings previously had, and the claims cannot again be presented except by a party properly entitled to the possession, and after due proceedings as in an original case.[1]

Calls for the relinquishment of land inadvertently patented.—A patent had inadvertently issued to the Central Pacific Railroad for certain land, the tract having been returned as mineral land by the Surveyor-General, and the return never having been disproved in the manner prescribed by law and the instructions. Afterward, an application was made for a mining patent for placer mining ground embracing this tract. The applicant submitted several affidavits, in which it was alleged that the applicant for patent and his grantors had held and worked the premises described in the application for nineteen years last past; that the value of labor and improvements upon the tract, made after the year 1861, was $35,000 in gold coin, and that the tract had been continuously worked as a placer claim from the latter date. It also appeared, from the certificate of the recorder and ex-officio auditor of the county, that the grantors of the applicant had paid taxes upon their mining claim, situate upon the tract, for the years 1868, 1869, 1870.[2]

The company was called upon to relinquish the land to the United States.

If a mine is erroneously patented, as, for instance, to a railroad company, and the latter does not relinquish the land when called upon, a patent will be granted for the mining claim as though no prior patent had been issued. Notwithstanding a patent covering the mine had issued to the Central Pacific Railroad Company, a placer claim was taken up for patenting, the company not having relinquished its patent.[3]

Where a party refuses to surrender a patent inadvertently and unlawfully issued, instructions will be issued to the United

---

Empire Mining Co. Decision of Commissioner, April 11th, 1871, Copp's U. S. Mining Decisions, 41.

[1] In re Kansas Lode, Decision of Commissioner, Feb. 27th, 1872, Copp's U. S. Mining Decisions, 79.

[2] Decision of Commissioner, October 23d, 1873, Copp's U. S. Mining Decisions, 227.

[3] In re Dutch Flat Cañon Placer Claim, 1 Copp's Land-owner, 2.

States Attorneys of the District to take proceedings to have the patent set aside and canceled. The adverse claimants may prosecute the suit.[1]

§ 92. Second patent — Entries of mineral lands by settlers and corporations.—While the office was held to have the power to issue a second patent, for the purpose of correcting a mistake or inadvertence, it was doubted whether it had that power in a case in which the first patent was obtained by artifice or fraud, upon a record regular upon its face. This is a right or power to decide upon questions of fraud, after the consummation of an entry and the execution and delivery of a patent thereon. Jurisdiction over questions of fraud more properly pertains to Courts of Equity; and, as they have the power to afford ample relief, parties are relegated to their remedies in the Courts; and therefore, instead of issuing a second patent in cases where the first has been obtained by fraud, to the injury of parties having a right or equity therein, and the facts are brought before the office, it will bring the matter to the attention of the Department of Justice, and ask that the party injured be permitted to use the name of the United States in the prosecution of proper proceedings.[2]

§ 93. Minerals discovered after patent to agricultural claimant.—The Commissioner held, in 1873, that mineral deposits discovered upon land after a United States patent therefor has issued to a party claiming under the laws regulating the disposal of agricultural lands and there being no reservations of mineral lands, pass with the patent, and the General Land Office has no further jurisdiction in the premises.[3]

§ 94. Setting aside patent.—The Land Office will do all in its power to set aside patents erroneously issued. If a patent has erroneously and inadvertently issued, the Land Office holds that it is proper to recite that fact, and issue another patent on the premises. But where it was contended that a second patent

---

[1] In re Wyoming Mine, Opinion Assistant Attorney-General U. S. January 14th, 1873; Copp's Mining Decisions, 152.
[2] Decision Commissioner, July 26th, 1873, Copp's U. S. Mining Decisions, 213.
[3] Ibid. July 10th, 1873. Ibid. 208.

could not properly issue, and as there is no question about the right to proceed in equity, and in the name of the United States to set aside a patent improperly granted, the Attorney-General was requested to institute such suits in behalf of the proper parties.[1]

§ 95. **Number of patents.**—One person may secure title to several mining claims. The statute does not restrict the number of patents, but gives the right to proceed to procure Government title to as many valid mining claims as he may have the possessory right to under local laws, and upon which the necessary amount has been expended in labor or improvements.[2]

§ 96. **Protests against issuance of patents—Status of protestants.**—Protests are not made by any party to the record in interest, but are made by a third party who stands in the light of *amicus curiæ*, and who has the right of showing only that the applicants have not complied with the law.[3] Parties who have not filed their adverse claims in time, and who stand in this relation, cannot take appeals from the General Land Office to the Secretary of the Interior.[4]

A protest has no such office to perform as that upon its being filed any right of intervention accrues save only in the nature of a challenge of the applicant's own showing, or that through its instrumentality any trial of unascertained rights may be authorized. It is held that for ascertaining the proper and necessary recital of a patent in a given case, the applicant is bound by the terms and disclosures of such filings, as conformably with the law, he rests his right to enter and purchase upon; and that for the further ascertainment and protection of rights, and as a

---

[1] Stark *v.* Starrs, 6 Wall. 402; Henshaw *v.* Bissell, 18 Ibid. 264; Wandering Boy Mine *v.* Highland Chief Mine, 2 Copp's Land-owner, 2; In re Prince of Wales, Antelope, Wandering Boy, Highland Chief, and Wellington Mines, Utah, Decision of Secretary of Interior, April 1st, 1875, 2 Copp's Land-owner, 2; 1 Copp's Land-owner, 43.

[2] Decision Commissioner, September 21st, 1872, Copp's U. S. Mining Decisions, 145.

[3] In re Kempton Mine, Decision of Secretary, January 2d, 1875, 1 Copp's Land-owner, 178.

[4] Application of Lambard, In re Mt. Pleasant Mine and Earl Mine, Decision Acting Secretary of Interior, Feb. 17th, 1877, 3 Copp's Land-owner, 194; Decision Acting Secretary, March 24th, 1876, In re Boston Quicksilver Mine.

duty on the part of the United States, it is held that the examination of the General Land Office should, whether protest be filed or not, proceed beyond the papers filed by the applicant and into those general records of the Office which evidence the final disposition made of the public domain; and if upon examination it is found that any part of the premises applied for have been previously disposed of, that express exception thereof should be inserted in the subsequent patent.[1]

Where a protest was not filed until after the expiration of the period allowed for that purpose, it was not permitted to suspend proceedings on an application for a placer location. The rule in the Flagstaff case was applied.[2]

A protest must be sworn to before an officer authorized to administer oaths in the land district where the claim is situated. Adverse claimants, notwithstanding default in making the claim, may be considered as parties to the contest for the purpose of showing from the records that the claimants have not complied with the law.[3]

§ 97. An illegal location invalidates subsequent proceedings.

A location being illegal and void, the subsequent proceedings, even if in due form, are also invalid, especially where the preliminary proceedings are insufficient to give the Land Office jurisdiction. In such cases, the applicants for patent can only protect their rights by the commencement of new proceedings, after a full and complete abandonment of the mine by prior occupants not holding the fee-simple title.[4]

§ 98. Location by a minor.

If a location is made by a person under twenty-one, he doing business for himself, and in his own name, and the location being in his own name, he has

---

[1] In re Hercules Lode and Seven-thirty Lode, Decision of Commissioner, August 17th, 1874, 1 Copp's Land-owner, 82.

[2] Weske v. Leet, Decision of Acting Secretary, May 11th, 1872, Copp's U. S. Mining Decisions, 95. See Flagstaff Lode; Highland Chief Lode, Copp's Mining Decisions, 61.

[3] McMurdy v. Streeter, 1 Copp's Land-owner, 34; In re Northern Light and Fairview Mines.

[4] In re Santa Rita del Cobre Mine, Decision of Comr. April 15th, 1873; Copp's U. S. Mining Decisions, 188; Decision of Acting Secretary, Nov. 6th, 1873; Ibid. 191.

the right to dispose of whatever he acquires by virtue of the location. His conveyance is not null and void. Nor can he for that reason interpose successfully an adverse claim to the application of his grantee, especially if he asserts the adverse claim alone and not with or by his guardian.[1]

§ 99. Applications for several lodes and a mill-site— Claim partly in one district and partly in another.—Where a party applied for fourteen quicksilver mines and a mill-site, it was held necessary for the company to file fourteen separate and distinct applications for patents.

An application for patent can embrace but one lode or vein, except in cases where placer claims embrace, within their exterior boundaries, several lode-claims.[2] If a mill-site is claimed with a mine, the application for patent may embrace the mill-site. Applications for mines must be made in the districts in which they lie.

In all cases where mining claims lie partly in one land district and partly in another, applications for patents therefor should be filed in that district where the principal workings of the claim are situated, as shown by the plat and field-notes; and the diagrams and notices should be posted near to such workings. A copy of the notice and of the diagram should be posted in the register's office in each district.

The notice posted in the office of the register, where the *application for patent is not filed*, should state where the application for patent for the premises therein described has been filed, and the date of filing of such application.

To the end that the applicants may be to as little expense as possible in the matter of publishing the notices required by law, the publication may be made of the several notices of intention to apply for patents for the mines in one advertisement, wherein will be accurately described the premises embraced by each of the applications which may be filed in the office.

This advertisement will of course be published for the period of time required by law. If published in a weekly paper, the

---

[1] In re Zella Lode, Decision of Acting Commissioner, June 9th, 1873, Copp's U. S. Mining Decisions, 202.
[2] Rev. St. 2333. Sec. 11, Act of May 10th, 1872.

advertisement must be inserted in ten consecutive issues of the paper; but if published in a daily paper, sixty days must elapse between the first and last insertions of the advertisement.[1]

§ 100. Delaying action at request of Congressional committee.—Where a mining company was prepared to establish, to the satisfaction of the Department, their claim, under the Statute of 1866, to receive a patent for certain lands in California, they were held to have the legal right to have the question of their claim to such patent passed upon. The Department is bound to consider and determine the same, notwithstanding requests from Congressional committees for a suppression of action.

If, under the law, parties have the title and are prepared to furnish the proper proofs of it, the law gives them the right to a patent, and the issuance of a patent is not made discretionary with the executive officers of the Government. When a right is created by law and a duty devolved upon an executive department under the same law, the enjoyment or enforcement of such right cannot be suspended at the request of a committee of Congress, and probably could not be by the action even of both Houses of Congress, by any means short of a change in the law itself. The Department can only pay attention to such requests when it affects a discretionary power.[2]

§ 101. The affidavit—Proper party to make it.—When the original locators make the application for patent, then one of them must make the affidavit required by the statute, but when the original locators have assigned their interests, and the application is made by the assignees, then the assignees are the claimants, and one of them may make the affidavit.[3]

It was so held in the case of the Kempton mine, where it was objected that there was no affidavit of the *proper party*

---

[1] In re Lake Quicksilver Mining Co. Decision of Commissioner, Nov. 12th, 1875, 2 Copp's Land-owner, 130.

[2] In re New Idria Mining Co. Opinion of the U. S. Attorney-General, June 22d, 1869, 13 Opinions of the Attorneys-General, 112.

[3] In re Kempton Mine, Decision of Secretary of Interior, January 2d, 1875, 1 Copp's Land-owner, 178; Rev. Stats. 2325.

that the plat and notice were posted in a conspicuous place on the claim during the period of publication. The sixth section provides that "at the expiration of the sixty days of publication *the claimant* shall file his affidavit showing that the plat and notice have been posted in a conspicuous place on the claim during said period of publication." It was argued that the claimant referred to was one of the original locators. But this was held not to be necessarily the case, and the doctrine above stated was applied, and the applicant held to be a claimant within the purview of the law, although not an original locator, but an assignee.[1]

§ 102. Verification of affidavit.—Section 2335 of the Revised Statutes is as follows: " All affidavits required to be made under this chapter may be verified before any officer authorized to administer oaths within the land-district where the claims may be situated, and all testimony and proof may be taken before any such officer, and, when duly certified by the officer taking the same, shall have the same force and effect as if taken before the register and receiver of the land-office. In cases of contest as to the mineral or agricultural character of land, the testimony and proofs may be taken as herein provided on personal notice of at least ten days to the opposing party; or if such party cannot be found, then by publication of at least once a week for thirty days in a newspaper, to be designated by the register of the land office as published nearest to the location of such land; and the register shall require proof that such notice has been given."[2]

There must be an application *under oath*. An application simply signed by one as president and another as secretary of a company, and not sworn to, is not sufficient.[3]

Neither the register nor receiver has authority to deputize

---

[1] In re Kempton Mine, Decision of Secretary of Interior, January 2d, 1875, 1 Copp's Land-owner, 178.

[2] Rev. Stat. 2335; Sec. 13, Act 1872, 17 U. S. Stat. 95. See Rev. Stats. 2321. Sec. 14 of the Act of 1870, 16 U. S. Stat. 217; read Sec. 14: That all *ex parte* affidavits required to be made under this act, or the act of which it is amendatory, may be verified before any officer authorized to administer oaths within the land district where the claims may be situated.

[3] Jefferson M. Co. v. Penn M. Co.; In re Penn Quartz Mine, Decision of Commissioner, July 21st, 1874, 1 Copp's Land-owner, 66.

any person to administer oaths, and papers sworn to before any person purporting to act as deputy for either register or receiver cannot be received as evidence.[1]

§ 103. **The location notice—Its sufficiency.**—In the inquiry into the regularity of the proceedings prior to the application for a patent, the location notice is one of the first papers demanding the supervision of the Department.

In the case of the Prince of Wales Lode[2] the Secretary of the Interior took occasion to remark upon the latitude allowed in the construction of such notices, and said: "It should be borne in mind that the discovery of lodes, and the preparation of location notices for the same, are generally made by unlettered men, and it would be productive of great hardship, and perhaps generally result in an entire loss of their valuable discoveries, if they were held to technical accuracy in their notices of location. Accordingly, it has been uniformly held by the courts and the Department that extreme liberality should be shown to these notices, and if they were sufficiently certain to put an honest inquirer in the way of ascertaining where the lode was, that was sufficient."

Accordingly, many location notices, neither very certain nor regular in form, have been held sufficient by the Department, especially in the absence of adverse claims.[3] But the sanction

---

[1] Decision of Acting Commissioner January 27th, 1876, 2 Copp's Land-owner, 162.

[2] Decision of Secretary of the Interior, April 1st, 1875, 2 Copp's Land-owner, 2.

[3] *What is a sufficient notice of location*—In the case of the Prince of Wales Lode, (Decision of Secretary of the Interior, April 1st, 1875, 2 Copp's Land-owner, 2) a notice of location in the following form was held sufficient:

"The Prince of Wales Lode.

"Discovered by Thomas E. Owens, August 1st, 1870. We, the undersigned, in company and undivided, claim 1,200 feet on the above lode or mass of ore, or whatever it may contain, 200 feet for discovery and 1,000 feet for location along this vein, wherever it may run, together with all dips, spurs, angles, and variations, with all the privileges granted by the laws of the district and the Congressional laws of the United States. This lode is situated on the right-hand fork of the creek known as Silver Fork, within about 200 feet in a southeasterly direction of the lode called the 'Antelope,' in Big Cottonwood cañon, and now supposed to run in a southwesterly and northeasterly direction.

"Discovery—Thomas E. Owen, 400; H. W. Bishop, 200; T. Robinson, 200; J. J. Dussain, 200; H. Burnette, 200;" See, also, 420 Mining Co. v. Bullion Mining Co. Decision of Secretary, March 22d, 1875, 2 Copp's Land-owner, 5.

Notices and sworn statements.—Where there was no adverse claim filed, and

no sworn statement disputing the sworn proof filed by the applicant, the following notice and sworn statement was held sufficient for a location previous to the passage of the Mining Acts:

"Notice is hereby given that we, the undersigned, have located, and claim ten (10) claims of two hundred (200) feet each on this ledge or lode of precious-bearing metals, running from this notice in a southerly direction, two thousand (2,000) feet. This company shall be known as the South Comstock Gold and Silver Mining Company. This is a re-location of the Lady Adams Company, dated March 28th, 1860."

Then follow the names of the locators, and the number of feet claimed by each.

On the 15th of May, 1872, the company was incorporated under the laws of the State of California, and on the following day one George W. Rodgers, who had purchased the entire interest of his co-locators, conveyed said mining premises to said Company.

In the sworn statement of the Superintendent of the South Comstock Gold and Silver Mining Company, accompanying said application for patent, he alleged that "said Company has become the owner of, and is now in the actual, quiet, and undisturbed possession of, and entitled to the possession of, two thousand (2,000) linear feet of the Comstock lode, vein, or deposit, bearing gold and silver, with surface ground for the convenient working thereof, as allowed by the local customs and rules of miners; said mineral claim, vein, lode, and deposit, and surface ground, being situate in the Devil's Gate Mining District, in the counties of Lyon and Storey, in the State of Nevada, and being more particularly set forth and described in the official field-notes of survey thereof herewith filed, dated the 7th day of April, 1874, and in the official plat of said survey." He also alleged that said "company and their ancestors, grantors, and predecessors in interest have held, occupied, and improved said claim, and maintained the actual undisputed possession thereof, from the date of said record of location, to wit: the 30th of April, 1872, to the present date."

In the sworn statement of G. W. Rogers, filed by the attorneys for the patentees in the office, with their argument in this case, he alleged " that said ledge has very large, heavy, and prominent croppings on the surface indicating its course or direction, said croppings being in some places at least 200 feet in width, the greatest width of the ledge so claimed by said South Comstock Gold and Silver Mining Company being shown by the croppings at 'Devil's Gate,' near the southern end of said claim, where the out-croppings are at least sixty feet high, and not less than 400 feet in width." (In re South Comstock Gold and Silver Mining Co., Decision of Commissioner, December 29th, 1875, 2 Copp's Land-owner, 146. See Decision of Secretary of the Interior, March 22d, 1875, 420 Mining Co. v. Bullion Mining Co. 2 Copp's Land-owner, 5; Decision of April 1st, 1875, In re Antelope, Prince of Wales, and Wandering Boy Mines, 2 Copp's Land-owner, 2; Decision of Commissioner, Jan. 18th, 1875, In re Red Pine Mine, affirmed by Secretary of the Interior, March 8th, 1875, 1 Copp's L. O. 135; 2 C. L. O. 56.

The case of "The Antelope Lode."—The location of this lode was made on the 15th of June, 1870, and recorded June 18th, 1870. It was as follows:

"The Antelope Lode, June 15th, 1870. Miners' Notice—We, the undersigned, claim three thousand (3,000) feet in this ledge or lode, with all its dips, angles, spurs, and variations, to be known as 'The Antelope Lode.' Also, two hundred (200) feet discovery, running one thousand (1,000) feet easterly, two thousand (2,000) feet westerly direction, situate at the head of the first south fork below mill known as 'Mill F,' in the right-hand fork of said fork."

DISCOVERY, (Signed by 15 locators.)

Application for patent was made on the 30th of December, 1873. Publication was made in the Salt Lake *Tribune,* commencing on the 4th of January, 1874. The owners of the Wellington Lode filed, on the 4th of March, 1874, an adverse claim.

On the 18th of August, 1874, the applicants for patent of the Antelope Lode filed in the General Land Office an abandonment in writing of all that portion of their claim covered by the adverse claim of the Wellington. The commissioner thereafter treated the Wellington claim as out of the case, and thereupon informally decided that the applicants were entitled to patent; and on the 26th of August, 1874, a patent was issued for the Antelope Lode, excluding the premises claimed by the Wellington. No notice of this decision was given to the Wellington or their attorneys. They claimed that they should have had notice, and that they had the right of appeal to the Department, which right had been cut off by the neglect to give them notice, and by the issuance of the patent. They alleged that they desired to appear and protest against the issuing of the patent to the claimants. They claimed that the location-notice did not describe the claim as minutely as the local laws required, in that it did not name the starting point, and did not show that the locators marked their claim with stakes or hillocks, with the names of the claimants on a distinctly written notice; that they had failed to show that the locators had done twenty-five dollars worth of work within ten days after recording their claim; that they failed to show that one of the locators, not one of the applicants, was a citizen of the United States; that the required amount of improvement and expenditure was not done on the claim, but was done on another—the Prince of Wales; that the notice and diagram were posted on the Prince of Wales, instead of the Antelope; that they failed to show that the publication-notice was given in a paper designated by the register; that no final survey of the claim as patented was made; that the claim was floated; and above all, that the filing of the adverse claim required that *all proceedings* should be suspended until after the judgment of the Court had been rendered.

The Antelope Lode, as originally located, covered a portion of the premises afterwards included in the patent to the Prince of Wales Lode. Those portions of the claim not so included, and perhaps others, were patented to the Antelope claimants.

The question was presented, whether the irregularities were such as to require the Department to institute proceedings to set aside the patent.

The Secretary said: "If there are no adverse interests, then it seems to me that there is no good ground for interference with the patent. There is no pretense that any adverse interests have been injuriously affected, except those of the Wellington claimants."

Requirements of re-location certificate.—In Philadelphia Lode *v.* Pride of the West Lode, Colorado, a location was sustained under the following state of facts: It was shown, by certified copies of the notices of location, that the Pride of the West lode was located June 10th, 1874, and record made of such location on the 19th of the same month, and again located August 7th, 1874, and recorded on the same day.

On the 18th day of September, 1874, an additional certificate of location was recorded. In this notice it was stated that they claimed the Pride of the West lode according to the survey made the 16th September, 1874, by T. M. Trippe, U. S. Deputy Surveyor, as follows, viz: "Running from the discovery tunnel S. 20 deg. 30 min. E. (Mag. Var. 14 deg. 30 min.) 922 feet, thence S. 6 deg. 30 min. W. (Mag. Var. 14 deg. 30 min.) 578 feet, being 1,500 feet linear and horizontal measurement along the surface of the lode, with 150 feet in width on each side of the center line thereof."

The first two location notices recorded did not give the courses along the line of the premises claimed, but the last one did.

§ 103   PATENTS TO MINERAL LANDS.   157

The act of the Colorado legislature concerning mines, which was approved February 13th, 1874, and which went into force June 15th, 1874, provides in the thirteenth section thereof that "if at any time the locator of any mining claim heretofore or hereafter located, or his assigns, shall apprehend that his original certificate was defective, erroneous, or that the requirements of the law had not been complied with before filing, or shall be desirous of changing his surface boundaries, or taking in any part of an overlapping claim which has been abandoned, * * * such locator or his assigns may file an additional certificate subject to the provisions of this act," etc.

It appeared that the locators of the Pride of the West lode made an additional certificate, based upon an actual survey executed by Deputy Surveyor Trippe, and filed their additional certificate dated 18th September, 1874, for record in the office of the County Clerk and Recorder of La Platte County, Colorado, September 18th, 1874.

The Deputy Surveyor in his sworn statement alleged that when he made the survey of September 16th, 1874, he "placed six posts upon the claim, in full compliance with all the requirements of the law."

It was urged that, as the additional location certificate did not state the land or mining district, the county, State, or Territory, in which the claim was located, and failed to state the date of the location, the same was void, and not made in accordance with law.

This location notice was signed by all the locators, was dated September 18th, 1874, and recited the fact that it was a *re-location* notice, and that the claim was situated on a certain mountain. This notice was signed by all the original locators, was recorded the same day that it was dated, in the same book of records that the two former locations of the claim were recorded.

In both of the former notices, the mining district, county, and Territory were stated, also the date of the location of the claim.

This re-location certificate, based upon an actual survey made by a U. S. Deputy Mineral Surveyor, was held to be made in conformity with the provisions of the local laws and Congressional enactments.

The objection to the *location* of said claim was accordingly overruled.

Philadelphia Lode *v.* Pride of the West Lode, Decision of Commissioner, Aug. 28th, 1876, 3 Copp's Land-owner, 82; Decision of June 10th, 1876, Ibid.

Re-locations in Colorado.—The Act of Colorado, approved February 13th, 1874, declares the manner in which re-locations may be made in that Territory. Where re-locations are made under the Colorado act, by parties who have the possession and the right of possession to a mining claim, for the purpose of changing the surface boundaries, increasing the width of surface ground, or other reasons, the parties who apply for patents for such mines should file a copy of the original notice of location, an abstract of title tracing the record title from the original locators to the re-locators, a copy of the re-location notice, and an abstract of any transfers made of the re-location. Where parties make applications for patents for mines which have been re-located as *abandoned*, they should file with their application a copy of the re-location notice, and an abstract of all the transfers thereunder. They should also file proof, *full, positive,* and *complete*, in regard to the abandonment of prior location, setting forth the facts necessary to show such abandonment. Where a party applies for a patent for a mine to which he claims the right of possession by reason of the fact that co-claimants have failed to contribute their *pro rata* share of the amount required by law to be expended annually, the applicant must file with his application for patent and other proofs, a copy of the original notice of location, an abstract of all transfers thereunder, and proofs that provisions of the fifth section of the Mining Act of May 10th, 1872, were fully and strictly complied with by the

given to these informal notices was generally in cases of locations prior to the passage of the mining acts. Since these acts there are certain requirements that *must* be complied with.[1]

§ 104. Parol evidence admissible to aid the notice of location.—That parol evidence is admissible to aid in the location of a mining claim, and define what tract is embraced in a location, appears to be well settled. Where the testimony of four deputy mineral surveyors, and four others, their attendants, had been filed, and showed that they had made a careful survey of the premises, and found that the location-application and patent were for substantially the same premises, these persons, having the means of knowing, and having no motive for misstating the facts, an objection that the application for patent and the final survey and patent did not conform to the original location, was considered not supported by the weight of the evidence, and was overruled.[2]

§ 105. The plat must show accurately the boundaries of the claim.—A Surveyor-General of the State will not be instructed to correct the plat and field-notes of his survey of the mining premises claimed by the applicant, and for which a patent is requested, in order that said plat and field-notes may be intelligent and in conformity with law. If the parties have complied with the law, no corrections are necessary. If they have not, the application is invalid.

Where neither the plat nor field-notes showed the exterior boundaries of the claim, nor their length, it was held that they did not show "accurately the boundaries of the claim," especially as no evidence was furnished that the exterior boundaries

---

party or parties who had made the required expenditures and improvements, and given notice thereof in due form to the claimants who had failed to contribute. The proof in cases of abandonment and notice to co-claimants must be clear, positive, and in strict compliance with the statutes.

(Letter from the Acting Commissioner of the General Land Office, to the Register and Receiver at Central City, Colorado, dated April 21st, 1876. Application of Hazen Cheney, 3 Copp's Land-owner, 37.)

[1] See Ante, Sec. 57.

[2] Instructions, Commissioner General Land Office, November 20th, 1873; Kelly v. Taylor, 23 Cal. 11.

of the claim had been "distinctly marked by monuments on the ground."[1]

§ 106. **Surveys should show exterior boundaries.**—In all cases the plat and field-notes of survey should show the exterior boundary of the claim for which an application for patent is made. The width of the claim, as represented upon the plat and described in the field-notes, in no case should exceed the amount of surface ground allowed by local laws and customs.[2]

§ 107. **Specific surface ground.**—In the case of the 420 Mining Company v. The Bullion Mining Company,[3] it was claimed that the application was invalid because it failed to designate the specific surface ground claimed. The second section of the Act of 1866, under which the application was made, provided that the diagram filed should be " so extended, laterally or otherwise, as to conform to the local laws, customs, and rules of miners." The application, of which the diagram was in fact a part, must have alleged the claim as required by such local laws, customs, and rules, and if no surface ground was provided for, a failure or omission to state the amount claimed, by specific description, was not a defect in pleading. The Secretary of the Interior, in deciding the case on appeal, said:

" The claim under consideration was located June 23d, 1859, the entire premises claimed being then within the limits of the Gold Hill Mining District, having a regularly adopted code of laws. This district was subsequently, during the year 1859, divided, and the now Virginia District created therefrom. This last-mentioned district adopted a code of laws September 14th, 1859. The Bullion claim lies partly within each of these districts. "Upon a careful examination of the laws of both Gold Hill and Virginia Districts I am unable to find any provision whatever, giving surface ground to quartz or ledge claims. Sec. 13 of the Gold Hill laws is as follows, viz: 'All quartz claims shall not exceed 300 feet in length, including the dips and spurs.'

---

[1] Decision of Commissioner, January 6th, 1874, Copp's U. S. Mining Decisions, 340; Rev. Stat. 2325, Sec. 6, Act of 1872.
[2] Decision of Commissioner, Sept. 11th, 1873, Copp's U. S. Mining Decisions, 223.
[3] Decision of the Secretary of the Interior, March 22d, 1875, 2 Copp's Landowner, 5.

"Article 1 of the Gold Hill laws is as follows, viz: 'All quartz claims hereafter located shall be 200 feet on the lead, including all its dips and angles.' There are no other provisions whatever relating to the quantity or extent of lode claims, and no provisions relating in any way to surface ground, except such as are evidently intended to apply to placer locations. It is clear, however, that it was the custom or rule in these districts to take as surface for quartz claims all the ground lying between the two walls of the lode.

"The application under consideration expressly alleged that the local laws, customs, and rules did not permit any surface ground to be occupied except the surface of the vein or lode; that the walls of the lode were at that time unascertained and unascertainable, but that the surface of the premises claimed was the surface of said lode, estimated at 111 acres. The lode itself, so far as known, was accurately described in both application and diagram.

"I think this application describes the claim as the law required it should be described. It literally follows your instructions issued under the act then in force. That act required the diagram to be so extended as to conform to local laws, and the local laws allow just such a claim as is described, and none other. "I am not entirely clear that to have been more explicit would have made the claim liable to the very objection now urged."

§ 108. Posting on claim, proof of.—The applicant must post a copy of the plat, together with a notice of the application for a patent, in a conspicuous place on the land embraced in such plat, previous to the filing of the application for a patent, and shall file a copy of the notice in the land office. The plat must be accompanied with the notice of intention to apply for a patent therefor; the notice must give the date of posting, the name of the claimant, the name of the claim, mine, or lode, the mining district and county, whether the location is of record, and if so, where the record may be found; the number of feet claimed along the vein and the presumed direction thereof; the number of feet claimed on the lode in each direction from the point of discovery, or other well-defined place on the claim:

the name or names of adjoining claimants on the same or other lodes; or if none adjoining, the names of the nearest claims.[1]

In the case of the Kempton Mine, it was objected that there was no sufficient proof that the plat and notice were posted in a conspicuous place on the claim. The objection was not that there was not a plat and notice posted on the claim, but that there was no proof *what plat and notice* were thus posted. There were on file, however, the affidavits of numerous persons that the plat and notice were posted in a conspicuous place on the claim, and so remained during the entire period of publication. They did not specify particularly what plat and notice, because at the time they were taken there was no controversy about their contents; but they did show that a plat and notice, which they all understood to be in due form, were properly posted. The objections were overruled.[2]

§ 109. Publication of the notice.

—The register of the land office, upon the filing of the application, plat, field-notes, notices, and affidavits, shall publish a notice that such application has been made, for the period of sixty days, in a newspaper to be by him designated as published nearest to such claim; and he shall also post such notice in his office for the same period.[3]

The publication is at the expense of the claimant. In all cases sixty days must intervene between the first and last insertion of the notice in the newspaper. The notices published and posted must be full and complete, and embrace all the data given in the notice posted upon the claim. The greatest care should be exercised in the preparation of these notices, inasmuch as upon their accuracy and completeness will depend, in a great measure, the regularity and validity of the whole proceeding. After the sixty days' publication has expired, the claimant will file his affidavit showing that the plat and notice remained conspicuously posted upon the claim sought to be patented during the sixty days of publication.[4]

---

[1] Rev. Stat. 2325. Instructions Feb. 1st, 1877, Subdivision 29.

[2] In re Kempton Mine, Decision of Secretary, Jan. 2d, 1875, 1 Copp's Landowner, 178.

[3] Rev. Stat. 2325.

[4] Instructions Feb. 1st, 1877, Subdivisions 34–40.

W. C.—11.

Published notices must be numbered to correspond with the record of applications, and instead of being headed "Lode Notice," etc., the words "Mining Application No. —" should be used, inserting the number of the application.[1]

The notice must be published with the knowledge of the register, and in a newspaper designated as published nearest the claim, or the application will be rejected.[2]

The publication of notices may be in newspapers published weekly, but must be for the full period of sixty days. A publication in a weekly paper for nine successive weeks (nine insertions) is not a publication "for the period of sixty days." The publication must, moreover, be in only one newspaper for the prescribed period.[3]

An objection that the publication is not according to law in point of time, should be made before patent. It is considered too late afterwards. And so with an objection that the proof of posting the notice and diagram on the claim did not show when, where, or for what period the same was posted.[4]

A clerical error, as issuing the final certificate of entry to Thomas Butterfield, and issuing the patent to Thomas Butterwood, the true name, and the one mentioned in the application and notice, does not invalidate the patent.[5]

Character of the register as agent for the applicant.—For the purpose of preparing and publishing the notice, the register acts as the agent for the applicant, and it is the latter's duty to see that the officer substantially complies with the law. The object of notice is to advise those who may have adverse interests of the pending of a claim that may affect their interest, and it will not answer to say that the applicant shall not be responsible for the notice, or that he shall receive his patent, although the officer neglected to do his duty.[6]

---

[1] Decision of Acting Commissioner, March 7th, 1876, 2 Copp's Land-owner, 180.
[2] In re Cascade Lode, 1 Copp's Land-owner, 50.
[3] McMurdy v. Streeter, 1 Copp's Land-owner, 34; In re Northern Light and Fair View Mines, 1 Copp's Land-owner, 34; In re Secret Cañon Quartz Mine, Decision of Commissioner, November 12th, 1873, Copp's Mining Decisions, 234.
[4] Prince of Wales Lode, Decision of Secretary, April 1st, 1875; 2 Copp's Land-owner, 2.
[5] Ibid.
[6] In re Flagstaff Lode, Opinion of Assistant Attorney-General, November 24th, 1871, Copp's U. S. Mining Decisions, 70, 71, 72.

§ 110. Time of publication.—The first day should be excluded, and the last included in the computation. Where an affiant showed that the notice of intention to apply for a patent was published "from January 7th to March 7th, 1874, inclusive," the notice being published twenty-four days in the month of January, twenty-eight days in February, and seven days in March, making fifty-nine days in all, it was held insufficient.[1]

A case arose where the Register directed the notice to be published in a newspaper called the *Daily Herald* for sixty days. The notice was published in said paper only on the 28th, 29th, 30th and 31st of December—four insertions. The notice was then discontinued in the *Daily Herald* by instruction of the applicant, and inserted in the *Weekly Herald*. In the *Weekly Herald* the notice was inserted from the 6th of January to the 2d of March, 1876.

In the case of the Jenny Lind Mining Co. et al. *v.* Eureka Mining Co., the Secretary of the Interior held that in estimating the sixty days of publication required by the act of May 10th, 1872, the first day of publication should be excluded and the last included.

In the case of J. H. McMurdy et al. *v.* E. S. Streeter et al., the Secretary of the Interior held that "the time elapsing between the first and last insertions must include the full period of sixty days." From the 6th of January to the 2d of March, excluding the first day, being only fifty-six days, the publication, therefore, in the *Weekly Herald* was held not sufficient, even though the notice had been inserted therein by direction of the register.

The applicants, therefore, did not give sufficient notice by publication, having published the notice for four days only in the paper designated by the officer to whom the law has delegated the power to authorize the publication of notices in case of applications for patent for mining claims.

The second notice was inserted in the *Weekly Herald* without authority of the register, and for fifty-six days only.

---

[1] Decision of Commissioner, July 21st, 1874, Jefferson Mining Co. *v.* Pennsylvania Mining Co. In re Penn. Mine, 1 Copp's Land-owner, 66; Jenny Lind Mining Co. *v.* Eureka Mining Co. Decision of Secretary of the Interior; Ibid. Prince of Wales Lode. Decision of Secretary, April 1st, 1875, 2 Copp's Land-owner, 2.

The statute having in this material requirement been disregarded, the publication as made, and all subsequent proceedings founded upon it, were held irregular and invalid, and the application for patent was accordingly rejected.[1]

§ 111. Counting the sixty days for publication of notice.—The notice of intention to apply for a patent must be published for the period of time required by law.

In computing the time for the sixty days' publication, the date of the paper as given thereon governs, even though, as a matter of fact, the paper may have been previously issued and put in circulation.

The first day of publication is excluded, and the last included in the computation. This much-vexed question in regard to the inclusion and exclusion of the first day, appears to be settled as follows: When the computation is to commence from an *act done*, the day on which the act is done is excluded.[2]

The cases also establish the proposition, that where there is a doubt as to whether the day in which an act is done should be included or excluded, that construction should be adopted which will support a contract or deed, rather than that which would destroy it; that which will prevent a forfeiture, rather than create one—and in cases of statutory enactment, that which will be most favorable to the party for whose benefit the statute was enacted. The provision that there should be a publication of sixty days was made for the benefit of adverse claimants, and for the purpose of giving them an opportunity to assert

---

[1] Decision of Acting Commissioner, April 29th, 1876, 3 Copp's Land-owner, 18.
[2] In re Eureka, Montana, Excelsior, King David, and May Henrietta Lodes, Jenny Lind Mining Co. v. Eureka Mining Co. Decision of Secretary, Nov. 24th, 1873, Copp's U. S. Mining Decisions, 169. Opinion of Assistant-Attorney-General, Sept. 30th, 1873, Ibid, 170. See, also, Griffith v. Bogert, 18 How. U. S. 162; 4 Kent, 105, note, 11th Ed.; 2 Parson's Cont. 663, note; Pope v. Headen, 5 Ala. 433; Lyon v. Hunt, 11 Ala. 295; Lang v. Phillips, 27 Ala. 311; Kimm v. Osgood, 19 Mo. 60; 25 Miss. 48; Bigelow v. Willson, 1 Pick. 485; State v. Schwerle, 5 Pick. 279; Wiggin v. Peters, 1 Met. 127; Farwell v. Rogers, 4 Cush. 460; Weeks v. Hull, 19 Conn. 376; Carleton v. Byington, 16 Iowa, 588; Carothers v. Wheeler, 1 Oregon, 194; Judd v. Fulton, 10 Barb. 117; Bissell v. Bissell, 11 Barb. 96; Cornell v. Moulton, 3 Denio, 12; Barr v. Lewis, 6 Texas, 76; State v. Gasconade Co. Ct. 33 Mo. 102; Cann v. Warren, 1 Houston, Del. 188; Gorham v. Wing, 10 Mich. 486; Sheets v. Selden, 2 Wall. 177; Page v. Weymouth, 47 Maine, 238; Walsh v. Boyle, 30 Md. 262; Thorne v. Moshor, 20 N. J. Eq. 257; Gorst v. Lowndes, 11 Sim. 434; Wilkinson v. Gaston, 9 Queen's B. 141.

their adverse claims; and in case of doubt as to whether the first day of publication should be included or excluded, that doubt should be decided in their favor.[1]

Where notice of the application is published in one paper a portion of the sixty days, and in another paper for the balance of the time, the notice is not published according to law. The law must be strictly complied with. In such case the applicant must commence *de novo*, although no new survey is necessary.[2]

§ 112. **Proof of publication.**—Where it was objected that the proof of publication did not state the last day of publication—an affidavit of the editor of the paper in which the notice was published stated that "the attached notice was published in the Salt Lake *Review* for a period of ninety days, commencing August 15th, 1871." This was held sufficient prima facie proof. And where it was objected that the notice and diagram were not posted on the claim until five days after they were filed in the Land Office, and five days after the publication had been commenced, the proof showed that they were posted on the claim for more than ninety days, it was held that they should have been posted before the publication; but the omission was an irregularity only, and was not fatal.[3]

§ 113. **The newspaper in which the notice is to be published.**—The register must publish the notice in a "newspaper to be by him designated as published nearest to such claim." His duty is to designate it, and it should be the paper published nearest to the claim. The public have a right to look to the paper published nearest the claim as the one in which a notice of application for a patent should appear. If two or more papers of repute are published equidistant, or very nearly so, from the claim, the register must designate the one in which the notice shall appear; but in other cases the paper published nearest the claim must be designated, provided the same is a reputable newspaper of general circulation. If such

---

[1] See cases cited Ante.

[2] In re Secret Cañon Quartz Mine, Decision of Commissioner, November 12th, 1873, Copp's Mining Decisions, 234.

[3] Wandering Boy Lode, Decision of Secretary, April 1st, 1875, 2 Copp's Landowner. 2.

a paper should be published within two miles of the claim, and another should be published in a town six miles from the claim, the register has no right or discretion to choose the latter.[1]

In the case of the Omaha Gold Quartz Mine[2] it was objected that the notice was not duly published, having been published in a newspaper called the *Nevada Transcript*, Nevada City, California, instead of in the *Grass Valley Union*, published at Grass Valley, California, and that the notice and diagram were not posted *conspicuously* upon the claim. By the affidavits of the superintendent of the mine, and eighteen other persons, it was shown that the notice and diagram were posted in the most conspicuous place upon the claim, near the center thereof, upon a prominent point, about eight feet south of the main traveled trail leading across the mine, and that the notice and diagram could be seen at a distance of more than five hundred feet. The evidence upon this point was considered satisfactory. The notice in the case was published in the *Nevada Transcript*, a weekly paper published at Nevada City, California, by direction of the register, whose duty, under the law, it is to publish the notice " in a newspaper to be by him designated as published nearest to said claim." " It is true," said the acting Commissioner, " that Grass Valley is nearer to said mine than Nevada City is. Both towns are situated, however, in the same township, and but a few miles apart. By the sworn statement of Samuel Bethel, U. S. Deputy Mineral Surveyor, residing in Nevada County, California, it appears that the *Transcript* is the official newspaper of Nevada County, and has a large circulation in Grass Valley, and that nearly all official notices are published therein. The publication of notice in this case is deemed satisfactory, and for the following reason, viz: It was published in the paper designated by the register, and in a newspaper of general circulation published near to said mine." The objections urged against the sufficiency and formality of posting and publishing the notices were overruled by the Commissioner, but sustained by the Secretary, on appeal, and the doctrine enunciated at the head of this section adopted.

---

[1] Decision of Secretary, December 1st, 1876, 3 Copp's Land-owner, 163; In re Omaha Quartz Mine.

[2] In re Omaha Gold Quartz Mine, Decision of Acting Commissioner, May 12th, 1876, 3 Copp's Land Owner, 36.

It is suggested that all departures from the strict letter of the law in this matter of publication are dangerous, and that a strict compliance with the terms of the act is the only proper course for the register or the applicant.[1]

§ 114. **Defects in the published notice.**—Where the published notice failed to give the bearing from the meridian, and left it uncertain whether that bearing should be east or west, but so far as it went agreed with the application, and the notice stated that the company had filed in the office a diagram, together with a notice of an intention to apply for a survey, etc., and this diagram and notice contained a true description, it was held that these defects were not of so material a character as to require the published notice to be set aside.[2]

§ 115. **Discrepancies between final survey and patent and the original application and published notice.**—The object of requiring notice to be given by publication is to inform all parties, who may have an adverse interest, of the premises sought to be acquired, so that they may appear and assert their rights. If the notice describes premises in which others have no interest, then such other persons may safely neglect to appear and set up any claim. They are bound by the notice, and

---

[1] *Paper printed partly in one district and partly in another.*—There is, it is held, no objection or impropriety in a newspaper proprietor's issuing his paper with one side of it entirely blank, or filled with matter printed in another city or State. In case one side of the paper is printed, and the paper is published in a given town in the district, the notice should be published in the newspaper "published nearest to such claim." Many of the papers published in sparsely inhabited parts of the country are printed on one side in another city or State, while the other side is filled with local news or advertisements. The object of publishing the notice is to notify all whose rights might be prejudiced by the issuance of a patent as applied for, in order that they may present their objections. This can best be accomplished by publishing the notice in a paper *published* nearest the claim, and in a paper of general circulation in that vicinity. Where there were two papers published in the district, and each was printed on one side in the city of St. Louis, Mo. while the other side of each issue was printed in the district, the one published nearest to the claim was pointed out as the proper paper, and it was held sufficient, though part of it was published out of the State. Decision of Commissioner, January 4th, 1877, 3 Copp's Landowner, 196.

[2] In re Flagstaff Lode, Decision Attorney-General, November 24th, 1871, Decision Secretary Interior, December 5th, 1871, reversing Decision Commissioners, November 10th, 1871, on this point. Copp's U. S. Mining Decisions, 61, 70, 71.

if they neglect it, they must do it at their peril; but the moment they find that the notice does not ask for anything in which they have an interest, that moment they may safely sleep, if they please. They are not bound, and should not be bound, to look after subsequent proceedings for fear that there may be a subsequent claim set up to their property. There can be no subsequent claim that varies materially from the original one, which is embodied in the application and publication. The law must be followed. The proceeding is a special statutory proceeding, and all the provisions of the law must be carefully, and, as some authorities say, strictly pursued. Actual notice without publication will not answer. Written notice would not be sufficient, because the statute says that there must be notice by publication.

If the published notice described certain premises, none other can be afterwards claimed and appropriated without a new application and new published notice, and if there should be a subsequent effort to include premises other than those included in the original application and notice, and an adverse claimant should appear and assert his claim to the new premises thus sought to be appropriated, and should fail in maintaining his claim, either by reason of not filing the same in time or for defect in form, he would not be thereby in any worse position than he would have been if he had not appeared at all. In this class of cases consent cannot give jurisdiction. It is a substantial compliance with the statute which alone can give jurisdiction.

In the case of the Prince of Wales v. The Highland Chief Mine, Utah,[1] it was claimed that the final survey and patent of the Highland Chief did not follow the original application and notice, and that the claim was floated to the eastward so as to include the discovery and works of the Prince of Wales.

The Highland Chief was located September 12th, 1870. In the location notice the lode was described as "commencing at the discovery stake and running 600 feet in a *southerly* direction, and 600 feet in a *northerly* direction therefrom. * * * Situate about five or six hundred feet *westerly* from the Young

---

[1] Decision of Secretary of Interior, April 1st, 1875, 2 Copp's Land-owner, 2; Ibid. 43.

Columbia and Wandering Boy lodes, Big Cottonwood District, Utah Territory."

A location 500 or 600 feet westerly from the Wandering Boy lode, would have excluded the premises in controversy.

The diagram of the Highland Chief, attached to its application for a patent, represented the Prince of Wales and Wandering Boy lodes as lying to the east of the premises claimed by the Highland Chief. The application and publication notice both alleged that, "from discovery shaft the lode extends northeasterly six hundred (600) feet and southwesterly therefrom six hundred (600) feet. There were no known adjoining claimants at either end; the nearest known claims being the Prince of Wales and Wandering Boy Mines, on *the easterly side of said lode*."

Under such an application and published notice, it was considered clear that the applicants had no right to go to the eastward, so as to take in and appropriate the mines which they alleged were on the "easterly side" of their lode.

The location, application, and published notice of the Highland Chief severally excluded the premises of the Prince of Wales Mine. It further appeared, from the testimony on file, that the owners of the Highland Chief, in the early stages of their proceedings for patent, did not intend to include the Prince of Wales Mine.

The surveyor who made their original diagram, testified that he was instructed to avoid the Prince of Wales Mine, and that he did so. There was nothing in the case that indicated any intention on their part to appropriate it, until after they discovered that the final survey might be construed to include it, and they had succeeded in excluding its adverse claim. It was held an error to include it in their patent.

§ 116. Discrepancies between the published notice and diagram filed.—Where a notice as published was consistent with the application as far as it went, but failed to state the courses and distances in full, and omitted one of the bearings from the meridian, yet it stated that the company had, on a certain day, filed in the office "a diagram of the same, together with a notice of intention to apply for a survey," etc., and the diagram and notice referred to contained a true description of the premises, the maxim of *id certum est, quod certum reddi potest* was applied, and the reference to the diagram and notice filed was held to cure the defects of the published notice.

It may be questioned whether this doctrine will not be found dangerous as a precedent. And as in the particular case it was not necessary to declare such a doctrine, inasmuch as the application was rejected on other grounds, it is much in the nature of a *dictum*.[1]

The notice is in the nature of a summons, by which opposing claimants are notified that proceedings have been initiated under the law to obtain a patent from the Government for the land therein described and specified, and that if they fail to answer or file their adverse claim within the period fixed by law, their right to appear is barred; and it would certainly contravene all analogies of the law to render judgment in favor of plaintiff for premises other than those for which parties have been summoned to defend, and then rule out all defendants whose rights might be thus jeopardized, on the ground that their right to appear had become barred by statutory limitations.[2]

§ 117. Discrepancies between the published notice, the diagram and posted notice.

—In a case which presented the following peculiar state of facts, it was held that the notice was sufficient:

Each one of these papers described the claim as commencing "at a point south 49 deg. west from the shaft upon the Winnebago Lode, at the distance of 56½ feet"; the *courses* agreed in all these papers. The application for patent, the notices posted and published, all gave the length of the claim as 1,400 feet.

The published notice, after giving the length of the claim as 1,400 feet, described the premises as commencing 56½ feet S. 49 deg. W. from the Winnebago shaft. Thence S. 49 deg. W.

---

[1] In re Flagstaff Lode, Decision of Secretary, December 5th, 1871; Opinion of Assistant Attorney-General, November 24th, 1871, reversing on that point decision of Commissioner, November 10th, 1871, Copp's U. S. Mining Decisions, 61, 70, 71. But see Decision of Acting Secretary, New Idria Claim, Ibid. 47.

[2] In re Flagstaff Lode, Decision of Commissioner, November 10th, 1871; Copp's U. S. Mining Decisions, 61.

eighteen hundred and seventy-two feet. Thence S. 51 deg. W. 350 feet. Thence S. 54 deg. W. 871½ feet to " western boundary, embracing a surface claim of 70,000 square feet, and is more fully described upon the diagrams and notices thereof filed this day in this office, and to be posted upon the claim itself."

The sum of the distances as given above, along the vein, to wit: 1872, 350, and 871½ feet, was *three thousand and ninety-three and a half feet*, although in the same notice it was stated that the claim was only *fourteen hundred feet in length.*

It was held that no one could have been misled by this notice. If in any doubt in regard to the length of the claim, a party could satisfy himself upon this matter by calling at the local office; for the same notice which contained this discrepancy stated "that the claim is more fully described upon the diagrams and notices thereof, filed this day in this office." (The local land office.)

The diagram and notice posted in the register's office were posted on the same sheet of paper. The sum of the distances along the vein as shown upon the diagrams, to wit: 178½, 350, and 871 feet, was 1,399½ feet. The notice stated that the claim was "1,400 feet in length," and gave the courses and distances along the vein. The sum of the distances given in the notice, to wit: 187½, 350, and 871½ feet, was *fourteen hundred and nine feet.*

The smallest number of feet called for in either the notice, diagram, or published notice, was *thirteen hundred and ninety-nine and one-half feet.*

The claim as finally surveyed along the center line was *thirteen hundred* and ninety-nine, and four hundred and seventy-five thousandths feet, or twenty-five thousandths of a foot less than the smallest number of feet called for in either of the documents.[1]

§ 118. Discrepancies between final survey and patent and the application.—In the case of the Prince of Wales Mine *v.* the Highland Chief Mine, Utah Territory, it was claimed

[1] In re Equator Lode, Decision of Commissioner, October 26th, 1875, 2 Copp's Land-owner, 114.

that the patent for the Highland Chief did not follow the final survey in this: that its final survey did not include any of the surface ground of the Prince of Wales, while it was conceded that the patent did include all the surface ground where the Highland Chief crossed the Prince of Wales Lode and its discovery shaft, and many of its valuable works.

The field-notes of this survey, made October 5th, 1871, upon this point were as follows: "From post N. 2, I run No. 53 deg. E. 919 (feet) to Prince of Wales claim 1,200 (feet); leave Prince of Wales claim." And again: "From post No. 4, I run S. 53 deg. W. 258 (feet) to Prince of Wales claim, 495 (feet); leave Prince of Wales claim."

The natural construction of this language was considered to be, that the spaces between the 919 and 1,200 feet on one side, and 258 and 495 feet on the other side, were omitted. If they were, the description was correct. If there was doubt whether they were omitted or not, it was held proper to explain that doubt by the testimony of experts in surveying. A deputy United States mineral surveyor, who made this survey, testified that he did omit the surface premises of the Prince of Wales, and that he intended so to do. The Secretary said: "He found the Prince of Wales Company in the actual occupancy of this surface ground. He saw that it had its discovery shaft and valuable mining works upon it; and he probably knew that the Prince of Wales Company was the first locator, inasmuch as he was a surveyor, and familiar with the mines in that location. He would, therefore, very naturally pass over the premises, and exclude them from his survey, unless he had directions from his employers to do otherwise. I do not think that he had any such instructions, and my reason for so thinking will appear when I come to consider another branch of this subject. It is true that the surveyor, in making up the area of his survey, did not exclude from such area the surface ground of the Prince of Wales, amounting to 24-100 of an acre. It probably escaped his recollection when he came to make his plat. In my judgment the weight of the evidence shows that the surface ground was excluded from the survey. It should, therefore, have been excluded from the patent, and it was error to include it." [1]

---

[1] Decision Sect'y Interior, April 1st, 1875, 2 Copp's Land-owner, 2; 1 Ibid. 43.

In reference to the objections that the location and application for patent and final survey do not agree, it appears that reference may be had to parol evidence to determine the location of the claim. Where four deputy mineral surveyors and their four attendants testified that the location, application for patent, and final survey were for substantially the same premises, and the testimony moreover corroborated these statements, the objections on this ground were overruled.[1]

§ 119. **New survey pending another application.**— After an application has been made for patent for a given mining claim, such claim is virtually withdrawn from market; pending the final disposition of the case, and no survey, as the basis of a patent, should receive the approval of the Surveyor-General for the same tract, until the first application has been disposed of. Parties may, however, have the *field work* of a survey of their claim made at any time, and if executed by a duly appointed mineral surveyor, such survey may receive the approval of the Surveyor-General at any time when no application for patent is pending for the same mine, if it is found upon examination that the survey is correct and made in accordance with law.[2]

§ 120. **Discrepancies between the survey and diagram filed.**—The final survey must substantially follow the claim described in the application. In a case where they did not correspond, but there was a variation in the description of 81° 39′ the application for patent was rejected, and the survey was not approved, but proceedings were required to be commenced *de novo*.[3]

§ 121. **Discrepancies between the survey and notice —Matters of description.**—Where, upon comparing the final

---

[1] In re Wandering Boy Lode, Decision Secretary, April 1st, 1875, 2 Copp's Land-owner, 2.

[2] In re Crown Point Lode, Decision of Commissioner, November 5th, 1874, 1 Copp's Land-owner, 133.

[3] In re Flagstaff Lode, Decision of Secretary, December 5th, 1871; Opinion of Assistant Attorney-General U. S., November 24th, 1871, affirming on this point a decision of Commissioner, November 10th, 1871, Copp's U. S. Mining Decisions, 61, 70, 71.

survey with the original notice and diagram, it is found that a discrepancy exists between them, as, for instance, a difference of ten degrees and twenty minutes between the tract of land for which the applicants gave legal notice that they would apply for a patent, and the tract which they had surveyed and platted by the United States Surveyor; to proceed to grant title on the survey would be equivalent to issuing a patent for a claim for which no notice had ever been given, and is unauthorized by the mining acts. In such cases no patent will be issued until the plat and field-notes of a corrected survey are received, describing the premises substantially as set forth in the diagram and notice. The Surveyor-General will be ordered to direct his deputy who executed the survey to proceed to correctly survey the claim, without additional charge to the applicant.[1]

§ 122. **Errors in survey.**—In the case of the Philadelphia Lode v. The Pride of the West Lode,[1] an error was made in the survey made as the basis of the last recorded notice of location, the posts at the southerly end of the claim having been placed about three feet too far south, the course between the posts at the angles on the easterly and westerly sides, and the posts established at the southeasterly and southwesterly corners of the claim, given as S. 6 deg. 30 min. W., instead of S. 6 deg. 42 min. W., the actual course between the points. With these exceptions the description given in the plat and field-notes agreed with the description contained in the last recorded notice of the location. The principle that courses and distances must give way when in conflict with fixed objects and monuments was applied.[2]

In the same case it was also urged that, as the discovery of the Philadelphia Lode was in reality outside of the boundaries of the Pride of the West claim instead of within such boundaries as represented upon the plat, the survey was erroneous.

It was not claimed that the courses and distances between the

---

[1] Application of International Mining and Exchange Company; In re Hercules Lode, April 19th, 1872, Decision of Commissioner, Copp's U. S. Mining Decisions, 90.

[2] Decision of Commissioner, August 28th, 1876, 3 Copp's Land-owner, 82; Decision of June 10th, 1876.

several posts described in the plat and field-notes of the Pride of the West were erroneously given.

The Commissioner said: "The fact that the discovery shaft of the Philadelphia Lode is represented upon said plat as lying within the exterior boundaries of the Pride of the West survey, while in reality it lies five feet to the east of the easterly boundary of said survey, will not prejudice the right of the Philadelphia claimants in any respect; as the patents in all cases of applications arising under the mining act follow the description of the premises as given in the field-notes of survey thereof.

"It is urged that no patents can issue upon said application, as the end lines of the claim as surveyed are not parallel to each other, as required by the last clause of the fifth section of the Act of May 10th, 1872.

"The course along the northerly end line of said survey is N. 69 deg. 80 min. E., while the course along the southerly line is N. 86 deg. 18 min. W. These end lines are perpendicular to the side lines but are not parallel to each other, there being an angle in the side lines between the northerly and southerly ends thereof. It might be questioned whether there has been a failure to comply with the *spirit* and *intent* of that provision of said section which requires that 'the end lines of each claim shall be parallel to *each* other.' But as a claimant may at any time abandon the whole or any part of his application for patent, *a strict compliance with the letter of the law* in regard to end lines may be secured by the applicants filing an abandonment to so much of the premises embraced by their application as may be necessary to render the end lines parallel, and having an amended survey filed.

"Should such abandonment be filed, the rights of no parties other than the applicants would be affected thereby, as there is no adverse claimant to that portion of the premises embraced in said survey which it would be necessary to abandon to make the end lines parallel."

Errors in survey.—Where any material error occurs in the survey, so as to mislead parties who may have the right to file adversely, or not to apprise them of the exact boundaries, extent, nature and location of the claim, the applicant must commence *de novo* by filing with the local land officers a plat and field-notes "showing accurately the boundaries of the claim," and publish a notice accurately describing the claim; for the patent when issued must conform to and agree with the description given in the plat and field-notes.[1]

§ 123. When applications for patent will be rejected—Errors and defects in patent and application.—If the record title is found defective the application will be rejected, and so if a previous application has been made for the same ground and withdrawn, pending a suit in court commenced by adverse claimants, it will be denied.[2]

Recalling patent.—After a patent has once been issued, it is contrary to the fixed policy of the Department to recall the same, unless it be shown that an error has been committed in the description of the tract, or a mistake made in the name of patentee.[3]

§ 124. Sworn statement.—It is sufficient if the sworn statement of all the applicants shows that they have the possessory right to the claim by virtue of a compliance by themselves and their grantors with the mining laws. If it be alleged that a notice was posted at the point of discovery of the lode, giving the names of claimants, number of feet claimed, and the general direction of the premises claimed, that the notice was recorded, and that the amount of labor required by law has been performed, and the claim is described in the location notice with such a degree of accuracy that parties can easily ascertain its exact locality, it is held that not giving the exact *course and distance* between the lode and a natural or artificial monument is not fatal to the application, where the local law provided that in making a record of location of any claim "the same shall be definitely described with reference to some natural or artificial monument."[4]

---

[1] Decision of Com., April 17th, 1873, Copp's U. S. Mining Decisions, 193.
[2] Brown v. Lewis, In re Cascade Mine, 1 Copp's Land-owner, 50.
[3] In re Washington Lode, Decision Acting Commissioner, April 5th, 1872, Copp's U. S. Mining Decisions, 88.
[4] In re King of the West Lode, City Rock and Utah Claimants v. Pitts, 1 Copp's Land-owner, 116; Decision of Commissioner, December 14th, 1874.

§ 125. Approval of survey—Jurisdiction of Surveyor-General.—The approval of a survey of a mining claim by the Surveyor-General is merely an indorsement thereon, over his own signature, that the survey is correct, and that it has been made in accordance with law and instructions, and until he has actually affixed his signature approving such survey, no appeal lies to the land office, as an appeal cannot lie from a *proposed* action or decision. If, however, a protest is filed against a given survey, the plat and field-notes of survey are to be transmitted to the General Land Office, together with all the papers which may have been filed with the case, that such action may be taken as the law and the facts may warrant. The Surveyor-General has no jurisdiction in the matter of deciding the respective rights of parties in cases of conflicting claims. Each applicant for a survey is entitled to a survey of the entire mining claim *as located*, if held by him in accordance with the local laws and Congressional enactments. If, in running the exterior boundaries of a claim, it is found that two surveys conflict, the plats and field-notes should show the extent of the conflict, giving the area which is embraced in both surveys, and also the distances from the established corners at which the exterior boundaries of the respective surveys intersect each other. If parties desire to protect their interests, which would be adversely affected by the issuance of a patent for the claim as surveyed, they must file an adverse claim against such application in the manner and form prescribed by the statute, for in no other way can their alleged adverse rights be adjusted.[1]

§ 126. Proof of citizenship only required of applicants.—Where it was alleged that the patent issued without proof that the original locators were citizens of the United States, it was not claimed that there was proof that the applicants for the patent were not citizens. It has not been the practice of the Land Office to require proof that the original locators were citizens, except in those cases where they were applicants for patent. It will not be presumed that they were not citizens in

---

[1] In re Crown Point Lode, Decision of Commissioner, November 5th, 1874, 1 Copp's Land-owner, 133.

the absence of an allegation or objection to that effect, before the issuing of patent. After patent has actually issued, it is held too late to make such an objection.[1]

§ 127. Miscellaneous.

Bona fide application for patent.—In the absence of any adverse claim, a *bona fide* application for patent under the Act of 1866 was considered such an appropriation of the premises embraced therein as takes them out of the application of the local laws.[2]

A portion of a claim, uncontested, may be patented, the parties having complied with the requirements of the law.[3]

Exemplified copies of patents are furnished only to parties in interest.[4]

Assignment of patents.—There are no rules or regulations governing the assignment of patents issued by the Land Office. Such patents are conveyances of the title to certain lands, previously existing in the United States, and if these parties desire to transfer to others the title thus acquired, they must conform to the laws of the *locus rei sitæ* relating to the conveyance of realty.[5]

Refunding purchase-money.—The money paid for a mining claim will not be refunded when a decision is made reducing the extent of a claim, except for so much of the superficies as is not included in the reserves necessary to cause the claim to conform to the local laws and customs of the miners. When, however, a decision is rendered by which a claim, erroneously extended, is reduced in size, the purchase-money will be returned, to the extent necessary to make the payment meet the requirements of the law.[6]

---

[1] Kempton Case, In re Wandering Boy Lode, Decision of Secretary, April 1st, 1875, 2 Copp's Land-owner, 2.

[2] Daney G. & S. M. Co. v. Sapphire M. Co., Decision Secretary, June 29th, 1875, 2 Copp's Land-owner, 66, 67.

[3] Decision of Commissioner, February 27th, 1872, Copp's U. S. Mining Decisions, 78.

[4] In re Daniel Peters Lode, Decision of Commissioner, January 2d, 1872, Copp's U. S. Mining Decisions, 76.

[5] Decision Commissioner, January 21st, 1869, Copp's U. S. Mining Decisions, 18.

[6] Decision Commissioner, September 14th, 1870, Copp's U. S. Mining Decisions, 32. See, generally, as to nature of United States patents, and the title conveyed,

Patterson v. Tatum, 3 Sawyer C. C. 164; Wilcox v. Jackson, 13 Pet. 499; Bagnell v. Broderick, 13 Pet. 436; Hooper v. Scheimer, 23 How. 235; Johnson v. Towsley, 13 Wall. 72; Samson v. Smiley, Ibid. 91; Davenport v. Lamb, 13 Wall. 418; White v. Cannon, 6 Wall. 443; Galloway v. Finley, 12 Pet. 264; Dredge v. Forsyth, 2 Black, 563; Schedda v. Sawyer, 4 McL. 181; Ballance v. Forsyth, 13 How. 18; S. C. 6 McL. 562; Gregg v. Tesson, 1 Black, 150; Mann v. Wilson, 23 How. 458; Lafayette's Heirs v. Kenton, 18 How. 197; Stoddard v. Chambers, 2 How. 285; Field v. Seabury, 19 How. 223, 333; Minter v. Crommeline, 18 How. 87; U. S. v. Arredondo, 6 Pet. 736; New Orleans v. De Armas, 9 Pet. 223; New Orleans v. U. S. 10 Pet. 662; Nelson v. Moon, 3 McL. L. 319; Reichart v. Felps, 6 Wall. 160; Stark v. Starr, 6 Wall. 402; Brush v. Ware, 15 Pet. 93; S. C. 1 McL. 533; Morgan v. Curtenius, 4 McL. 366; Beard v. Federy, 3 Wall. 479; U. S. v. Hughes, 11 How. 552; S. C. 4 Wall. 232; U. S. v. Stone, 2 Wall. 526; Hoofnagle v. Anderson, 7 Wheat. 212; McArthur v. Browder, 4 Wheat. 488. As to sales of public lands, generally, and the power of Congress to sell, see U. S. v. Gratiot, 14 Pet. 526; S. C. 1 McL. 454; Oliver v. Piatt, 3 How. 333; S. C. 1 McL. 295; Wilcox v. Jackson, 13 Pet. 498; Miller v. Kerr, 7 Wheat. 1; Root v. Shields, 1 Wool. 340.

CHAPTER IX.

ADVERSE CLAIMS—PROCEEDINGS IN COURT.

§ 128. Adverse claims.
§ 129. Adverse claims under Act of 1866.
§ 130. Adverse claims under statutes now in force—details of procedure.
§ 131. Who may file.
§ 132. Verification of adverse claim.
§ 133. Verification of adverse claims by agents of companies.
§ 134. Time of filing.
§ 135. Commencing second suit—Dismissal of former suit.
§ 136. What constitutes an adverse claim.
§ 137. Necessary allegations.
§ 138. What adverse claimant must show.
§ 139. Form of adverse claim.
§ 140. Prima facie adverse claim.
§ 141. Sufficient filing.
§ 142. Adverse claim must be accompanied by certified survey.
§ 143. The object of giving notice by publication.
§ 144. Jurisdiction of the Land Office over adverse claims.
§ 145. Notice of suit.
§ 146. Authority of register to dismiss.
§ 147. Proceedings in Court—proper party to commence suit.
§ 148. Possession as equivalent to adverse claim—parties to institute suit.
§ 149. What are Courts of competent jurisdiction.
§ 150. Contests in Court—jurisdiction.
§ 151. Jurisdiction of State Courts.
§ 152. Transfer of causes to United States Courts—jurisdicion of mining causes.
§ 153. Cancelation of entry pending suit.
§ 154. Stay of proceedings.
§ 155. Filing consent to judgment.
§ 156. Laches in bringing suit.
§ 157. Prosecution of suits—reasonable diligence.
§ 158. Abandonment of portion of adverse claim.
§ 159. Abandonment of surface ground.
§ 160. Cross-applications—delay.
§ 161. Fees on filing adverse claim.
§ 162. Miscellaneous.

§ 128. Adverse claims.—Section 2326 of the Revised Statutes of the United States reads as follows: "Where an adverse claim is filed during the period of publication, it shall be upon oath of the person or persons making the same, and shall show the nature, boundaries, and extent of such adverse claim, and all proceedings, except the publication of notice and making and filing of the affidavit thereof, shall be stayed until the controversy shall have been settled or decided by a Court of competent jurisdiction, or the adverse claim waived. It shall be the duty of the adverse claimant, within thirty days after filing his claim, to commence proceedings in a Court of competent jurisdiction, to determine the question of the right of possession, and prosecute the same with reasonable diligence to final judgment; and a failure so to do shall be a waiver of his adverse claim. After such judgment shall have been rendered, the party entitled to the possession of the claim, or any portion thereof, may, without giving further notice, file a certified copy of the judgment-roll with the register of the Land Office, together with the certificate of the Surveyor-General that the requisite amount of labor has been expended or improvements made thereon, and the description required in other cases, and shall pay to the receiver five dollars per acre for his claim, together with the proper fees, whereupon the whole proceedings and the judgment roll shall be certified by the register to the Commissioner of the General Land Office, and a patent shall issue thereon for the claim, or such portion thereof as the applicant shall appear, from the decision of the Court, to rightly possess. If it appears, from the decision of the Court, that several parties are entitled to separate and different portions of the claim, each party may pay for his portion of the claim, with the proper fees, and file the certificate and description by the Surveyor-General, whereupon the register shall certify the proceedings and judgment roll to the Commissioner of the General Land Office, as in the preceding case, and patents shall issue to the several parties according to their respective rights. Nothing herein contained shall be construed to prevent the alienation of the title conveyed by a patent for a mining claim to any person whatever.[1]

[1] Sec. 7, Act 1872, 17 U. S. Stats. 93, omitted the clause relating to proofs of citizenship, which is incorporated in Sec. 2321 Rev. Stats.; otherwise the sections are identical.

Sec. 6 of the Act of July 26th, 1866, 14 U. S. Stats. 252, was as follows: That whenever any adverse claimants to any mine, located and claimed as aforesaid, shall appear before the approval of the survey, as provided in the third section

§ **129. Adverse claims under Act of 1866.**—Under the Act of 1866 the opposing claimant was required to file his adverse notice with the register and receiver; and, in order that it might appear to those officers whether or not the adverse claim was such a one as was contemplated by the sixth section, they required the opposing claimant to present his affidavit, setting out in detail the nature of his adverse claim, stating when and how it originated—whether by purchase or by location—the names of all the original locators, with a certified copy of the original location from the mining recorder's office; and if he claimed as a purchaser, an abstract of title, certified by the recorder, tracing the title to the possession from the original locators to the claimant, should be furnished. Such affidavit and accompanying papers were to be carefully examined by the register and receiver, and if, in their judgment, an adverse claim was made out, they suspended all further action on the application for patent, until an adjustment was had in the local courts; if they found otherwise, they refused to suspend, but in either event the papers filed, both by the applicant for patent and the adverse claimant, were referred to the General Land Office for review, where the decision of the register and receiver was either affirmed or set aside, and all parties in interest notified of the result.[1]

In the case of placer claims upon surveyed lands, where no survey was required, the adverse claimant was required to appear before the entry was made; but if, from any cause, such adverse claimant was unable to appear within the time specified, but appeared before the patent was issued, the register nevertheless took his sworn statement, and transmitted it to the General Land Office, for such action as the Commissioner might deem proper. When the parties were notified that an adverse claim was made out, it became the duty of the adverse claimant immediately to commence action in Court, and to prosecute the same to final judgment or decree, by which the further proceedings of the office were governed. In default of such suit being

of this act, all proceedings shall be stayed until a final settlement and adjudication in the Courts of competent jurisdiction of the rights of possession to such claim, when a patent may issue as in other cases. (See Sec. 2325 Rev. Stats.)

[1] Instructions August 8th, 1870, Copp's U. S. Mining Decisions, 259.

instituted within a reasonable time, the original claim was dealt with as if no adverse interest had been asserted. Every facility was to be afforded to parties desiring to avail themselves of the privileges accorded by these enactments, and completed cases were to be promptly reported to the General Land Office. Monthly returns were to be made of all entries of lode and placer claims, with details specifically showing what lands were entered.[1]

§ 130. **Adverse claims under statutes now in force—Details of procedure.**—An adverse mining claim must be filed with the register of the same land office with whom the application for patent was filed, or in his absence, with the receiver, and within the sixty days' period of newspaper publication of notice. The adverse notice must be duly sworn to by the person or persons making the same before an officer authorized to administer oaths within the land district, or before the register and receiver: it must fully set forth the nature and extent of the interference or conflict; whether the adverse party claims as a purchaser for valuable consideration or as a locator; if the former, a certified copy of the original location, the original conveyance, a duly certified copy thereof, or an abstract of title from the office of the proper recorder, should be furnished, or if the transaction was a verbal one, he will narrate the circumstances attending the purchase, the date thereof, and the amount paid, which facts should be supported by the affidavits of one or more witnesses, if any were present at the time, and if he claims as a locator he must file a duly certified copy of the location from the office of the proper recorder.[2]

In order that the "boundaries" and "extent" of the claim may be shown, it is incumbent upon the adverse claimant to file a plat showing his claim, and its relative situation or position with the one against which he claims, and the extent of the conflict. This plat must be made from an actual survey by a United States Deputy Surveyor, who will officially certify thereon to its correctness; and in addition there must be attached to such plat of survey a certificate or sworn statement

---

[1] Instructions June 8th, 1870, Copp's U. S. Mining Decisions, 259.
[2] Instructions June 10th, 1872, Subdivisions 47, 48; February 1st, 1877, 44–52.

by the surveyor as to the approximate value of the labor performed or improvements made upon the claim by the adverse party or his predecessors in interest, and the plat must indicate the position of any shafts, tunnels, or other improvements, if any such exist upon the claim of the party opposing the application, and by which party said improvements were made.[1]

Upon the foregoing being filed within the sixty days as aforesaid, the register, or in his absence, the receiver, will give notice in writing to both parties to the contest that such adverse claim has been filed, informing them that the party who filed the adverse claim will be required, within thirty days from the date of such filing, to commence proceedings in a Court of competent jurisdiction, to determine the question of right of possession, and to prosecute the same with reasonable diligence to final judgment; and that should such adverse claimant fail to do so, his adverse claim will be considered waived, and the application for patent will be allowed to proceed upon its merits.[2]

When an adverse claim is filed, the register or receiver will indorse upon the same the precise date of filing, and preserve a record of the date of notifications issued thereon; and thereafter all proceedings upon the application for patent will be suspended, with the exception of the completion of the publication and posting of notices and plat, and the filing of the necessary proof thereof, until the controversy shall have been adjudicated in Court, or the adverse claim waived or withdrawn. The statute itself fully provides for proceedings after rendition of judgment by the Court.[3]

§ 131. **Who may file.**—If claimants are in fact incorporated as a company, their united interests constituting but one claim patentable on the required expenditure in labor and improvements, there is no question as to the right of such company to appear by an authorized agent or attorney, and no reason appears why any one of the several members may not be thus authorized. This is the usual practice in other States and Ter-

---

[1] Instructions June 10th, 1872, Subdivision 49, February 1st, 1877, 44–52; Decision of Secretary, In re Webster Lode, 2 C. L. O. 31.
[2] Instructions June 10th, 1872, Subdivision 50, February 1st, 1877, 44–52.
[3] Instructions June 10th, 1872, Subdivision 91, February 1st, 1877, 44–52; Decision of Secretary, In re Webster Lode, 2 C. L. O. 31.

ritories, and as the individuals composing these corporations are often widely scattered, a different rule would frequently render proceedings under the mining act impracticable. Where the interest is of such a mutual character as in the case of these companies, there would seem to be no occasion for requiring the personal appearance at the local office of each individual of a company. One member of a company therefore may file an adverse claim in behalf of the whole, and each member is not required to appear before the register and receiver.[1]

§ 132. **Verification of adverse claims.** — The adverse claim must be "upon oath of the person or persons making the same." An officer authorized to administer oaths within the land district may administer an oath to an adverse claim outside of the district but inside of the limits of his jurisdiction, if the latter extends within the land district where the claims are situated.

Where the facts were that the adverse claim was sworn to before a deputy clerk of the District Court in and for the county; and although there was no testimony showing the exact part of the county where the oath was administered, it was shown that the office and residence of the clerk were in a town in the county, and that it was probable that the affidavit was made at that place. The line between the district where the mine was situate and the district where the town was located ran so as to leave the town in one district and the mine in the other. The affidavit being administered in the district where the town was situate, it was held a sufficient compliance with Sec. 2335 of the Revised Statutes, it being within the jurisdiction of the clerk.[2]

An adverse claim will be rejected if not sworn to before an officer authorized to administer oaths within the land district where the claim is situated.[2] A protest verified before a commissioner of deeds for the State, but residing elsewhere, does not comply

---

[1] Decision of Commissioner, January 28th, 1869, Copp's U. S. Mining Decisions, 19.

[2] Corning Tunnel M. Co. v. Pell, In re Slide Lode, Decision of Secretary, February 17th, 1877, 3 C. L. O. 195, reversing S. C. Decision Commissioner, 3 C. L. O. 130, and distinguishing the case from that of Dardanelles Mining Co. v. Cal. M. Co., Copp's Mining Decisions, 161, Infra.

with the act. The instructions issued under the Act of 1866 required all affidavits to be made before the register and receiver, but the Acts of 1870 and 1872 authorize them to be made before any officer within the land district who has authority to administer oaths. The authority was limited to the district, so as to make it practicable to punish those guilty of perjury in making the oath.[1]

The jurat to the adverse claim must be made by the party and cannot be made by an attorney. The law does not provide that the adverse claim shall be verified upon the oath of an agent or attorney. Without statutory authority an attorney cannot make the oath for his client. An adverse claim, therefore, verified by an attorney, is not sufficiently verified under the act, and will be rejected.

But one of the adverse claimants may make an affidavit as the representative of the others making the same claim.[2]

Where several parties unite in the adverse claim, the jurat is sufficient if made by one of them.

The filing of an adverse claim with the register is a sufficient filing under the act. The official indorsement of the filing is prima facie evidence that they were filed as of that date.

§ 133. Verification of adverse claims by agents of companies.

An incorporated company must necessarily act through its officers or agents. The company, as a company,

---

[1] Corning Tunnel Mining & Reduction Co. v. Pell, In re Slide Lode, Decision of Commissioner, November 3d, 1876, 3 Copp's Land-owner, 130, 131; Rev. Stats. 2335, 2336; Decision of Acting Secretary, October 28th, 1873; Decision of Commissioner, March 7th, 1873; In re Dardanelles Mining Co. v. Bosphorus Lode, Copp's U. S. Mining Decisions, 160, 161; Decision of Secretary of Interior, McMurdy v. Streeter, April 30th, 1874.

Decision of Acting Secretary, October 28th, 1873, Copp's U. S. Mining Decisions, 161; Affirming Decision of Commissioner, March 7th, 1873, Copp's U. S. Mining Decisions, 160; In re Bosphorus Lode; In re Dardanelles Mining Co.; In re California Silver Mining Co.

[2] Decision of Commissioner, December 14th, 1874, In re King of the West Lode, City Rock and Utah Clamaints v. Pitts, 1 Copp's Land-owner, 146; Jenny Lind Mining Co. v. Eureka Mining Co., Decision of Secretary of Interior, November 24th, 1873; Opinion of Assistant Attorney-General in same case; In re Eureka, Montana, Excelsior, King David, and Mary Henrietta Lodes; Jenny Lind Mining Co. v. Eureka Mining Co., Decision of Secretary, November 24th, 1874, Copp's U. S. Mining Decisions, 169; Opinion of Assistant Attorney-General, September 30th, 1873, and November 22d, 1873. Ibid. 172, 175.

cannot make oath to the statements contained in an adverse claim presented by it. Such a company may therefore file an adverse claim sworn to by its agent or attorney. A distinction is to be drawn in this respect between incorporated and unincorporated companies or associations. In the latter case the act does not provide that the claim may be made upon the oath of an agent or attorney, and without statutory authority an attorney cannot make oath for his client, and in such cases adverse claims so verified will be rejected.[1]

§ 134. **Time of filing adverse claims.**—The rule excluding adverse claims not filed within the period prescribed after the commencement of the publication of notice, has been so often applied by the Department, that it must now be regarded as fully settled. No further time can be granted,[2] and no adverse claim can be considered, if filed after the expiration of the period of application.[3]

The words "and after the expiration of said period, if no adverse claim shall have been filed," seem clearly to require the adverse claims "to have been filed" prior to the expiration of the time for notice. The meaning is that the register shall give the notice required for the prescribed period, and that parties having adverse claims shall have the entire period in which to file their claims, and that upon its expiration, the very next day thereafter, if there have been no adverse claims filed, the claimant shall have the right to apply to the Surveyor-General for a survey, and upon its being approved, and the land paid for, and the proper papers forwarded to the Commissioner, he shall be entitled to his patent.[4]

§ 135. **Commencing second suit—Dismissal of former suit.**—The commencement of another suit against applicants

---

[1] Equator Mining & Smelting Co. v. Marshall Silver Mining Co., Decision of Acting Commissioner, October 26th, 1874, 1 Copp's Land-owner, 132; Decisions of Secretary and of Assistant Attorney-General, November 22d and November 24th, 1873, 1 Copp's Land-owner, 132.

[2] In re Unicorn Lode, Decision of Commissioner, April 18th, 1873, Copp's U. S. Mining Decisions, 194; Decision of Secretary, March 14th, 1872, Copp's U. S. Mining Decisions, 74; Seymour v. Woods, Decision of Commissioner, March 23d, 1877.

[3] In re Equator Lode, Decision of Commissioner, 2 Copp's Land-owner, 114.

[4] Ibid.

for a patent, after the time allowed by the Office to bring suit to adjudicate the right of possession to the mine has expired, will not be considered by the Office, and the mineral entry will be allowed.

Thus, where a suit was brought within the time, but afterwards dismissed on complainant's motion without prejudice, and another suit instituted after the time, the latter was not considered. In the same case, an adverse claim not filed within the proper time was likewise ignored.[1]

§ 136. What constitutes an adverse claim.—The adverse claim must show the nature, boundaries, and extent of the claim. An adverse claim cannot consist of a mere informal protest; such a one will not suspend proceedings. It is contrary to the spirit and letter of the law and the practice of the General Land Office, to permit one person, or association of persons, to file one protest against several applications for patents for separate and distinct lodes. Where there were three separate and distinct applications for patents for three separate and distinct lodes, each application was held an entirety, and to rest upon its own merits. As each application is for a separate and distinct portion of mineral land, parties who desire their adverse claims considered must file a separate and distinct adverse claim against each application separately. They must strictly comply with the law and the instructions, and file with the local land officers within the time prescribed by law, and in proper form, a separate and distinct adverse claim against each application which it is alleged conflicts with the premises owned by such adverse claimants. Where applicants for patents strictly comply with the law and the instructions, a like requirement will be imposed upon adverse claimants.[2]

---

[1] In re Pelican Lode, Decision Commissioner, July 8th, 1872, Copp's U. S. Mining Decisions, 126; Decision of Secretary of the Interior, July 5th, 1872, Copp's U. S. Mining Decisions, 127; Decision of Secretary of the Interior, November 16th, 1872, Ibid. 127.

[2] In re Zella Lode; In re Mountain Tiger Lode; In re Rockwell Lode; Decision of Acting Commissioner, June 9th, 1873, Copp's U. S. Mining Decisions, 202. See 420 Mining Co. v. Bullion Mining Co. 3 Sawyer C. C. 638; S. C. 9 Nevada.

§ 137. **Allegations of the adverse claim.**—If the adverse claimants properly *allege* that they are the *owners* of the claim, that is good pleading, and sufficient to notify the applicant for patent of what is claimed. The material thing is ownership, in accordance with the rules and regulations of miners. But where the adverse claimant failed to file with his adverse claim record evidence that he had title to the premises claimed by him, by purchase from the parties who had record "title to the mine," and the adverse claimant did not positively allege ownership, but only that he claimed as purchaser, and referred to certain deeds which showed that he had purchased from a party who had no right, the claim was rejected.

Parties are held to a full and unequivocal declaration as to the fact of ownership. Thus, in the case just adverted to, a party couched his claim of title in such language that, when coupled with the transactions he pleaded in the matter of certain conveyances under which he claimed, raised the presumption that his estate was rather one of trust than of fee or ownership, and the adverse claim was rejected. In this case the deeds referred to showed that he purchased from a party who had no right, title, or interest in the mine; a party who had previously conveyed to a foreign corporation. It was urged that he held the property in trust for the company, and that the latter was the party whose interests were, in fact, to be protected by the adverse filing. The claim, as trustee, was not considered sufficient, and the doctrine just enunciated was applied.[1]

§ 138. **What the adverse claimant must show.**—The adverse claimant must show sufficient compliance with the mining laws, and the usages and customs of mining districts, to make it clear that he is acting in good faith.

Where there was nothing in the protest or affidavit to show that the adverse claimants had complied with the laws, or that it was a valid subsisting claim at the time the grantors of the applicant made their location, and there was nothing to show

---

[1] In re King of the West Lode, City Rock and Utah Claimants v. Pitts, 1 Copp's Land-owner, 146; Decision of Commissioner, December 14th, 1874; Jenny Lind Mining Co. v. Eureka Mining Co., Opinion of Assistant Attorney-General of the United States.

that the adverse claimant or his grantors performed labor or made improvements, or was in possession at the time of filing application, and moreover, the extent and boundaries of the adverse claim were indefinite, it was rejected.[1]

Where it was not shown that a party filing a protest had any authority to file it, and give notice of an adverse claim, except the statement in his affidavit that he had the authority; nor did it appear by his affidavit or otherwise that he was a member of the company in whose behalf he appeared, it was held that a mere allegation of authority is not sufficient—the authority to act must be shown.

In the case of the Eureka Mining Company v. the Jenny Lind Mining Company et al., decided Nov. 24th, 1873, it was held "that the jurat to the adverse claim required by the seventh section of said act, (Act of May 10th, 1872) must be made by the party, and cannot be made by an attorney."[2]

When, however, the party in interest is an incorporated company which can only act through an agent from the necessity of the case, this rule must be somewhat modified. In such case the company may verify its protest by the oath of its president or other executive officer; or it may, by letter of attorney, appoint some proper person to act for it, who would then be clothed with sufficient power to make an affidavit. Without such authority a person not an executive officer of the company would have no right to act, and whatever he might do in its behalf would be invalid, and should be so regarded.[3]

§ 139. **Form of adverse claim.**—It must be made out in the form prescribed by the Act of May 10th, 1872, and the Revised Statutes and the instructions issued thereunder. If it is not so made out it will be rejected.

The record must show that a survey was made of the premises claimed adversely. The diagram or plat filed must be "made from an actual survey by a United States deputy," or other

---

[1] Hawley Consolidated Mining Co. v. Memnon Mining Co.; In re Sheridan Lode, Decision of Secretary of Interior, Feb. 12th, 1876; 2 Copp's Land-owner, 180. Affirming Decision of Commissioner, Ibid.

[2] Copp's U. S. Mining Decisions, 169.

[3] Hawley Consolidated Mining Co. v. Memnon Mining Co.; In re Sheridan Lode, Decision of Secretary of Interior, Feb. 12th, 1876, 2 Copp's Land-owner, 180.

surveyor. A certificate or sworn statement must be attached to the plat or diagram, signed by a deputy or other surveyor as to the correctness thereof.

There must also be a "certificate or sworn statement by the surveyor as to the approximate value of the labor performed or improvements made upon the claim of the adverse party," either attached to said plat or on file with the case.

The adverse claimant must also, under oath, show the nature, extent, and boundaries of his adverse claim, and make out a case showing at least a formal conflict with the claim of the applicants for a patent. If this be not done, the proceedings for a patent will not be suspended.[1]

The adverse claimant should show that he has complied strictly with the local laws, and the nature and extent of the alleged conflict, and that the conflicting locations are on one and the same lode.

He should set forth the facts in detail upon which he bases his adverse claim, so that the office can apply the law. Allegations of conclusions of law are equally out of place in the Land Office as in pleadings.[2]

As adverse claims must comply with the law and the instructions, and be in the prescribed form, where the record did not show that any survey was made of the premises claimed adversely, and where the diagram or plat which was filed did not show that it was made from an actual survey by a United States deputy or other surveyor, and there was no certificate or sworn statement attached to the plat or diagram signed by a deputy or other surveyor, as to the correctness thereof, and no "certificate or sworn statement by the surveyor, as to the approximate value of the labor performed, or improvements made upon the claim of the adverse party," either attached to the plat or on file in the case, and, therefore, nothing to show the nature, extent, and boundaries of the adverse claim, it was not recognized, and such

---

[1] In re War Eagle Mine, Decision of Commissioner, May 1st, 1873, Copp's U. S. Mining Decisions, 195.

[2] In re Wandering Boy Lode; In re Porcupine Mine, Decision of Commissioner, May 6th, 1873, Copp's U. S. Mining Decisions, 197.

a claim is not sufficient to stay proceedings on application for a patent.[1]

And where an adverse claimant failed to file a copy of the original notice of location of his claim from the office of the proper recorder, to show the number of feet embraced by said location, the number of locators, or the number of feet acquired by purchase, and failed to produce evidence in regard to expenditures on the lode, and did not show the nature or extent of the alleged conflict, and did not assert that the two locations were on one and the same lode, the adverse claim was rejected.[2]

The adverse claimants must also set forth the facts in detail upon which they base their adverse claim, and how the premises described in the application for patent conflict, and the extent of the conflict, and must not state legal deductions or conclusions. For this reason, an allegation "that sufficient work and all acts and things were done according to the acts of Congress, the mining laws of the district and customs of miners, to hold and possess the same," is bad and insufficient.[3]

An actual survey must be made of the entire mineral claim. The claimant is not allowed to color a portion of the applicant's survey, and treat it as his entire adverse claim.[4]

It has been held that an omission to file an abstract should be treated as an irregularity only, and not as a defect that vitiates the adverse claim. A party claiming only an equitable right cannot be considered as an adverse claimant—especially where a party's asserted equities grow out of transactions having to do with the mining tract covered by that and not by some other but conflicting tract or claim.[5]

---

[1] In re War Eagle Mine, Decision of Commissioner, May 1st, 1873, Copp's U. S. Mining Decisions, 195.

[2] In re Wandering Boy Lode; In re Prince of Wales and Antelope Lodes; In re Porcupine Lode; Decision of Commissioner, May 6th, 1873, Copp's U. S. Mining Decisions, 197.

[3] Ibid.

[4] Bates v. Chambers, In re Daniel Webster and Homestead Mines, 1 Copp's Land-owner, 98.

[5] In re Mono Mine, Claimants v. Gisborn, Decision of Commissioner, March 27th, 1874, 1 Copp's Land-owner, 135; Opinion of Assistant Attorney-General, September 30th, 1873; Jenny Lind Mining Co. v. Eureka Mining Co., Decision of Secretary of the Interior, November 24th, 1873, Copp's Mining Decisions, 173.

The cause of action, the settlement of which is referred to the Courts, is not one created, nor is the remedy defined by the act. The subject-matter of the controversy to be determined is not whether an applicant shall have a patent from the United States for his location, but it is whether one party unlawfully withholds the possession of the premises, or any part of them, from the other; and upon that issue the unlawful entry or cause of action may be shown to have accrued to the plaintiff at any time within the running of the Statute of Limitations. A suit, therefore, commenced *before* the adverse claim is filed, suspends proceedings as well as one commenced within thirty days afterward.

An adverse claim should be so drafted as to inform a person that a portion of the mining claim he is seeking to obtain a patent for does not belong to him, but to the adverse claimant; and with such precision as to fairly advise him of the nature, boundaries, and extent of the adverse claim, so that he may prepare himself to establish his own, on the trial before the Courts and meet the adverse claim.[1]

---

[1] An adverse claim containing the following allegations has been approved by an Assistant Attorney-General of the United States, viz: That the adverse claimant is the "lawful owner and entitled to the possession of about 1,100 feet of the said Eureka Lode," that it is the "owner, by location of the persons composing said association, and in possession of the following named lodes or veins of quartz, or other rock in place, bearing silver and other metals, viz: The south extension of the Bullion Lode, the Queen Victoria Lode, and the Pride of the West Second Lode, situated, located, and recorded in the Tintic Mining District, Juab County, Utah Territory." That "on the 17th day of March, A. D. 1871, the several premises hereinafter described were mineral lands of the public domain, and each contained a vein or lode of quartz, and other rock in place, bearing and containing silver and other minerals, and said premises were entirely vacant and unoccupied, and were not owned, held, or claimed by any person or party as mining claims or otherwise, and that while the same were so vacant, unoccupied, and unclaimed, the persons (see Exhibit B) forming the association known as the Jenny Lind Mining Company, each and all being citizens of the United States at the time, did enter upon and explore and discover the south extension of the Bullion, containing 3,000 feet linear measurement, which was located March 20th, and recorded April 10th, 1871; the Queen Victoria Lode, containing 2,000 feet linear measurement, located March 17th, and recorded March 18th, 1871; and the Pride of the West Second Lode, containing 1,000 feet linear measurement, located June 21st, 1871, and recorded September 19th, 1871." "That the said Jenny Lind Company, and the persons composing the same, have continuously held and occupied and been in the actual possession of the mining premises and lodes since the date of location of the same, with the knowledge of the Eureka Company and its agents, and without any

The fact that an adverse claim is not accompanied by the plat of survey and field-notes, is not necessarily such an irregu-

opposition whatever from it (the Eureka Company). That the locators of said lodes, and the Jenny Lind Mining Company respectively, have in all respects complied with every custom, rule, regulation, and requirement of the mining laws of said mining district, and thereby became and are owners (except as against the paramount title of the United States) and the rightful possessors of said mining claim and locations." And "that the Vice-President of the Eureka Company, at the time of his filing the application therefor, well knew that the Jenny Lind Mining Company was the owner, in possession and entitled to the possession, of so much of said mining ground embraced within the survey and plat of said applicant as is hereinbefore stated; and the said Jenny Lind Mining Company is entitled to all the silver and other metals in said southern extension of the Bullion Lode, the Queen Victoria Lode, and the Pride of the West Second Lode."

It further appeared that the Eureka Company, on the 10th of October, 1872, entered into a written contract which was proffered to the Jenny Lind Company, but never executed by it, in which the Eureka offered to convey to the Jenny Lind, when patent should be issued to it, the said Bullion, Queen Victoria, and Pride of the West Lodes, in consideration that the Jenny Lind Company would refrain from filing an adverse claim to the application of the Eureka Company for patent. This writing contained the following clause: "And, whereas, the said party of the first part (the Eureka Company) has no claim to any part of said Queen Victoria, Pride of the West, and Bullion locations, their dips, angles, and spurs."

It was objected that the adverse claim was defective in this, that it alleged ownership by location of the south extension of the Bullion Lode, the Queen Victoria Lode, and Pride of the West Second Lode, while the record of location showed that they were made by persons some of whom were not members of the Jenny Lind Company, and that there was no allegation or proof that such persons ever assigned or conveyed their interest to the Company, and therefore it was urged the adverse claim did not "show the nature of the claim," as required by the act.

Though the exhibits did show that the persons who organized the Jenny Lind Company were not identical with some of the locators of the lodes, yet the objections were overruled, the Assistant Attorney-General remarking, in his opinion, to the Secretary of the Interior: "Suppose the adverse claim had alleged ownership by location and the exhibits had shown ownership by purchase, the claim would have undoubtedly been good. The material thing is ownership in accordance with the rules and regulations of miners. All that is alleged; and it is also alleged that the Eureka had full knowledge of the ownership and possession, and never asserted any claim to the contrary.

"The statement in the written agreement goes further, and admits that said Company had no claim to any part of the said lodes of the Jenny Lind Company. It is claimed that this admission should not be regarded in the case, because it was pending a treaty of compromise. Grant it: yet it is the admission of a fact made without any stipulation that it should be without prejudice and according to the American cases, is receivable as an admission against the Eureka Company."

Eureka Mining Co. v. Jenny Lind Mining Co., Decision of Secretary and Opinion of Assistant Attorney-General U. S., November 22d and 24th, 1873, reversing Decision of Commissioners, March 26th, 1873, Copp's U. S. Mining Decisions,

larity as will justify an exclusion of the claim, especially if reasonable means were used to procure the survey and field-notes, as if claimants prevented the protestants from obtaining them by obtaining control of the United States deputy surveyors, and preventing them from making the survey.

While the regulations issued by the Commissioner require the plat and field-notes, they were not intended to operate as a bar where an applicant in good faith has done all in his power to comply with them. So the fact that no abstract of title accompanies the adverse claim is not a fatal irregularity, if the adverse claimants allege that they are the owners of the claim.[1]

A statement in the adverse claim that the affiant is president of the company is prima facie evidence of the fact, and so is the official seal of a notary public as to the fact of his notaryship.[2]

§ 140. Prima facie adverse claim.

The question as to what constitutes a prima facie adverse claim was passed upon in the Land Department, in the case of Bullion Mining Company v. 420 Mining Company. On Nov. 6th, 1867, the Bullion Mining Company filed, under the Act of 1866, an application for patent for 1,200 linear feet of the Comstock Lode, Nevada, and also a diagram of the premises claimed. The notice of intention to apply for a patent was published on the 14th of Nov. 1867, and for a period of ninety days thereafter. Various adverse claims were filed and suits commenced thereon, but they were all either abandoned, dismissed, or settled, except the adverse claim of the 420 Mining Company, which was filed on the 4th of February, 1868, and within the ninety days' notice provided for by the Act of 1866, and was properly verified.

The claim stated that the 420 Mining Company was "a corporation duly organized, and now existing under the laws of the State of California; that the 420 Mining Company is the owner

---

166, 169, 175. See, also, Mount v. Bogart, Anthon, 190; Maucy v. Carter, 4 Conn. 633; Fuller v. Hampton, 5 Conn. 516; Sanborn v. Neilson, 4 N. H. 501; Delogey v. Rentoue, 1 Martin, 175; Marvin v. Richmond, 13 Den. 58; Cole v. Cole, 34 Me. 542.

[1] Eureka Mining Co. v. Jenny Lind Mining Co., Opinion Attorney-General, September 30th, 1873, Copp's U. S. Mining Decisions, 170–173.

[2] Ibid.

of and has for more than nine years last past, been in the possession of 420 feet of the lode known and called "The Comstock Lode" * * "that 420 feet of the north end of the mining ground claimed by the said Bullion Mining Company is the mining ground of the said 420 Mining Company," "that on or about the 16th day of Nov., 1865, the said Bullion Mining Company, as plaintiff, commenced an action against the 420 Mining Company, as defendant, in the District Court of the First Judicial District, Nevada, in and for Storey County, to recover from the said defendant the possession of the northern 420 feet of the said mining ground described by said notice * * * * * and that said suit is still pending in said Court and undetermined." Only the application for patent and a diagram of the premises claimed were filed by the Bullion Company previous to Oct. 22d, 1872, except the proof of the publication of notice, which was filed on the 2d of March, 1868. The Bullion Company's application was not under oath, and they failed to file with their application any evidence, record or otherwise, tending to show that they were in a condition to apply for a patent, or that they had any record or other title to the premises described in their application. They filed no proof that they had previously occupied and improved the premises in accordance with local customs and rules; that they had expended on the premises, in actual labor and improvements, a sum of not less than $1,000; and they, as it appeared, recognized the fact that there was a "controversy or opposing claim" in regard to part of the premises described, by commencing suit against the 420 Company to recover possession of the ground claimed by it.

On the 22d of October, 1872, the Bullion Company filed a certificate of incorporation, an abstract of title, and several affidavits in regard to possession, improvements, and the posting of the notice and diagram upon the claim. On the 15th of January, 1873, they filed a copy of the local mining laws, and further affidavits in regard to possession and compliance with the local laws, and on the 18th of January, 1873, were permitted to enter their claim.

It appeared by the clerk's certificate that, after the dismissal of the suit commenced by the Bullion Company, the 420 Mining

Company commenced suit, November 29th, 1872, against the Bullion Company, to adjudicate the right of possession. This was held to be within a reasonable time after the dismissal of the suit by the Bullion Company. A copy of the complaint was on file in the Land Office.

The attorney for the 420 Company filed a certified copy of the certificate of incorporation of said company, and the affidavit in regard to the adverse claim.

The question was presented on this state of facts, whether the 420 Company had presented such an adverse claim as was contemplated by the Act of 1866, and one which should be adjudicated in the local Courts before patent issued. It was acknowledged that both the application for patent and the adverse claim were incomplete; but it was held that the respective rights of the two companies should be adjudicated in a Court of competent jurisdiction before patent issued, and that a prima facie adverse showing had been made out by the 420 Company, inasmuch as the adverse claim was filed in due time, and was under oath; the premises had been in litigation for many years, and a suit in ejectment was pending between the companies in the local Courts at the time the application for patent was filed, and in regard to the premises in dispute; and that when the Bullion Company abandoned that suit, the 420 Company took the necessary steps to secure a decision in the Courts.[1] Where the Land Office has decided that an adverse claimant has made out a prima facie adverse showing, and the contest is transferred to the Courts of competent jurisdiction, and they decide that the adverse claimant has no right, title, or interest in the premises, the decision is final and binding on the office. The claimant cannot afterward question its mineral character.[2]

§ 141. Sufficient filing.—The Commissioner has not the power to make a regulation in conflict with the law. The Commissioner, in his regulations issued under the Act of 1866,

---

[1] Bullion Mining Company v. 420 Mining Company, Decision of Commissioner, August 19th, 1873, Copp's U. S. Mining Decisions, 219; see S. C. 9 Nevada, 240; 3 Sawyer C. C. 634.

[2] Evans v. Randall, Decision Secretary, March 23d, 1876, 3 Copp's Landowner, 2.

required that the adverse claim should be filed with the register. or, in his absence, with the receiver. But a filing in the office of the register is substantially a filing with the register and receiver within the meaning of the law. It is not necessary that both the receiver and register should mark documents filed, or that two copies should be filed.[1]

§ 142. **Adverse claims must be accompanied by a certified survey.**—The protest or adverse claim must conform strictly to the law and the instructions, and cannot otherwise operate as a bar to the issuance of a patent as applied for.

A case arose where there was nothing with the papers in the case to show that an "actual survey" was made of the premises claimed by the adverse claimants. A plat was filed, but no surveyor had officially certified thereon as to its correctness. There was no certificate or sworn statement by a surveyor, "as to the approximate value of the labor performed, or improvements made upon the claim of the adverse party," either attached to the plat or on file with the case. The adverse claimants alleged that they were prevented from having a survey made, "as the agent of the company refused to allow the deputy-surveyor whom the claimants had engaged to survey and plat the same, to go on to the plat and survey of said company for that purpose." But with the papers was found an affidavit of a person who swore that he was the deputy mineral surveyor, and that the adverse claimants never made application to him for an official survey of any part of the claim described.

No evidence was on file showing agency, nor that an oath was made to the adverse claim as required by law. The jurat being taken before the Clerk of a District Court of the United States, there was no seal to the jurat. The adverse claim was rejected and the entry allowed and approved.[2]

§ 143. **The object of giving notice by publication** is to afford an opportunity to appear and be heard against the ap-

---

[1] Eureka Mining Co. v. Jenny Lind Mining Co. Opinion of Attorney-General, September 30th, 1873, Copp's U. S. Mining Decisions, 170.

[2] Decision of Secretary of the Interior, December 11th, 1872, Approving Decision of Commissioner, July 17th, 1873, Copp's U. S. Mining Decisions, 337; Rev. Stat. 2326, Act of 1872, Sec. 7, 17 U. S. Stats. at Large, 92; Instructions June 10th, 1872, Subdivision 49.

plication, to all persons who may be injuriously affected by the issuance of a patent for the premises claimed. When a proper notice is given, parties in interest who fail to appear and object, do so at their peril. They cannot disregard the notice unless the advertisement covers claims in which they have no interest. If it does, they are not required to appear and watch the further progress of the case, lest premises should be substituted which were not contained in the notice, and which may include valuable mining interests of their own. If they have carefully examined the notice during the period of publication, and find that it does not describe premises in which they have an interest, they may safely dismiss the subject, and conclude that they cannot be prejudiced by any subsequent proceedings in the case. Should it appear that the parties and officers, after the publication of the notice, improperly and illegally change the description of the premises, so that the final survey covers premises not included in the advertisement, and in which third parties have an interest, then such third parties have the right to appear at any stage of the proceedings before patent, and call the attention of the tribunal having jurisdiction over the subject to the fact that the *record* shows that the applicant has no right to a patent for the premises included in his final survey, for the reason that he has not given public notice, for the period required by law, of his intention to apply for a patent for the premises, as the notice he gave described other premises.

A case arose where parties requested to be allowed to present a caveat against the issuing of a patent, and to submit testimony showing that the application for patent ought not to be granted. There was a material difference between the original application and notice, and the final survey, in accordance with which a patent had been issued by the Commissioner, before appeal to the Secretary of the Interior. The party appeared in time and made the proper objections to the sufficiency of the case of the applicant as shown by the record. Their objection, it was held, could not be considered, because patent had issued. The proper practice was said to be to allow the adverse claimant to show, when their application for patent reached the Office, that their right to the lodes in controversy, or either of them, or any part of them, was superior to that of the original applicant, and that the portion of the same so found superior had been included in the patent issued to the original applicant, and thereupon to issue a patent to the adverse claimant for the portion so found superior in right, reciting therein the fact that a former patent had inadvertently and erroneously issued for the same to the original applicant. This decision was under the Act of 1866.[1]

§ 144. Jurisdiction of the Land Office over adverse claims.—The discovery and location of a mining claim are the first steps taken to initiate a right thereto, the basis upon which rests all subsequent proceedings; and parties, whether applicants or adverse claimants, are bound, in asserting their claims before the office, to the surface ground which is embraced by the original location. The question of how far the General Land Office may extend its examinations into the sufficiency of an asserted adverse claim, does not seem to be fully settled. Under what state of facts, notwithstanding the timely filing in the local office of the requisite papers, followed by resort to the proper Court, the Commissioner may by law, and should ignore the adverse claim, and proceed to patent the tract applied for without waiting for the determination of the action in Court, is a question frequently presenting itself. That the mere presentation of an adverse claim, followed by proceedings in the Courts, does not oust the jurisdiction of the Land Office, is settled. In Jenny Lind Mining Co. v. Eureka Mining Co.,[2] after thorough examination, the adverse claims of certain lodes were rejected by the Department for insufficient verification. This, however, was a rejection for insufficiency of form, and did not settle the question as to whether or not the Office may consider the sufficiency of the substance of the claim as presented. It is held that it may to this extent, that if upon examination of the claim presented, treating it for the purposes of an examination as the pleading of the claimant, it is found to be bad on general demurrer, then it ought to be rejected.

"The nature, boundaries, and extent of such adverse claim"

---

[1] Tiernan v. The Salt Lake Mining Co., Decision of Secretary of Interior, April 28th, 1874, 1 Copp's Land-owner, 25; See S. C. Decision of Commissioner, December 16th, 1872, Copp's U. S. Mining Decisions, 153.

[2] Copp's Mining Decisions, 173.

are required to be shown. If upon that showing the party himself, notwithstanding his declaration of conflict, pleads a location, which, allowing reasonable latitude for want of care or technical knowledge in the locators or draughtsmen, of their notice of location, does not evidence a conflict in fact, the applicant for patent ought not to be delayed for the trial of an alleged fact, whose non-existence stands admitted in his opponent's case. Where this is the condition of things the adverse claim will be rejected.[1]

The Land Office will not take notice of the filing of a bill in equity to restrain the applicants for a patent from proceeding with their application, where no adverse claim has been filed in time.[2]

§ 145. **Notice of suit.**—As it is made the duty of the adverse claimant to commence suit within thirty days after filing his claim, and a failure to do so being deemed a waiver of his adverse claim, where the adverse claimant or his attorney neglects to file in the local land office evidence that suit has been commenced, as directed by the statutes, the register and receiver are bound to presume that the adverse claim is waived, and they will act accordingly.[3]

§ 146. **Authority of register to dismiss an adverse claim.**—The register and receiver have no authority to dismiss an adverse claim "nor to receive additional proof, either from the applicant for patent or the adverse claimant, after the time prescribed by law for publication has expired, and before the "controversy shall have been settled or decided by a Court of competent jurisdiction, or the adverse claim waived," unless such adverse claimant shall fail to commence proceedings in Court within the time required by law ,i. e., within thirty days after filing his claim; in which last event the application is allowed to proceed as if no adverse claim had been asserted.

Should the register and receiver decide that an adverse claim

---

[1] In re King of the West Lode, City Rock and Utah Claimants v. Pitts, Decision of Commissioner, December 14th, 1874, 1 Copp's Land-owner, 146.

[2] In re Red Pine Mine, Decision of January 18th, 1875, 1 Copp's Land-owner, 162.

[3] Decision of Commissioner, August 6th, 1875, 2 Copp's Land-owner, 82.

has been made out in proper form, and stay proceedings upon the application for patent, the applicants may appeal from such decision to the Commissioner of the General Land Office; and on the other hand, should the local land officers decide that no adverse claim made in the proper form had been filed, the adverse claimants have the like right of appeal. But in no event can additional proof of any kind be received upon such appeal.[1]

§ 147. Proceedings in Court—Proper party to commence suit—Time.—It is the duty of the adverse claimant, within thirty days after filing his claim, to commence proceedings in a Court of competent jurisdiction, to determine the question of the right of possession, and prosecute the same with reasonable diligence to final judgment.

The action must be commenced by the *adverse claimant* in order to entitle him to a stay of proceedings. The Act of 1872 expressly requires it to be done within thirty days from the filing of the adverse claim, and the Act of 1866, it has been held, required it within a reasonable time. In the case of the 420 Mining Company v. the Bullion Mining Company,[2] the adverse claim was filed February 4th, 1868, and no suit or action was commenced by the adverse claimants until over four years and a half thereafter, viz: on November 29th, 1872. This was held not to be within reasonable time, and the pendency of a suit commenced by the applicant against the adverse claimant to try the right of possession was held not to excuse the failure of the adverse claimant to bring suit himself. The Secretary of the Interior, in deciding the case on appeal, said: " The fact that an action was pending in the local Courts in which the Bullion Company was plaintiff and the 420 Company defendant, did not relieve the latter company from the obligation imposed by the statute. That proceeding was within the control of the plaintiff, and could at any time have been terminated by a withdrawal of the suit, or by submission to a non-suit. Under such circumstances there would have been no final adjudication of the rights of possession as required by the act. That particular

---

[1] Overman Silver Mining Co. v. Dardanelles Silver Mining Co., Decision of Commissioner, April 11th, 1873, Copp's U. S. Mining Decisions, 181.

[2] Decision of Secretary of Interior, March 22d, 1875, 2 Copp's Land-owner, 5.

suit would have ended, but no final adjudication would have been reached. And this is precisely what occurred in this case. The plaintiff withdrew its suit, and left the question as to the right of possession just where it was before. It had nothing to lose, and perhaps something to gain by this move, and by making it, simply and properly exercised a strict legal right. A similar result, working a practical defeat of the provisions of the statute, is liable to occur at any time in this class of cases, if it be held that the pendency of a suit against a protestant relieves him from the duty of making himself plaintiff in another suit.

"The evident intent of the statute was to stay proceedings only when the protesting party within reasonable time commences, and with reasonable diligence pursues, his remedy against the claimant. This construction of the act was adopted by your Office, and included in your instructions to the local officers, under date of June 25th, 1867, prior to the filing of this adverse claim; and in my opinion it is the only consistent construction of which the language is susceptible. I am, therefore, of opinion that the 420 Company have not complied with the law in this respect, and for this reason are not entitled to a further stay of proceedings."

§ 148. **Possession as equivalent to an adverse claim— Parties to institute suit—Protests.**—In Becker v. Citizens of Central City of Colorado,[1] the former claimed, under the Act of 1866,[2] 3,000 linear feet of a mineral deposit near Central City, Colorado, known as the Gunnell Extension, or White Lode. He alleged full compliance with the law and instructions, but his claim was opposed by citizens of Central City, Colorado, who, before the expiration of the ninety days provided in the third section, filed with the Commissioner of the General Land Office a remonstrance protesting against the issuing of the patent, representing that said Gunnell Extension, or White Lode, as claimed by Becker, extended to a considerable distance under town lots and improvements owned and occupied by them in said city.

[1] Opinion of Assistant Attorney-General, August 7th, 1871; Decision of Acting Secretary of the Interior, August 9th, 1871, 2 Copp's Land-owner, 98.
[2] 14 Stats. 251.

The Commissioner, in a letter to the register and receiver, May 6th, 1870, said: "Although such protests do not, in the opinion of the Commissioner, constitute such an adverse claim as would properly come within the purview of the sixth section of the Mining Act, yet in view of the magnitude of the interests represented to be involved, it is deemed but fair to have the rights of *all* the parties determined by the local tribunals, and you will accordingly notify *all* parties claiming adversely to said application of Becker, that they will be allowed sixty (60) days from the date of your notification, in which to institute proceedings in Court to adjudicate their respective rights in the premises."

Under this decision the petitioners were in doubt whether the duty of commencing proceedings in the Courts devolved under it upon them or the claimant. December 12th, 1870, the Commissioner further instructed the register and receiver that was the duty of the *town lot claimants* to commence such proceedings. From the decision of the Commissioner the town claimants appealed.

The Assistant Attorney-General, in his opinion, said:

"The case presents two questions for consideration:

"1st. Is the claim of the petitioners an adverse claim within the meaning of the sixth section?

"2d. Who must commence the proceedings in the local Courts?

"Possession is one of the elements of title, and is made by this statute a necessary subject of inquiry. If found to be in any one other than the claimant, it is a bar to the issuing of a patent, at least until adjudged wrongful in the manner pointed out in the sixth section.

"There can be no question about this, if the possession relates to the vein or lode, the mine itself; but it is said that it is otherwise if it relates to the surface of the land.

"In the present case, the application for a patent includes the surface and soil, as well as the mineral. I am of opinion that the persons in possession of this surface are adverse claimants, and have an adverse claim within the meaning of this law, and are entitled to be heard in the local Courts, before a patent is issued.

"*Second.* Who should commence the proceedings?

"As a general rule, the suit should be commenced by the party who sets up the adverse claim. I think this rule should apply to all cases except those in which the adverse claimant is in the evident and open possession of the premises, tract, lode, or vein, or a portion of the same. When thus in possession, an adverse claimant who attacks his right to possession should certainly be required to take the initiative. To hold otherwise would be against all the analogies of the law.

"In the case now under consideration, the adverse claimants are in the evident and open possession of the surface of the ground, or a portion thereof, and under the rule as above stated should be made defendants to the proceedings which Becker should be required to bring against them."

A decision of the Commissioner not in accordance with these views was reversed, and sixty days given to Becker after the receipt of notice within which to commence proceedings against the parties in possession.

§ 149. What are Courts of competent jurisdiction.—In

the case of the 420 Mining Company *v.* the Bullion Mining Company,[1] the case was tried in the District Court for the First Judicial District of Nevada, the Court having original jurisdiction in this class of cases. The Court found, as a matter of fact, that the Bullion Company had title to the land in contest; that the 420 Company had no title, and that the Bullion Company had been in exclusive possession since the year 1865. Judgment having been entered for the defendant, and appeal having been taken, the appellate tribunal, the highest in the State, unanimously affirmed the decision of the Court below. The Department, in 1874, considered this as a final adjudication by "Courts of competent jurisdiction." The Secretary, on appeal, said: "I see no good reason now for changing the opinion then expressed. I do not understand that the Supreme Court of the United States has jurisdiction over this class of cases upon writ of error. It certainly cannot change the facts found by the Court below. These facts conclusively establish the right of the Bullion Com-

---

[1] Decision of Secretary, March 22d, 1875, 2 Copp's Land-owner, 5. See 9 Nevada, 240.

pany to the possession of this lode under local laws, so far as that question can be considered by the Department in connection with a possible further stay of proceedings. The Department is only authorized to stay proceedings until the right of possession has been finally adjudicated in the Courts of competent jurisdiction. I think such rights have been so finally adjudicated, where facts are finally found which unmistakably control their disposition." And for the reason that the 420 Company had failed to commence suit within reasonable time after filing its adverse claim, and its suit, when brought, having, so far as the questions before the Department were concerned, been finally decided in favor of the Bullion Company, the former company was held not entitled to any further stay of proceedings.

§ 150. **Contest in Court—Jurisdiction.**—The meaning of Section 2326 of the Revised Statutes U. S. is, that all cases which may arise in the disposal of the mineral lands, shall be tried and determined, if tried at all, in a Court of competent jurisdiction; that the adjudication and determination of that Court shall be final, and a patent for the tract in controversy shall issue to the successful party or parties, upon showing further compliance therewith. It is equally clear that when the Court has acquired jurisdiction of the subject-matter in controversy, all other proceedings except those mentioned must be stayed until such determination is made, if the suit be prosecuted with reasonable diligence.

The only question which can ever rise is, whether the adverse claimant has complied with the terms of the act, so as to bring his case within it. He must file his claim during the period of publication, showing its nature, boundaries, and extent, and bring suit for the recovery of the possession of it within thirty days thereafter, or be deemed to have waived it. When he has done all this, according to law, it is only necessary for the Department to pass upon the regularity of the claim, leaving the rights of the parties to be determined by the Court.[1]

It is the duty of all the officers under whose notice an adverse

---

[1] C. T. M. Co. v. Pell, Decision of Secretary, Feb. 17th, 1877, 3 Copp's Landowner, 195.

claim properly comes, to examine it and determine whether the claimant has substantially set forth under oath, its "nature, boundaries, and extent"; but if a compliance with the law is shown in these particulars, and a suit has been instituted to determine the rights of the parties, the Land Office can proceed no further with the investigation. It is the duty of the Court in which the suit is pending, to determine all other questions relating to the controversy.

Where, therefore, the adverse claimant has complied with the act, has filed his claim under oath during the period of publication, showing the origin of his title, as well as the nature, boundaries, and extent of the claim, and has brought suit within the time prescribed to recover possession of the portion claimed by applicants, applications for patents will be suspended until the final adjudication and determination of the rights of the parties involved in the suits instituted in the Courts, or until it is shown that such suits have not been prosecuted with reasonable diligence. Objections that go to the merits and not to the form of the claim, are to be tried in the suits in Court, and are not to be further considered by the Land Office, until the final determination of the suit.

Where the objection was that the claim differed in point of description from the original location, and that the adverse claimant had no title to the tract claimed, or if he had, he held it in secret trust for a foreign corporation, and was, therefore, not entitled to present a claim, both of these objections were considered as going to the merits of the case, and not to the form of the claim, and therefore properly triable in the Courts; pending which trial the Land Office refused to enter into their investigation. Upon the institution of the suit in time, the jurisdiction is transferred to the Courts, and the Department has no further duties to perform until a final determination of the case.[1]

§ 151. Jurisdiction of State Courts.—The law provides that where an adverse claim is filed within the time and in the manner specified in the act, certain proceedings " shall be stayed

---

[1] Chambers v. Pitts, In re King of the West Lode, Decision of Secretary of the Interior, December 26th, 1876, 3 Copp's Land-owner, 162.

until the controversy shall have been settled or decided by a Court of competent jurisdiction, or the adverse claim waived. It is the duty of the adverse claimant, within thirty days after filing his claim, to commence proceedings in a Court of competent jurisdiction, to determine the question of the right of possession, and prosecute the same with reasonable diligence to final judgment, and a failure to do so shall be a waiver of his adverse claim."

The act further provides that after such judgment shall have been rendered, the party entitled to the possession of the claim * * may * * file a certified copy of the judgment-roll with the register of the Land Office." And upon compliance with this and other provisions in said act, "a patent shall issue thereon for the claim, or such portion thereof as the applicant shall appear, from the decision of the Court, to rightly possess."[1]

The Supreme Court of Nevada has held that Congress did not, by the passage of this act, nor by the previous Mining Acts, confer any additional jurisdiction upon the State Courts. The object of the law was understood to be, to require parties protesting against the issuance of a patent to go into the State Courts of competent jurisdiction, and institute such proceedings as they might, under the different forms of action therein allowed, elect, and there try the "rights of possession" to the claims.

The Mining Acts did not attempt to confer any jurisdiction not already possessed by the State Courts, nor to prescribe different forms of action. The State statutes regulating the mode of procedure, and the State Statutes of Limitation, were held to apply to all such controversies. An actual, exclusive, and uninterrupted adverse possession for the statutory period, constitutes a bar.

The pendency of a suit to recover possession of a mining claim does not estop the plaintiff, in case of a suit subsequently commenced against himself, from setting up the Statute of Limitations, and claiming rights and privileges thereunder.[2]

---

[1] 17 U. S. Stats. 1872, 91, Sec. 7; Rev. Stats. Sec. 2326.
[2] 420 Mining Co. v. Bullion Mining Co. 9 Nev. 240; 3 Sawyer C. C. 634.

§ 152. **Transfer of causes to the United States Courts —Jurisdiction of mining causes.**—In cases where the only questions to be litigated in suits to determine the right to hold mining claims are, as to what are the local laws, rules, regulations, and customs by which the rights of the parties are governed, and whether the parties have in fact conformed to such local laws and customs, the Courts of the United States, it has been held, have no jurisdiction of the cases under the provisions of the Act giving jurisdiction in suits "arising under the Constitution and laws of the United States," etc., and entitled "An Act to determine the jurisdiction of Circuit Courts of the United States, and to regulate the removal of causes from State Courts, and for other purposes." Approved, March 3d, 1875. (18 U. S. Stats. 470.)[1]

---

[1] An Act to determine the jurisdiction of Circuit Courts of the United States, and to regulate the removal of causes from State Courts, and for other purposes. —*Be it enacted by the Senate and House of Representatives of the United States of America in Congress assembled,* That the Circuit Courts of the United States shall have original cognizance, concurrent with the Courts of the several States, of all suits of a civil nature at common law or in equity, where the matter in dispute exceeds, exclusive of costs, the sum or value of five hundred dollars, and arising under the Constitution or laws of the United States, or treaties made, or which shall be made, under their authority, or in which the United States are plaintiffs or petitioners, or in which there shall be a controversy between citizens of different States, or a controversy between citizens of the same State claiming lands under grants of different States, or a controversy between citizens of a State and foreign states, citizens or subjects; and shall have exclusive cognizance of all crimes and offenses cognizable under the authority of the United States, except as otherwise provided by law, and concurrent jurisdiction with the District Courts of the crimes and offenses cognizable therein. But no person shall be arrested in one district for trial in another, in any civil action before a Circuit or District Court. And no civil suit shall be brought before either of said Courts against any person by any original process or proceeding in any other district than that whereof he is an inhabitant, or in which he shall be found at the time of serving such process or commencing such proceeding, except as hereinafter provided; nor shall any Circuit or District Court have cognizance of any suit founded on contract in favor of an assignee, unless a suit might have been prosecuted in such Court to recover thereon if no assignment had been made, except in cases of promissory notes negotiable by the law merchant and bills of exchange. And the Circuit Courts shall also have appellate jurisdiction from the District Courts under the regulations and restrictions prescribed by law.

Sec. 2. That any suit of a civil nature, at law or in equity, now pending or hereafter brought in any State Court where the matter in dispute exceeds, exclusive of costs, the sum or value of five hundred dollars, and arising under the Constitution or laws of the United States, or treaties made, or which shall be made, under their authority, or in which the United States shall be plaintiff or petitioner, or in which there shall be a controversy between citizens of different

In the case of Trafton v. Nougues, in the Circuit Court of the United States of the Ninth Judicial Circuit in and for the Dis-

States, or a controversy between citizens of the same State claiming land under grants of different States, or a controversy between citizens of a State and foreign states, citizens or subjects, either party may remove said suit into the Circuit Court of the United States for the proper district. And when in any suit mentioned in this section there shall be a controversy which is wholly between citizens of different States, and which can be fully determined as between them, then either one or more of the plaintiffs or defendants actually interested in such controversy may remove said suit into the Circuit Court of the United States for the proper district.

SEC. 3. That whenever either party or any one or more of the plaintiffs or defendants, entitled to remove any suit mentioned in the next preceding section, shall desire to remove such suit from a State Court to the Circuit Court of the United States, he or they may make and file a petition in such suit in such State Court before or at the term at which said cause could be first tried, and before the trial thereof, for the removal of such suit into the Circuit Court to be held in the district where such suit is pending, and shall make and file therewith a bond, with good and sufficient surety, for his or their entering in such Circuit Court, on the first day of its then next session, a copy of the record in such suit, and for paying all costs that may be awarded by the said Circuit Court, if said Court shall hold that such suit was wrongfully or improperly removed thereto, and also for there appearing and entering special bail in such suit, if special bail was originally requisite therein, it shall then be the duty of the State Court to accept said petition and bond, and proceed no further in such suit, and any bail that may have been originally taken shall be discharged; and the said copy being entered as aforesaid, in said Circuit Court of the United States, the cause shall then proceed in the same manner as if it had been originally commenced in the said Circuit Court; and if in any action commenced in a State Court the title of land be concerned, and the parties are citizens of the same State, and the matter in dispute exceed the sum or value of five hundred dollars, exclusive of costs, the sum or value being made to appear, one or more of the plaintiffs or defendants, before the trial, may state to the Court, and make affidavit, if the Court require it, that he or they claim and shall rely upon a right or title to the land under a grant from a State, and produce the original grant, or an exemplification of it, except where the loss of public records shall put it out of his or their power, and shall move that any one or more of the adverse party inform the Court whether he or they claim a right or title to the land under a grant from some other State, the party or parties so required shall give such information, or otherwise not be allowed to plead such grant, or give it in evidence upon the trial ; and if he or they inform that he or they do claim under such grant, any one or more of the party moving for such information may then, on petition and bond, as hereinbefore mentioned in this act, remove the cause for trial to the Circuit Court of the United States, next to be holden in such district; and any one of either party removing the cause shall not be allowed to plead or give evidence of any other title than that by him or them stated, as aforesaid, as the ground of his or their claim, and the trial of issues of fact in the Circuit Courts shall, in all suits, except those of equity and of admiralty and maritime jurisdiction, be by jury.

SEC. 4. That when any suit shall be removed from a State Court to a Circuit Court of the United States, any attachment or sequestration of the goods or estate of the defendant, had in such suit in the State Court, shall hold the goods

trict of California, Sawyer, Circuit Judge, rendered a decision February 5th, 1877, in which he very fully considered the

or estate so attached or sequestered to answer the final judgment or decree, in the same manner as by law they would have been held to answer final judgment or decree had it been rendered by the Court in which such suit was commenced; and all bonds, undertakings, or security given by either party in such suit, prior to its removal, shall remain valid and effectual, notwithstanding said removal; and all injunctions, orders, and other proceedings had in such suit, prior to its removal, shall remain in full force and effect until dissolved or modified by the Court to which such suit shall be removed.

Sec. 5. That if, in any suit commenced in a Circuit Court, or removed from a State Court to a Circuit Court of the United States, it shall appear to the satisfaction of said Circuit Court, at any time after such suit has been brought or removed thereto, that such suit does not really and substantially involve a dispute or controversy properly within the jurisdiction of said Circuit Court, or that the parties to said suit have been improperly or collusively made or joined, either as plaintiffs or defendants, for the purpose of creating a case cognizable or removable under this act, the said Circuit Court shall proceed no further therein, but shall dismiss the suit, or remand it to the Court from which it was removed, as justice may require, and shall make such order as to costs as shall be just; but the order of said Circuit Court, dismissing or remanding said cause to the State Court, shall be reviewable by the Supreme Court on writ of error or appeal, as the case may be

Sec. 6. That the Circuit Court of the United States shall, in all suits removed under the provisions of this Act, proceed therein as if the suit had been originally commenced in said Circuit Court, and the same proceedings had been taken in such suit in said Circuit Court as shall have been had therein in said State Court prior to its removal

Sec. 7. That in all causes removable under this Act, if the term of the Circuit Court to which the same is removable, then next to be holden, shall commence within twenty days after filing the petition and bond in the State Court for its removal, then he or they, who apply to remove the same, shall have twenty days from such application to file said copy of record in said Circuit Court, and enter appearance therein; and if done within said twenty days, such filing and appearance shall be taken to satisfy the said bond in that behalf; that if the clerk of the State Court in which any such cause shall be pending, shall refuse to any one or more of the parties or persons applying to remove the same, a copy of the record therein, after tender of legal fees for such copy, said clerk so offending shall be deemed guilty of a misdemeanor, and, on conviction thereof in the Circuit Court of the United States, to which said action or proceeding was removed, shall be punished by imprisonment not more than one year, or by fine not exceeding one thousand dollars, or both, in the discretion of the Court.

And the Circuit Court to which any cause shall be removable under this Act, shall have power to issue a writ of certiorari to said State Court, commanding said State Court to make return of the record in any such cause removed as aforesaid, or in which any one or more of the plaintiffs or defendants have complied with the provisions of this Act for the removal of the same, and enforce said writ according to law; and if it shall be impossible for the parties or persons removing any cause under this Act, or complying with the provisions for the removal thereof, to obtain such copy, for the reason that the clerk of said State Court refuses to furnish a copy, on payment of legal fees, or for any

whole question, and commented on the difficulties to be met with in the construction of the act.

other reason, the Circuit Court shall make an order requiring the prosecutor in any such action or proceeding to enforce forfeiture or recover penalty as aforesaid, to file a copy of the paper or proceeding by which the same was commenced, within such time as the Court may determine; and in default thereof, the Court shall dismiss the said action or proceeding; but if said order shall be complied with, then said Circuit Court shall require the other party to plead, and said action or proceeding shall proceed to final judgment; and the said Circuit Court may make an order requiring the parties thereto to plead *de novo;* and the bond given, conditioned as aforesaid, shall be discharged so far as it requires copy of the record to be filed as aforesaid.

SEC. 8. That when in any suit, commenced in any Circuit Court of the United States, to enforce any legal or equitable lien upon, or claim to, or to remove any incumbrance or lien or cloud upon the title to real or personal property within the district where such suit is brought, one or more of the defendants therein shall not be an inhabitant of, or found within the said district, or shall not voluntarily appear thereto, it shall be lawful for the Court to make an order directing such absent defendant or defendants to appear, plead, answer, or demur, by a day certain to be designated, which order shall be served on such absent defendant or defendants, if practicable, wherever found, and also upon the person or persons in possession or charge of said property, if any there be; or where such personal service upon such absent defendant or defendants is not practicable, such order shall be published in such manner as the Court may direct, not less than once a week for six consecutive weeks; and in case such absent defendant shall not appear, plead, answer, or demur, within the time so limited, or within some further time, to be allowed by the Court in its discretion, and upon proof of the service or publication of said order, and of the performance of the directions contained in the same, it shall be lawful for the Court to entertain jurisdiction, and proceed to the hearing and adjudication of such suit in the same manner as if such absent defendant had been served with process within the said district; but said adjudication shall, as regards said absent defendant or defendants without appearance, affect only the property which shall have been the subject of the suit and under the jurisdiction of the Court therein, within such district. And when a part of the said real or personal property against which such proceeding shall be taken shall be within another district, but within the same State, said suit may be brought in either district in said State: *Provided, however,* That any defendant or defendants not actually personally notified as above provided, may, at any time within one year after final judgment in any suit mentioned in this section, enter his appearance in said suit in said Circuit Court, and thereupon the said Court shall make an order setting aside the judgment therein, and permitting said defendant or defendants to plead therein, on payment by him or them of such costs as the Court shall deem just; and thereupon said suit shall be proceeded with to final judgment according to law.

SEC. 9. That whenever either party to a final judgment or decree which has been or shall be rendered in any Circuit Court, has died or shall die before the time allowed for taking an appeal or bringing a writ of error has expired, it shall not be necessary to revive the suit by any formal proceedings aforesaid. The representative of such deceased party may file in the office of the clerk of such Circuit Court a duly certified copy of his appointment, and thereupon may enter an appeal or bring writ of error as the party he represents might have

He said: "I have had no little difficulty in satisfactorily construing this act. In the broad sense claimed by some, nearly all cases relating to the title to lands would be swept into the National Courts; for, in the new States, in every action of ejectment involving a question as to the real title, one party or the other goes back to a patent or other grant under the laws of the United States. Since the passage of the Act of Congress of 1866, and subsequent acts upon the same subject, expressly declaring the public lands to be free and open to exploration and occupation for mining purposes, subject to the local laws, regulations, and customs of miners; also, authorizing a sale and patent to parties establishing a right under such local laws, regulations, and customs, it seems to be claimed on this broad principle that all suits relating to disputes about mining claims may be transferred to the National Courts. But, clearly, the great majority of such cases only involve a litigation of precisely the same questions as were litigated in those classes of cases for the many years since the acquisition of California, prior to the passage of those acts of Congress; and they turn upon no disputed construction of the Constitution or the statutes of the United States. In fact, where a patent is authorized to be issued to the possessor under these acts in a contested case, the statute refers the parties to the ordinary tribunals of the country, to determine under the local laws and customs, irrespective of the acts of Congress, which party is entitled to the mining claim, and the patent issues to the party so determined to have the right. (The 420 Mining Company v. The Bullion Mining Company, 3 Sawyer, 634.) Thus, the rights of the parties are determined the laws, regulations, and customs of the locality outside the acts of Congress, without any discussion or controversy as to the construction of those acts. Since some of this class of cases transferred to this Court were retained, but with no little hesitation, the Supreme Court of the United States has decided

done. If the party in whose favor such judgment or decree is rendered has died before appeal taken or writ of error brought, notice to his representatives shall be given from the Supreme Court, as provided in case of the death of a party after appeal taken or writ of error brought.

SEC. 10. That all acts and parts of acts in conflict with the provisions of this Act are hereby repealed.

Approved March 3d, 1875.

several cases which afford a rule for the future, and which, it seems to me, exclude jurisdiction in many cases which the Bar appears to have supposed could be transferred. The case of McStay *v.* Friedman, 92 U. S. R. 724, was a case in which one of the parties relied: First, on the Statute of Limitations. Second, on the title acquired through the city of San Francisco, under the well-known Van Ness Ordinance, and the act of the legislature confirming it. On a writ of error to the State Court, it was sought to sustain jurisdiction of the United States Supreme Court, on the ground that the title derived through the city depended upon the Act of Congress of 1866, (14 St. 4) granting the land to the city in trust for those who held under the ordinances of the city, State Statutes, etc.

"The Court say: 'At the trial no *question was raised* as to the validity or operative effect of the act of Congress.' * * * 'The city title was not drawn in question. The real controversy was as to the transfer of that title to the plaintiffs in error; and this did not depend upon the 'Constitution, or any treaty statute of, or commission held, or authority exercised under, the United States.' Romie *v.* Casanova, 91 U. S. R. 380, is a similar case. At the present term of the Supreme Court, in a case which was *actually transferred from the State Court to this Court*, under section two of the Act of 1875, the same ruling was made. One party claimed certain lots in San Francisco by virtue of possession, in pursuance of the provisions of the Van Ness Ordinance and the Statutes of the State and of the United States confirming said title, while the city claimed the same as being a part of the public squares reserved and set apart for public purposes, in pursuance of the same ordinances and statutes. After the transfer a demurrer was interposed to the jurisdiction of this Court, on the ground that it presented no question arising under the act of Congress, the rights of the parties depending upon the construction of the ordinances of the city and the State statutes alone. On the other hand, it was earnestly urged that it was necessary to construe the act of Congress, in order to find out who the beneficial grantee intended by the act of Congress was. The Court, however, held that the act of Congress referred the question, as to who was entitled to the land, to the city ordinances and the statutes of the State upon the subject; and that their rights must be determined by a construction of those ordinances and statutes. The Supreme Court affirmed this ruling at the present term, thus holding that the same principle adopted in relation to the section providing for writs of error to the State Courts, is also applicable to cases of transfer from the State to the National Courts, under section two of the Act of 1875; that is to say, that unless there is some contest as to the construction of the act of Congress, there is no jurisdictional question in the case.

"So with reference to mining claims, the act of Congress grants certain rights to those who discover, take up, and work mining claims. But it refers the parties to the local laws of the States and Territories, and to the rules, regulations, and customs of miners of the district where the mines are situated, for the measure of their rights. If a dispute arises, as in the cases referred to, the act of Congress refers the parties to the ordinary tribunals, to determine it by the local laws and customs, and not by the act of Congress. Upon the trial of the rights to a mining claim, precisely the same questions are tried, and they are determined by the same laws and customs that were invoked as the measure of the rights of the parties before the act of Congress had been passed. Clearly, the great mass of these cases cannot involve the discussion or any dispute as to the construction of any act of Congress; and when they do not, under the decisions cited, this Court is without jurisdiction so far as this provision of the act is concerned. Where the controversy is upon matters other than the consideration of the Constitution or an act of Congress, the 'correct decision' of *such controversy* cannot possibly '*depend upon the right construction of either.*' No controversy can possibly arise under the Constitution or an act of Congress, when all parties agree as to its construction. There may be a contest as to other matters, but not as to the Constitution or laws in such cases.

"This action was brought in the State Court in Placer County to recover for trespass upon a gravel gold-mining claim, and seeking an injunction restraining the working of the claim by the defendant. There is no fact alleged, either in the complaint or the petition for transfer, indicating that there is any question involved other than those that usually arise in the trial of a right to a mining claim. And it affirmatively appears from the views stated in the petition that such are, in fact, the questions to be tried. It is alleged in the petition, it is true, that defendant located and held his claim under the several acts of Congress relating to the subject. But this is no more than can be said, in a general sense, of all mining claims, since the passage of the several acts referred to. But, as we have seen, that does not necessarily, nor even ordinarily, in this class of cases, involve any question of disputed construction of the act, or any right or question which is not to be determined by the local laws, rules, and customs, without reference to the acts of Congress, precisely as they were before there was any such act in existence

"The only other allegation is, that the 'right to said mining ground by plaintiff depends upon the laws of Congress, and the right or title of defendant to said mining ground aforesaid must also be determined by the acts of Congress under which defendant and petitioner claim title; and that the rights of the plaintiff as against defendant must be determined under the laws of Congress of the United States.' This is, in substance, two or three times repeated; but it is only the statement of a legal conclusion rather than a fact; and a conclusion manifestly founded upon the general idea that all mining claims are so held, that an action relating thereto, involving the rights of the parties to the mine, necessarily arises under the acts of Congress within the meaning of the act giving jurisdiction to this Court—an erroneous conclusion, if I am right in the views before expressed. These allegations express merely the opinion of the petitioner that a jurisdictional question will arise. In my judgment such averments are insufficient to justify a transfer, or retaining the case when brought here. The precise facts should be stated, out of which it is supposed the jurisdictional question will arise; and what the question is, and how it will arise, should be pointed out, so that the Court can determine for itself whether the case is a proper one for consideration in the National Courts. Otherwise the administration of justice will be greatly obstructed, and intolerable inconvenience be the result. Under the fifth section of the act, it is made the imperative duty of the Court *at any stage of the proceedings*, when it appears that 'such suit does not *really* and *substantially* involve a *dispute* or *controversy* properly within its jurisdiction,' to stop the proceeding and remand the case. Where a suit presents no disputed construction of an act of Congress—where there is no contest at all as to what the act means, or what rights it gives—where the only questions are as to what are the local mining laws, rules, and customs, and as to whether the parties have in fact performed the acts required by such local laws, rules, and customs, how can it be said, in any just sense, that such a suit '*really* and *substantially* involves a *dispute* or *controversy*' arising under an act of Congress? The location of the mine involved in the case is more than one hundred and fifty miles from San Francisco, where the Court is held; and many other cases may arise in this State, Nevada, and Oregon, in regard to claims lying from three to five hundred miles distant from the places where the National Courts are held, and between which places the means of communication are by no means easy or cheap. Generally in this class of cases the testimony rests mainly in parol, and there is a multitude of witnesses. The expense of prosecuting or defending such suits, at a large distance from the location of the mines, would be enormous. If the Court should accept a petition containing a bare statement of the opinion of the petitioner that the rights of the parties are derived under an act of Congress, as in this case, the result in most cases would be that the Court would not be able to determine whether the case 'really and substantially involves a dispute or controversy properly within the jurisdiction of the Court,' until the close of the testimony, when it would be necessary to remand the case at last. Such results would largely obstruct the due administration of justice, and work an intolerable inconvenience to honest suitors. Besides, it would encourage transfers of cases over which the Court has no jurisdiction, by unscrupulous parties, for the very purpose of deterring the adverse party from pursuing his rights by reason of the delays, inconvenience, and enormous expense of prosecuting an action of this class at a great distance from home. These difficulties would be especially onerous in cases relating to mining rights, where time is often as important as the right, in the several large States of the Pacific Coast and interior of the continent, and where a Court is held at but one point. A single State, in some instances, it must not be forgotten, contains more territory than all the Middle and New England States together.

"In view of these, in my judgment, weighty considerations, therefore, I think it of the highest importance to the rights of honest litigants, and to the due and speedy administration of justice, that a petition for transfer should state the exact facts, and distinctly point out what the question is, and how and where it will arise, which gives jurisdiction to the Court, so that the Court can determine for itself, from the facts, whether the suit does really and substantially involve a dispute or controversy properly within its jurisdiction.

"Whenever, therefore, the record fails to distinctly show such facts in a case transferred to this Court, it will be returned to the State Court, and under the authority given by Section 5, at the cost of the party transferring it. If I am wrong in my construction of the act, and of the recent decisions of the Supreme Court, the statute (Section 5) happily affords a speedy remedy, by writ of error, upon which this decision and the order remanding the case may be reviewed without waiting for a trial, and the question may as well be set at rest in this case as in any other. It is of the utmost importance that a final decision of the question be had as soon as possible. If counsel desire, I will order the clerk to delay returning the case till they have an opportunity to sue out and perfect a writ of error."

An order was entered returning the case to the State Court from whence it came, with costs against the party at whose instance it was brought to the United States Court.

§ 153. Cancelation of entry pending suit.—Where adverse claimants commence suit within the prescribed time, and the suit is pending and undetermined, no entry should be permitted by either party until a final decree of the Court. If one is improperly made under these circumstances, it will be canceled.[1]

§ 154. Stay of proceedings.—The pendency of a suit commenced on an adverse claim, and in proper time, operates as

---

[1] In re Hidden Treasure Lode; In re Saco Lode, Decision of the Commissioner, Oct. 23d, 1873, Copp's U. S. Mining Decisions, 228.

a stay of all proceedings before the Department on the application for patent until the same is determined.[1]

A suit commenced after the expiration of the thirty days cannot operate as a bar to the issuance of a patent.[2]

New trial as ground of suspension of proceedings.—Where it was not shown that an application for a new trial in the case of an adverse contest in the Courts had been perfected in accordance with the requirements of the local law, there being as conditions precedent to the granting of a new trial the payment of costs, and the vacating of the judgment rendered by the Court on the former trial, it was held necessary for the adverse claimants to show that their motion for a new trial had been granted without conditions. The Office would not recognize their simple application for a new trial as of sufficient force to warrant a further suspension of the case. A new trial must be granted unconditionally to warrant such suspension.[3]

§ 155. Filing consent to judgment.

—A party may waive his claim to the premises adversely claimed, and debar himself from asserting his right to the same in the future. If an applicant files a formal disclaimer of his right, title, and interest to the premises described in the complaint of the adverse claimant, and consents that the plaintiff may have judgment according to his prayer, the plaintiff obtains all he seeks, and the suit is virtually ended and the controversy settled. No reason then exists why a patent should not issue for the tract. The abandonment of the surface ground, or of the entire premises in controversy before the Department, and the continued prosecution of the suit involving the same premises before a Court of competent jurisdiction, are not justified by a correct interpretation of the law; but when the applicant, defendant in a suit by an adverse claimant in a Court of competent jurisdiction, waives his

---

[1] Application of Lambard, In re Earl Mine & Mt. Pleasant Mine, Decision Assistant Secretary, Feb. 17th, 1877, 3 Copp's Land-owner, 194; Ibid. Dec. 26th, 1876; King of the West v. City Rock Lodes, and Ibid. Jan. 3d, 1877; In re Last Chance Mine, No. 2.

[2] Melton v. Lambard, January Term, 1876, Supreme Court of California, Decision of Acting Secretary, Feb. 17th, 1877, 3 Copp's Land-owner, 194; Morse v. Streeter, Copp's U. S. Mining Decisions, 127.

[3] In re Bank of Commerce Lode, Decision Acting Commissioner, Nov. 18th, 1872, Copp's U. S. Mining Decisions, 149.

claim, confesses judgment, and thus acknowledges the plaintiff's superior right to the tract in dispute, he has done all that can be required of him in thus ending the controversy, and should be no longer deprived of a patent to premises to which he has shown himself legally entitled, and which are not embraced within the limits of the adverse claim.[1]

§ 156. **Laches in bringing suit.**—Suits must be commenced within the time prescribed after notification. Laches will be fatal to the claim. Where seven months elapsed after the attorney received information that a decision had been rendered directing the adverse claimant to bring suit, and no suit was commenced against the applicant for patent, the Office declined to further delay proceedings upon the application.[2]

The time for filing adverse claims will not be extended, nor can they be filed *nunc pro tunc*.[3]

Where a party makes an application for patent and shows compliance with the statute, his application for patent will not be indefinitely suspended at the instance of parties who show no desire to have their alleged adverse interests finally determined by the Courts. Where the application had been suspended nearly four years at the instance of the adverse claimants, and no suit was pending which was commenced within the time allowed, the only suit pending having been commenced more than eighteen months after a decision was rendered, the application was allowed to proceed.

The adverse claimants had also entered their discontinuance of the suit; whereupon judgment was rendered for defendant afterward, but not till eighteen months after the time allowed for another suit.[4]

---

[1] In re Application of Lambard, Decision Acting Secretary, Feb. 17th, 1877, 3 Copp's Land-owner, 194; Ibid. Dec. 26th, 1876; King of the West v. City Rock Lodes, and Ibid. Jan. 3d, 1877; In re Last Chance Mine, No. 2.

[2] In re Montana Fluming and Mining Company, Decision of Commissioner, Aug. 16th, 1873, Copp's U. S. Mining Decisions, 216.

[3] In re Jones & Matteson Lode, Decision of Commissioner, Aug. 19th, 1873, Copp's U. S. Mining Decisions, 218; In re Unicorn Lode, Decision of Commissioner, Copp's U. S. Mining Decisions, 194, April 18th, 1873.

[4] Wood v. Hyde, Decision of Commissioner, July 24th, 1874, 1 Copp's Land-owner, 67.

§ 157. Prosecution of suits by adverse claimants—Diligence.—The law not only requires an adverse claimant to *commence proceedings* in a Court of competent jurisdiction, but also to *prosecute the same with reasonable diligence to final judgment.*

A failure on the part of an adverse claimant to comply with either of these requirements is held to be a waiver of his adverse claim. Where more than three years had elapsed since a suit was commenced, and one special term and six regular terms of the Court had been held, and no trial of the cause had, the only orders entered being those of continuance, the applicants were required to furnish a certificate of the clerk of the Court showing at whose instance the several continuances were made. And in case it should appear from such certificate that the several continuances were granted at the instance and request of the adverse claimants, the applicants were ordered to be allowed to make entry of their claims, should no appeal be taken from the decision of the Commissioner within sixty days from the date of the notification to all parties in interest.[1]

§ 158. Abandonment of portion of adverse claim.—Applicants may abandon and file an abandonment of that portion of a claim claimed adversely, and which is represented and described in the plat filed with the adverse claim; the former will be permitted to receive, after survey, a patent for the remainder of the premises described in their application, as in such case no conflict exists.[2]

If a party files an adverse claim to an application, and for any reason concludes not to prosecute the same, he may file with the register and receiver a written statement of the fact that he does not intend to longer contest the right of the applicant, in which event all the papers filed by the applicant and the adverse claimants are to be transmitted to the General Land Office after the entry has been made. This abandonment must be filed before suit is commenced.

Papers filed by adverse claimaints must be received by

---

[1] Clark v. Calkins, Decision of Commissioner, 3 Copp's Land-owner, 98.
[2] In re Fairmount Lode and Mill-Site; In re Fenian Star Lode, Decision of Commissioner, Aug. 4th, 1874; 1 Copp's Land-owner, 82.

the register, and when papers have once been filed with the register they become part of the record, and can neither be withdrawn nor returned, but must be transmitted to the General Land Office with the other papers in the case.[1]

§ 159. Filing an abandonment of surface ground pending conflict, not a termination of contest.—Upon filing an adverse claim, the provision of the law is explicit that all the proceedings, except the publication of notice and the making and filing of the affidavit thereof, shall be stayed until the final adjudication of the case by the authorized tribunal, or a waiver of the adverse claim. The provision of the law that in case two lodes intersect, the prior location shall be entitled to the ore or mineral contained within the space of intersection, does not release the Department from the duty of abstaining from all further proceedings in the case, nor justify the issuing of a patent embracing the premises in controversy, with the exception of immaterial portions abandoned by the applicants as the surface ground. It is considered clear that it was the intention of Congress to refer all questions arising from a conflict of claims, where a suit is duly commenced, to a Court of competent jurisdiction, in the possession of the power necessary to ascertain the truth and facts relating to the same, a power not possessed by the Department; and it is therefore held to be the duty of the Department to refrain from any act that would in any manner interfere with the adjudication of the controversy.

Where, therefore, the Commissioner held that by reason of abandonment, "no necessity exists for a further suspension of proceedings upon the application for patent," his decision was reversed upon the ground that the adverse claim was filed; that the possession of the surface ground in dispute might be of the least importance, a mere incident, that other and far more important questions might be involved, (the location of the lode, for example) and to allow the defendants to obtain the advantage to be derived from the possession of a patent from the Government simply by filing in the office an abandonment of the sur-

[1] Jefferson Mining Co. v. Pennsylvania Mining Co.; In re Pennsylvania Quartz Mine, 1 Copp's Land-owner, 66; Decision of Commissioner, July 21st, 1874.

face ground, was considered an evasion of both the intent and letter of the law.[1]

§ 160. **Cross-applications—Delay in adverse claim.**—In the case of the Prince of Wales Mine v. Highland Chief Mine, Utah Territory,[2] there were cross-applications and conflicting adverse claims. The questions were solved by applying the doctrine of prior location. It appeared that the Prince of Wales Mine was first located and recorded, and that it made the first application for patent, and that the Highland Chief filed an adverse claim thereto after the period of publication had expired. The Secretary of the Interior said:

"If this application and adverse claim had been forwarded to the Commissioner by the local officers, as they were bound to do under the instructions, the adverse claim would have been rejected because not filed within the period of publication. The fault was not that of the Prince of Wales, and it ought not to suffer by the neglect of duty of any official. (Railroad v. Smith, 9 Wall. 99.)

"The Highland Chief afterwards made application for patent while that of the Prince of Wales was pending. The Prince of Wales filed an adverse claim after the period of publication had expired, and the Highland Chief for that reason caused its rejection. In other words, the Highland Chief, by the decision of this Department, struck out and got rid of the adverse claim of the Prince of Wales, for the very reason which should have excluded its adverse claim to the Prince of Wales application. The Prince of Wales had the prior right and the prior location, and it was manifest error in this Department to allow the Highland Chief to transpose the condition of the parties, and thereby materially change the rights of the contending parties."

§ 161. **Fees on filing adverse claim—Claim filed without payment of fees, how treated.**—Section 2238 of the

---

[1] Ayers v. Foley, Decision of Secretary, January 3d, 1877, 3 Copp's Land-owner, 196; Reversing Decision of Commissioner S. C., 3 Copp's Land-owner, 66, sub nom.; Sacramento Mining Co. v. Last Chance No. 2 Mining Co., reversing also on that point case of Antelope Lode, Decision of Secretary, April 1st, 1875, 2 Copp's Land-owner, 2, and approving King of the West Lode, Decision of Secretary, December 26th, 1876.

[2] Decision of the Secretary of Interior, April 1st, 1875, 2 Copp's Land-owner, 2.

Revised Statutes provides that the fees for filing and acting upon each adverse claim shall be five dollars for the register, and a like amount for the receiver. The eighty-ninth paragraph of circular instructions from the Land Office under the statute, provides that the fees shall be paid at the time of filing the adverse claim. An adverse claim cannot be considered as filed until the party who desires to assert an adverse claim against an application for patent has performed all the acts required of him by the Statute. The local officers are required to report to the Commissioner of the General Land Office the amounts received for filing and acting upon adverse claims, and to place said sums to the credit of the United States, and they have no authority of law to receive and place on file any adverse claims, until the legal fees for such filing have been paid in full. Parties who fail to comply with the plain and positive requirements of the law in asserting their adverse claims, cannot thereby prejudice the rights of applicants who strictly comply with the statute. Where, after the papers had been received, and on the succeeding day, the adverse claimant was telegraphed that the papers had been received without the fees, and he was instructed to send the fees, or the adverse claim could not be filed. Two days still remained within which the adverse claimants might have completed their case; but the required fees were not transmitted until the fifth of November, five days after the expiration of the sixty days' notice by publication, and after the period within which adverse claims must be filed; for this reason said papers were not considered as an adverse claim. Such a filing can only be considered as a protest made for the purpose of showing that the applicant has failed to comply with the mining act.[1]

§ 162. Miscellaneous.

Amendment of adverse claim.—An adverse claim cannot be amended after filing, so as to embrace a larger portion of the premises applied for than that described in the original adverse claim.[2]

---

[1] In re Omaha Gold Quartz Mine, Decision of Acting Commissioner, May 12th, 1876, 3 Copp's Land-owner, 36. See Rev. Stats. Sec. 2238; Instructions Feb. 1st, 1877, Subdivision 89.

[2] Decision of Com. Jan. 14th, 1873, Copp's U. S. Mining Decisions, 156.

Evidence of adverse claim.—An affidavit alleging that the contestants had been owners of certain portions of the claim for more than three years, and that they had worked the ground for several years, is not sufficient. It should state in detail the nature of the adverse claim, where and how it originated, whether by purchase or location, etc.[1]

Withdrawal of protest by cotenant.—Where one cotenant has made out a prima facie adverse showing to an application, he cannot be denied his right and privilege of having his adverse rights adjudicated in a Court of competent jurisdiction, by reason of other cotenants having declared their intention to make no further contest. In such a case, time will be given to institute proceedings in a Court of competent jurisdiction, to determine the rights of possession to the premises.[2]

Questions presented for adjudication by the Courts.—In order to ascertain which party is entitled to a patent, it is only necessary to determine which party, at the time of its issue, was the rightful owner of the mining claim in question, as against everybody but the United States, under the laws, rules, customs, and decisions of Courts in force at the time in the locality embracing it. The party who can maintain his right to the claim in the Courts of the country, as against any person but the United States, under those local laws, is the party upon whom Congress intended to confer the right to purchase, no matter how that right originated, if under those laws he has the present right. The object of the suit is simply to ascertain the party who has the right to the claim under the laws of the State and local rules and customs, for that person, when found, is the party upon whom the law confers the privilege—the right to purchase.[3] In suits to ascertain this inquiry, local statutes of limitation apply, and are recognized by the act.

Papers to be filed.—The adverse claimant should file with the other papers which go to make up his adverse claim, either an

---

[1] Thomas v. Richards, Decision of Secretary, March 19th, 1872, Copp's U. S Mining Decisions, 81.
[2] In re Harris Lode, Decision Commissioner, February 12th, 1873, Copp's U. S. Mining Decisions, 158.
[3] 420 Mining Co. v. Bullion Mining Co. 3 Sawyer C. C. 634. See further, as to when a judgment in a suit to try the right of possession of a mining claim is *res adjudicata*, S. C. 9 Nevada, 240.

abstract of the title to the premises claimed, together with a copy of the original notice of location, or certified copies of the original notice of location and the deeds of conveyance, tracing the right of possession from the original locators to such adverse claimant. Where an abstract of title is furnished instead of copies of the original deeds, such abstract should be full and complete, attested by the seal of the recorder.[1]

Applicants for different lodes may become adverse claimants to each other, where there is a claim that the two lodes are identical.[2]

Negligence.—A case having been once suspended and carried to the Courts for adjudication of adverse claims, and having been there dismissed for want of attention and presentation on the part of the adverse claimants, cannot be stayed a second time for such purpose, but must proceed upon the application for patent.[3]

An adverse claim which does not claim the mining ground included in the application for a patent, but simply states that they have a right to construct a dam, ditch, and bedrock flume, through, on, or across it, to connect with their dumping ground, is not an adverse claim within the meaning of the act.[4]

A party having no interest in the mine, and no authority to represent parties who had, is in no position to assert an adverse claim.[5]

Caveat against issuing patents.—Whatever objections third parties desire to make to the issuance of a patent for a mining claim, must be filed with the register and receiver within the prescribed time. At the expiration of that time, if no adverse claim has been filed, the matter is solely between the United States and the applicant for the patent.[6]

---

[1] Decision of Commissioner, October 31st, 1873, Copp's U. S. Mining Decisions, 232.

[2] In re Ajax or Big Indian Lode, Decision Commissioner, September 21st, 1869, Copp's U. S. Mining Decisions, 22.

[3] In re Mountain City Lode, Decision Commissioner, November 17th, 1869, Copp's U. S. Mining Decisions, 23.

[4] In re Application of Taylor & Smith, Decision of Commissioner, April 16th, 1871, Copp's U. S. Mining Decisions, 42.

[5] In re Alger Lode, Decision Commissioner, March 4th, 1872, Copp's U. S. Mining Decisions, 80.

[6] In re Flagstaff Case, December 16th, 1872, Copp's U. S. Mining Decisions, 153.

In the matter of proof of citizenship in setting up adverse claims, the law is complied with if the citizenship is properly alleged, and the fact is not controverted.[1]

A public highway is not an adverse claim where there is no claim to the mine. Should a patent be issued upon the application, the rights of all parties to the use of highways are as secure under the law as if the title had remained in the Government.[2]

Where suit has been decided.—When the register and receiver has been directed to suspend proceedings awaiting the final determination of a suit commenced on an adverse claim asserted against the application for a patent, and the suit has been decided in favor of the applicant, a copy of the decree filed with the register and receiver, and a certificate of the clerk of the Court that no suit is pending against said applicant, brought by the adverse claimant, bringing into question the title to said property, should be filed. Upon the filing of these papers with the register and receiver, they will allow the entry to be made.[3]

Rights of foreign corporations.—A foreign corporation purchasing a patent issued to citizens of the United States, takes all the rights and is entitled to all the privileges that would have accrued to the original patentees, had they retained their interest in the mine. An agent of such foreign corporation is to be treated precisely as would the patentee, so far as rights are concerned, under the United States patent.[4]

---

[1] Magnolia M. Co. v. Magnolia East & West Co., Decision of Commissioner November 27th, 1874, 1 Copp's Land-owner, 135; Decision of Secretary, July 28th, 1875, 2 Copp's Land-owner, 68; Eureka Co. v. Jenny Lind Co., Decision of Secretary, Copp's U. S. Mining Decisions, 169, 173, 177, 178; Kempton Case, Decision of Secretary, January 2d, 1875.

[2] Decision of Commissioner, December 29th, 1871, Copp's U. S. Mining Decisions, 76.

[3] In re Alger Lode, Decision Commissioner, October 30th, 1873, Copp's U. S. Mining Decisions, 232.

[4] In re Searle Lode, Decision of Commissioner, October 8th, 1875, 2 Copp's Land-owner, 115; Rev. Stat. Sec. 2326, last clause.

CHAPTER X.

PLACER CLAIMS—SURVEY, ENTRY, AND PATENT—DIMENSIONS OF CLAIMS—SUBDIVISIONS OF TEN-ACRE TRACTS—EVIDENCE OF POSSESSION—MODE OF OBTAINING PATENT.

§ 163. Conformity of placer claims to surveys—Limits and boundaries.
§ 164. Subdivision of ten-acre tracts—Extent of placer locations.
§ 165. Survey of placer claims—Limitations.
§ 166. Evidence of possession—Sufficient to establish right to patent.
§ 167. Proceedings for patent for placer claims.
§ 168. Details of procedure.
§ 169. Description in the notice.
§ 170. Entry and survey of placer claims under the Act of 1866.
§ 171. Survey of placer claims under the Acts of 1833, 1870.
§ 172. Survey and entry under the Act of 1870.
§ 173. Quantity of placer ground subject to location.
§ 174. Proofs necessary to establish possessory rights.
§ 175. Placer ground located after May 10th, 1872.
§ 176. Conflicting claims—Placer and lode claims.
§ 177. Miscellaneous provisions.

§ 163. Conformity of placer claims to surveys—Limits and boundaries.—Sec. 2329 of the Revised Statutes reads as follows: "Claims usually called 'placers,' including all forms of deposit, excepting veins of quartz, or other rock in place, shall be subject to entry and patent, under like circumstances and conditions, and upon similar proceedings, as are provided for vein or lode claims: but where the lands have been previously surveyed by the United States, the entry in its exterior limits shall conform to the legal subdivisions of the public lands."[1]

§ 164. Subdivision of ten-acre tracts—Extent of placer locations.—Sec. 2330 of the Revised Statutes reads: "Legal subdivisions of forty acres may be subdivided into ten-acre tracts; and two or more persons, or association of persons, having contigu-

---

[1] Rev. Stats. 2329; Sec. 12, (first clause) Act 1870, 16 U. S. Stats. 217; See Secs. (Rev. Stats.) 2319, 2330, 2331, 2334.

ous claims of any size, although such claims may be less than ten acres each, may make joint entry thereof; but no location of a placer claim, made after the 9th day of July, 1870, shall exceed 160 acres for any one person or association of persons, which location shall conform to the United States surveys; and nothing in this section contained shall defeat or impair any bona fide pre-emption or homestead claim upon agricultural lands, or authorize the sale of the improvements of any bona fide settler to any purchaser."[1]

§ 165. Survey of placer claims—Limitation of.—Sec. 2331 of the Revised Statutes is as follows: "Where placer claims are upon surveyed lands, and conform to legal subdivisions, no further survey or plat shall be required, and all placer mining claims located after the 10th day of May, 1872, shall conform as near as practicable with the United States system of public land surveys, and the rectangular subdivisions of such surveys, and no such location shall include more than twenty acres for each individual claimant; but where placer claims cannot be conformed to legal subdivisions, survey and plat shall be made as on unsurveyed lands; and where by the segregation of mineral land in any legal subdivision a quantity of agricultural land less than forty acres remains, such fractional

---

[1] Rev. Stats. 2330.

Sec. 12 of the Act of 1870, 16 U. S. Stats. 217, read: "That claims, usually called 'placers,' including all forms of deposit, excepting veins of quartz, other rock in place, shall be subject to entry and patent under this act, or under like circumstances and conditions, and upon similar proceedings, as are provided for vein or lode claims; *Provided*, That where the lands have been previously surveyed by the United States, the entry in its exterior limits shall conform to the legal subdivisions of the public lands, no further survey or plat in such case being required, and the lands may be paid for at the rate of two dollars and fifty cents per acre; *Provided further*, That legal subdivisions of forty acres may be subdivided into ten-acre tracts; and that two or more persons or association of persons, having contiguous claims of any size, although such claims may be less than ten acres each, may make joint entry thereof; *And provided further*, That no location of a placer claim, hereafter made, shall exceed 160 acres for any one person or association of persons, which location shall conform to the United States surveys; and nothing in this section contained shall defeat or impair any bona fide pre-emption or homestead claim upon agricultural lands, or authorize the sale of the improvements of any bona fide settler to any purchaser."

portion of agricultural land may be entered by any party qualified by law, for homestead or pre-emption purposes."[1]

§ 166. Evidence of possession sufficient to establish right to patent.—Section 2332 of the Revised Statutes reads: "Where such person or association, they and their grantors,[2] have held and worked their claims for a period equal to the time prescribed by the Statute of Limitations for mining claims of the State or Territory where the same may be situated,[3] evidence of such possession and working of the claims for such period shall be sufficient to establish a right to a patent thereto under this chapter, in the absence of any adverse claim; but nothing in this chapter shall be deemed to impair any lien which may have attached in any way whatever to any mining claim or property thereto attached prior to the issuance of a patent."[4]

---

[1] Rev. Stats. 2331.

Sec. 10 of the Act of 1872, 17 U. S. Stats. 94, was as follows: "That the act entitled, 'An Act to amend an act granting the right of way to ditch and canal owners over the public lands, and for other purposes,' approved July 9th, 1870, shall be and remain in full force, except as to the proceedings to obtain a patent, which shall be similar to the proceedings prescribed by Secs. 6 and 7 of this act, for obtaining patents to vein or lode claims; but where said placer claims shall be upon surveyed lands, and conform to legal subdivisions, no further survey or plat shall be required, and all placer mining claims hereafter located shall conform as near as practicable with the United States system of public land surveys, and the rectangular subdivisions of such surveys, and no such location shall include more than twenty acres for each individual claimant, but where placer claims cannot be conformed to legal subdivisions, survey and plat shall be made as on unsurveyed lands; *Provided*, That proceedings now pending may be prosecuted to their final determination under existing laws; but the provisions of this act, when not in conflict with existing laws, shall apply to such cases; *And provided also*, That where, by the segregation of mineral land in any legal subdivision a quantity of agricultural land less than forty acres remains, said fractional portion of agricultural land may be entered by any party qualified by law, for homestead or pre-emption purposes." (See, also, Rev. Stats. 2329, 2334.)

Sec. 16 of the Act of 1870, 16 U. S. Stats. 217, read: "That so much of the Act of March 3d, 1853, entitled, 'An Act to provide for the survey of the public lands in California, the granting of pre-emption rights, and for other purposes,' as provides that none other than township lines shall be surveyed where the lands are mineral, is hereby repealed. And the public surveys are hereby extended over all such lands; *Provided*, That all subdividing of surveyed lands into lots less than 160 acres, may be done by county and local surveyors at the expense of the claimants; *And provided further*, That nothing herein contained shall require the survey of waste or useless lands."

[2] Rev. Stats. 2332.

[3] See Sec. 2324, Rev. Stats.

[4] Sec. 13 of the Act of 1870, 16 U. S. Stats. 217, reads: "Sec. 13.—That where

§ 167. Proceedings for patent for placer claims.—Section 2333 of the Revised Statutes is in the following language: "Where the same person, association, or corporation, is in possession of a placer claim, and also a vein or lode included within the boundaries thereof, application shall be made for a patent for the placer claim, with the statement that it includes such vein or lode, and in such case [1] a patent shall issue for the placer claim, subject to the provisions of this chapter, including such vein or lode, upon the payment of five dollars per acre for such vein or lode claim, and twenty-five feet of surface on each side thereof. The remainder of the placer claim, or any placer claim not embracing any vein or lode claim, shall be paid for at the rate of two dollars and fifty cents per acre, together with all costs of proceedings; and where a vein or lode, such as is described in Sec. 2320, is known to exist within the boundaries of a placer claim, an application for a patent for such placer claim which does not include an application for the vein or lode claim shall be construed as a conclusive declaration that the claimant of the placer claim has no right of possession of the vein or lode claim; but where the existence of a vein or lode in a placer claim is not known, a patent for the placer claim shall convey all valuable mineral and other deposits within the boundaries thereof."[2]

§ 168. Details of procedure.—The provisions and regulations for obtaining patents to veins or lodes apply with some

said person or association, they and their grantors, shall have held and worked their said claims for a period equal to the time prescribed by the Statute of Limitations for mining claims of the State or Territory where the same may be situated, evidence of such possession and working of the claims for such period shall be sufficient to establish a right to a patent thereto, under this act, in the absence of any adverse claim: *Provided, however,* that nothing in this act shall be deemed to impair any lien which may have attached, in any way whatever to any mining claim or property thereto, attached prior to the issuance of a patent."

Sec. 13 of the Act of 1870 applied as well to lode as to placer claims, and lessened the amount of proof required to establish a right to a patent. Instructions, Aug. 8th, 1870, Copp's U. S. Mining Decisions, 253.

[1] Sec. 11, Act of 1872, 17 U. S. Stats. 94, was the same as above, with the addition of the following words in parenthesis: (Subject to the provisions of this act and the act entitled, "Act of 1870," instead of "subject to the provisions of this chapter.")

[2] Rev. Stats. 2333. See Sec. 11, Act of 1872, 17 U. S. Stats. 94; and also, Sec. 2325, Rev. Stats.

slight modifications in the notice, etc., to placer claims, regard being had to the different nature of the two classes of claims, placer claims being fixed, however, at two dollars and fifty cents per acre, or fractional part of an acre.[1]

Where placer claims are upon surveyed lands, and conform to legal subdivisions, no further survey or plat is required, and all placer mining claims located after May 10th, 1872, must conform as nearly as practicable with the United States system of public land surveys, and the rectangular subdivisions of such surveys, and no such location shall include more than twenty acres for each individual claimant; but where placer claims cannot be conformed to legal subdivisions, survey and plat must be made as on unsurveyed lands. But where such claims are located previous to the public surveys, and do not conform to legal subdivisions, survey, plat, and entry thereof may be made according to the boundaries fixed by local laws.[2]

By Section 2330 of the Revised Statutes, authority is given for the subdivision of forty-acre legal subdivisions into ten-acre lots, which is intended for the greater convenience of miners in segregating their claims, both from one another and from intervening agricultural land.

The proper construction is held to be, that these ten-acre lots in mining districts should be considered and dealt with, to all intents and purposes, as legal subdivisions; and that an applicant having a legal claim which conforms to one or more of these ten-acre lots, either adjoining or cornering, may make entry thereof, after the usual proceedings, without further survey or plat.[3]

§ 169. **Description in the notice.**—In cases of this kind, however, the notice given of the application must be very specific and accurate in description, and as the forty-acre tracts may be subdivided into ten-acre lots, either in the form of squares of ten by ten chains, or of parallelograms five by twenty chains, so long as the lines are parallel and at right angles with

---

[1] Instructions, June 10th, 1872, Subdivision 54. February 1st, 1877, Subdivisions 53, 54.

[2] Rev. Stats. 2331. Instructions February 1st, 1877, Subdivision 53.

[3] Instructions, June 10th, 1872, Subdivisions 56, 57, 58. February 1st, 1877, Subdivisions 55, 56.

the lines of the public surveys, it is necessary that the notice and application state specifically what ten-acre lots are sought to be patented, in addition to the other data required in the notice. Where the ten-acre subdivision is in the form of a square, it may be described, for instance, as the " S. E. ¼ of the S. W. ¼ of N. W. ¼," or if in the form of a parallelogram it may be described as the " W. ½ of the W. ½ of the S. W. ¼ of the N. W. ¼," or the " N. ½ of the S. ½ of the N. E. ¼ of the S. E. ¼ of Section ——, Township ——, Range ——," as the case may be ; but in addition to this description of the land, the notice must give all the other data that are required in a mineral application, by which parties may be put on inquiry as to the premises sought to be patented.[1] The proof submitted with applications for claims of this kind must show clearly the character and extent of the improvements upon the premises.

The proceedings necessary for the adjustment of rights, where a known vein or lode is embraced by a placer claim, are clearly defined in the eleventh section of the Act of 1872, Rev. Stats. 2333.

When an adverse claim is filed to a placer application, the proceedings are the same as in the case of vein or lode claims already described.[2]

§ 170. Entry and survey of placer claims under the Acts of 1866-70.

An applicant for a patent for a placer claim was required by the Act of 1866 to come within the same conditions applicable to claimants of veins or lodes, and the proceedings prior to the survey were the same in both instances.[3]

After the expiration of the ninety days' notice given in such cases, proof of which was to be made to the satisfaction of the register, the placer mining claimant, where the subdivision of a forty-acre tract was necessary, might engage, under private contract, either a United States deputy or a county or local surveyor to perform the work at the expense of the claimant; such forty-acre tract to be invariably laid off into four lots of equal area to suit the circumstances of the case, the survey to be ex-

---

[1] Instructions, June 10th, 1872, Subdivisions 59, 60, 61, 62 ; Ibid., February 1st, 1877, Subdivisions 53-60.

[2] See Ante, p. 180.

[3] Instructions, August 8th, 1870, Copp's U. S. Mining Decisions, 253, 265.

ecuted and sworn to, and plat and field-notes filed. Upon which the latter were transmitted by the register and receiver to the Surveyor-General for verification and approval, who, if he found the work to have been correctly executed, would give such ten-acre lot, where the same constituted the entire claim, its appropriate numerical designation in the order of surveyed mineral claims in the township; and where several of these ten-acre lots were contiguous, and constituted one claim, they would not receive separate numbers for each lot, but the whole would receive one number in the order of mineral claims in the township.

The Surveyor-General then marked such claims upon the original township plat on file in his office, and sent an authenticated copy of the plat and field-notes of the survey to the register of the proper local land office, and to the General Land Office, as in the case of vein or lode surveys.

Thereafter, if no adverse claim was presented, an entry was allowed of such claims at the rate of two dollars and fifty cents for each acre, or fractional part of an acre, embraced in the survey; the local land officers preserving an unbroken, consecutive series of numbers for all mineral entries, both lode and placer, and then reporting to the General Land Office in the usual manner.

These directions applied only to those placer claims which were upon surveyed land, and could not be entered into forty-acre legal subdivisions without interference with the rights of other bona fide mineral or agricultural claimants in the same tract; and in all cases, testimony was required as to whether or not such other claimants to such forty-acre tract existed, and where such were found, the applicant was required, at his own expense, to cause the survey into ten-acre lots, so as to segregate his claim from the remainder; and where there were no such other claimants to any portion of the forty-acre tract, the entry was required to conform in its exterior limits to such forty-acre legal subdivision.

Where there were several placer claims within the same subdivision, their occupants had the option of making joint entry of the land, or of having such smaller subdivisions made at their own cost, and receiving separate patents. Where the placer claim sought to be patented was upon unsurveyed land, a survey and plat thereof had to be made by a United States mineral deputy surveyor, under conditions similar to those applicable to surveys of veins or lodes.[1]

§ 171. Survey of placer claims under Acts of 1866-70.

—In making the survey where placer mines existed upon such forty-acre tracts, the subdivision was required to be invariably into ten-acre lots, in the form either of squares, one side of which should be ten chains, or in the form of parallelograms, one side of which should be five and the other twenty chains, as might the better embrace such placer claim. But the lines of these surveys were not allowed to run diagonally to those of the regular surveys, but were required to be parallel and at right angles therewith, so as to avoid confusion in the description of the remainder of the land.

In case there existed a vein or lode-claim upon such forty-acre tract, the subdivision into ten-acre lots was not imperative, and the survey in such case might be executed in such manner as would segregate the portion of land actually containing the mine, and used as surface ground for the convenient working thereof, from the remainder of the tract, which remainder would be patented to the agriculturist to whom the same might have been awarded, subject, however, to the condition that the land might be entered upon by the proprietor of any vein or lode for which a patent had been issued by the United States, for the purpose of extracting and removing the ore where found to penetrate or intersect the land so patented as agricultural, as provided for in the act.

Such survey when executed was to be properly sworn to by the surveyor, either before a notary public, officer of a Court of Record, or before the register or receiver, the deponent's character and credibility to be properly certified to by the officer administering the oath.

Upon the filing of the plat and field-notes of such survey duly sworn to, the same was transmitted to the Surveyor-General for his verification and approval; who, if he found the work correctly performed, properly marked out the same

---

[1] Instructions, May 6th, 1871, Copp's U. S. Mining Decisions, 261.

upon the original township plat in his office, and furnished authenticated copies of such plat and description both to the proper local land office and to the General Land Office, to be affixed to the duplicate and triplicate township plats respectively.

In cases where a portion of a forty-acre tract was awarded to an agricultural claimant, and he caused the segregation thereof from the mineral portion, such agricultural portion was not given a numerical designation, as in the case of surveyed mineral claims, but was simply described as the "Fractional —— quarter of the —— quarter of section ——, in township ——, of range —, — meridian, containing —— acres, the same being exclusive of the land adjudged to be mineral in said forty-acre tract." The surveyor was to correctly compute the area of such agricultural portion, which computation was to be verified by the Surveyor-General.

After the authenticated plat and field-notes of the survey were received from the Surveyor-General, the General Land Office issued the necessary order for the entry of the land, and in issuing the receiver's receipt and register's patent certificate, the latter officers were invariably to be governed by the description of the land given in the order from the General Land Office.[1]

§ 172. **Survey and entry under Act of 1870.**—Under the twelfth section of the Act of 1870 the Surveyors-General were authorized to have such subdivisions into ten-acre tracts made by their deputies, when applied for by claimants, numbering each ten-acre tract with consecutive numbers of claims in the township, as in the case of other mineral surveys; and if the service was performed by county and local surveyors, as authorized by the sixteenth section, it was the duty of the Surveyor-General to verify the surveys so executed, and if found correctly done, to adopt the same and certify the fact, appending his approval, as in cases of surveys made under his own direction. The expense of such subdividing was required to be defrayed by the mining claimants.[2]

---

[1] Instructions May 6th, 1871, Copp's U. S. Mining Decisions, 261.
[2] Instructions August 8th, 1870, Copp's U. S. Mining Decisions, 253.

§ 173. Quantity of placer ground subject to location.—

Sections 2330 and 2331 of the Revised Statutes, (see Ante, Secs. 164, 165) are construed to mean that, after the ninth day of July, 1870, no location of a placer claim can be made to exceed 160 acres, whatever may be the number of locators associated together, or whatever the local regulations of the district may allow; and that from and after May 10th, 1872, no location made by an individual can exceed twenty acres, and no location made by an association of individuals can exceed 160 acres, which location of 160 acres cannot be made by a less number than eight bona fide locators; but that whether as much as twenty acres can be located by an individual, or 160 acres by an association, depends entirely upon the mining regulations in force in the districts at the date of the location; it being held that such mining regulations are in no way enlarged by the statutes, but remain intact and in full force with regard to the size of locations, in so far as they do not permit locations in excess of the limits fixed by Congress; but that where such regulations permit locations in excess of the maximums fixed by Congress, they are restricted accordingly. The regulations as to the manner of marking locations on the ground, and placing the same on record, must be observed in the case of placer locations, so far as the same are applicable; the law requiring, however, that where placer claims are upon surveyed public lands, the locations must hereafter be made to conform to legal subdivisions thereof, as near as practicable.[1]

Placer claims before the Act of 1870.—It was held that in mining districts over which the lines of the public surveys had not been extended, a placer claim held and occupied according to the district regulations, upon which not less than $1,000 had been expended, might, in the absence of an adverse claimant, and after the usual proceedings, be surveyed, entered, and patented, whatever might be its shape or area, provided that such claim was located at a date prior to the passage of the Act of July 9th, 1870, which interdicted, after that date, the location of a claim by any person or association of persons, in extent

---

[1] Instructions, June 10th, 1872, Sudivisions 65, 66; Ibid. February 1st, 1877, Subdivisions 61-70.

exceeding 160 acres, whatever the mining regulations might prescribe.

But, upon lands which had been surveyed, no lot or claim smaller than ten acres could be patented to any person or association of persons, under said Act; the subdivision of forty-acre tracts into ten-acre legal subdivisions to be effected in the manner prescribed by the law and the instructions.[1]

The size of placer claims located prior to the Act of 1870 was regulated and controlled by the local law. Subsequent to July 9th, 1870, and prior to May 10th, 1872, no location of a placer claim could exceed 160 acres. From and after the passage of the Act of May 10th, 1872, no individual location can exceed twenty acres, and no location made by an association can exceed 160 acres. There is nothing in the mining acts of Congress forbidding one person, or an association of persons, *purchasing* as many *separate and distinct locations* as he or they may desire, and embracing in one application for patent the entire claim to which they have the possession and the right of possession, by virtue of compliance with the local laws and Congressional enactments. The law does not require an expenditure of $500 upon each *location* of a placer claim embraced in an application for patent, where the locations are contiguous and constitute one claim. Where an application embraces two or more distinct tracts of placer mining ground, the required amount—viz: $500—should be expended upon each tract, and a copy of the diagram and notice posted upon each tract, to entitle the claimant to make entry thereof.[2]

Placer claims on surveyed lands under Acts of 1866 *and* 1870.—In regard to placer claims on surveyed lands, where the claimant applied to enter 160 acres in legal subdivisions, no survey and plat of the claim were required; the entry being allowed to be completed at the local land office as soon as satisfactory proof had been made after the expiration of ninety days' notice and publication, provided that no adverse claimant had, in the meantime, appeared. Where the claimant of a

---

[1] Decision Commissioner, March 1st, 1871, Copp's U. S. Mining Decisions, 40.
[2] The laws do not limit the number of locations one person may make in a mining district. (1 Copp's Land-owner, 91; Decision of Commissioner, Nov. 21st, 1874, 1 Copp's Land-owner, 131; See Decision of Commissioner, July 10th, 1873, Copp's U. S. Mining Decisions, 211.)

placer mine desired the subdivision of a quarter section, the service might be performed by county and local surveyors, at the expense of the claimant, as required by law.[1]

§ 174. Proofs necessary to establish possessory right to placer claim.—Section 2332 of the Revised Statutes, by its provisions, greatly lessened the burden of proof, more especially in the case of old claims located for many years, the records of which, in many cases, have been destroyed by fire, or lost in other ways during the lapse of time, but concerning the possessory right of which all controversy or litigation has long been settled. The applicant is not required to produce evidence of location, copies of conveyances, or abstracts of title, as in other cases, but is required to furnish a duly certified copy of the Statute of Limitations of mining claims for the State or Territory, together with his sworn statement giving a clear and succinct narrative of the facts as to the origin of his title, and likewise as to the continuation of his possession of the mining ground covered by his application, the area thereof, the nature and extent of the mining that has been done thereon; whether there has been any opposition to his possession or litigation with regard to his claim, and if so, when the same ceased; whether such cessation was caused by compromise or by judicial decree, and any additional facts within the claimant's knowledge having a direct bearing upon his possession and bona fides which he may desire to submit in support of his claim.[2]

There should likewise be filed a certificate, under seal of the Court having jurisdiction of mining cases within the judicial district embracing the claim, that no suit or action of any character whatever involving the right of possession to any portion of the claim applied for is pending, and that there has been no litigation before said Court affecting the title to said claim or any part thereof, for a period equal to the time fixed by the Statute of Limitations for mining claims in the State or Territory, other than that which has been finally decided in favor of the claimants. The claimant should support his narrative of facts rela-

---

[1] Instructions, August 8th, 1870, Copp's U. S. Mining Decisions, 259.
[2] Instructions, June 10th, 1872, Subdivisions 68, 69; Ibid, February 1st, 1877, Subdivisions 61-70.

tive to his possession, occupancy, and improvements, by corroborative testimony of any disinterested person or persons of credibility, who may be cognizant of the facts in the case, and are capable of testifying understandingly in the premises. It is to the advantage of claimants to make their proofs as full and complete as practicable.[1]

§ 175. Placer mining claims located after May 10th, 1872, must conform, as nearly as practicable, with the public surveys. In other words, the location of a placer mine upon surveyed land, made after May 10th, 1872, should embrace legal subdivisions of the public lands, where this can be done without interfering with the rights of other bona fide mineral, agricultural, or other claimants in the same tract. Where placer mines are situated upon unsurveyed land, or where, by reason of some other bona fide claimant, a legal subdivision of surveyed land cannot be embraced in an application for patent, survey, plat, and entry must be made of the premises for which a patent is sought, in accordance with the boundaries fixed by local laws.[2]

§ 176. Conflicting claims—Placer and lode claims.— The premises described in an application for a quartz lode, embraced a portion of the premises described in an application for certain placer mining ground. No mention was made by the applicant for the placer ground that any vein or lode claim existed within the exterior boundaries of the premises described in their application for patent, and hence, in the language of the law it was a "conclusive declaration that the claimant of the placer claim has no right of possession of the vein or lode claim."

The applicant for the lode claim was allowed to proceed with his application for patent, and make entry of the premises described in his application, upon full compliance with the law and instructions.

---

[1] Instructions, June 10th, 1872, Subdivisions 70, 71, 72; Ibid, February 1st, 1877, Subdivisions 61–70.
[2] Decision Acting Commissioner, May 19th, 1873, Copp's U. S. Mining Decisions, 200; Instructions, June 10th, 1872; Ibid, 275. February 1st, 1877, Subdivision 53.

Applicants for placer mining claims are required to furnish proof that the premises described in their said applications do not contain any known veins or lodes of quartz, or other rock in place, bearing gold, silver, cinnabar, lead, tin, or copper.[1]

§ 177. Miscellaneous provisions.—*Cinnabar and copper deposits.*

—As these deposits are found "in rock in place," rather than in the form of placers, it is held by the Land Office that parties desiring to obtain patents for lands, valuable on account of the deposits of cinnabar or copper, must enter the same as lode-claims.[2]

Publication.—One notice may include a description of all such tracts, giving an accurate description of each parcel separately. This rule is, however, confined to placer claims in the same neighborhood, and not to claims situate at wide distances from each other in different land or mining districts.[3]

Liens are fully protected by the act, and the parties, after a patent, are in even a better condition to enforce their liens than if the question of titles was undetermined.[4]

Surveyed lands.—Placers must be regarded as being on unsurveyed land, until the township plat, approved by the Surveyor-General, is filed in the local office.[5]

Placer claims embracing five-acre lots must be surveyed when application is made for a patent, as the smallest legal subdivision of the public lands is a ten-acre tract.[6]

Certificates of improvements in case placer claims embrace legal subdivisions.—Where a placer claim is situate upon surveyed land, and conforms to legal subdivisions thereof, no survey or plat is required of the claim, and proof of improvements may consist of affidavits of parties who are familiar with the claim, and who can testify understandingly in regard to the character and amount of improvements.[7]

---

[1] In re Maryland Quartz Mine, Decision of Commissioner, October 17th, 1873, Copp's U. S. Mining Decisions, 226.
[2] Decision of Commissioner, Aug. 26th, 1871, Copp's U. S. Mining Decisions, 60.
[3] In re Franklin Lode of Colorado, Decision Commissioner, June 19th, 1871, Copp's U. S. Mining Decisions, 45.
[4] In re Powell Claim, Decision Commissioner, March 22d, 1871, Copp's U. S. Mining Decisions, 41.
[5] Decision of Commissioner, Aug. 27th, 1873, Copp's U. S. Mining Decisions, 222.
[6] Ibid. Oct. 23d, 1873, Ibid. 229.   [7] Ibid. Nov. 20th, 1873, Ibid. 235.

CHAPTER XI.

PUBLIC SURVEYS OVER MINERAL LANDS—SURVEYS OF MINING CLAIMS—DUTIES OF SURVEYOR-GENERAL—APPOINTMENT OF DEPUTIES.

§ 178. Appointment of surveyors of mining claims by Surveyor-General.
§ 179. Public surveys extended over mineral lands.
§ 180. Description of vein claims on surveyed and unsurveyed lands.
§ 181. Appointment of deputies.
§ 182. Charges for surveys and publications.
§ 183. Special instructions to deputies.
§ 184. Authority of deputies outside the district.

§ 178. Appointment of surveyors of mining claims by Surveyor-General.—"The Surveyor-General of the United States may appoint in each land district containing mineral lands, as many competent surveyors as shall apply for appointment to survey mining claims. The expenses of the survey of vein or lode claims, and the survey and subdivision of placer claims into smaller quantities than 160 acres, together with the cost of publication of notices, shall be paid by the applicants, and they shall be at liberty to obtain the same at the most reasonable rates, and they shall also be at liberty to employ any United States deputy surveyor to make the survey. The Commissioner of the General Land Office shall also have power to establish the maximum charges for surveys and publication of notices under this chapter; and in case of excessive charges for publication, he may designate any newspaper published in a land district where mines are situated, for the publication of mining notices in such district, and fix the rates to be charged by such paper; and to the end that the Commissioner may be fully informed on the subject, each applicant shall file with the register a sworn statement of all charges and fees paid by such applicant for publication and surveys, together with all fees and money paid the register and the receiver of the Land Office, which statement shall be transmitted, with the other papers in the case, to the Commissioner of the General Land Office."[1]

§ 179. Public surveys extended over mineral lands.— Sec. 2406 of the Revised Statutes reads: "There shall be no further geological survey by the Government, unless hereafter authorized by law. The public surveys shall extend over all mineral lands; and all subdividing of surveyed lands into lots less than 160 acres may be done by county and local surveyors at the expense of claimants; but nothing in this section contained shall require the survey of waste or useless lands."[2]

§ 180. Description of vein claims on surveyed and unsurveyed lands.—Section 2327 of the Revised Statutes is as follows: "The description of vein or lode claims, upon surveyed lands, shall designate the location of the claim with reference to the lines of the public surveys, but need not conform therewith; but where a patent shall be issued for claims upon unsurveyed lands, the Surveyor-General, in extending the surveys, shall adjust the same to the boundaries of such patented claim, according to the plat or description thereof, but so as in no case to interfere with or change the location of any such patented claim."[3]

§ 181. Appointment of deputy mineral surveyors.— Under Section 2334 of the Revised Statutes, the Surveyors-

---

[1] Rev. Stats. 2334.
See Sec. 12, Act 1872, 17 U. S. Stats. 95.
Fees of Registers and Receivers, see Rev. Stats. 2238.
See, also, Rev. Stats. Secs. 2330, 2331, 2406.
Sec. 12 of the Act of 1872, 17 U. S. Stats. 95, was the same as Rev. Stats. 2334, with the following words added: "The fees of the register and receiver shall be five dollars each for filing and acting upon each application for patent or adverse claim filed, and they shall be allowed the amount fixed by law for reducing testimony to writing, when done in the Land Office, such fees and allowances to be paid by the respective parties; and no other fees shall be charged by them in such cases. Nothing in this act shall be construed to enlarge or affect the rights of either party in regard to any property in controversy at the time of the passage of this act, or of the Act (of 1866); nor shall this act affect any right acquired under said act. And nothing in this act shall be construed to repeal, impair, or in any way affect the provisions of (The Sutro Tunnel Act)."

[2] Rev. Stats. 2406; Sec. 9, Act of July 9th, 1870, 16 U. S. Stats. 218. See Sec. 2334 Rev. Stats.

[3] Sec. 8, Act 1872, 17 U. S. Stats. 94. See Sec. 2325, Rev. Stats.

General of the several districts are required to appoint in each land district as many competent deputies for the survey of mining claims as may seek such appointment; it being distinctly understood that all expenses of these notices and surveys are to be borne by the mining claimants, and not by the United States; the system of making deposits for mineral surveys, as required by previous instructions, being revoked as regards field work, the claimant having the option of employing any deputy surveyor within such district to do his work in the field.[1]

With regard to the platting of the claim, and other office work in the Surveyor-General's office, that officer will make an estimate of the cost thereof, which amount the claimant will deposit with any Assistant United States Treasurer, or designated depository, in favor of the United States Treasurer, to be passed to the credit of the fund created by "individual depositors for surveys of the public lands," and file with the Surveyor-General duplicate certificates of such deposit, in the usual manner. The Surveyor-General is instructed to appoint mineral deputy surveyors as rapidly as possible, so that one or more may be located in each mining district for the greater convenience of miners. The usual oaths are required of these deputies and their assistants as to the correctness of each survey executed by them.[2]

§ 182. **Charges for surveys and publications.**—The law requires that each applicant shall file with the register and receiver a sworn statement of all charges and fees paid by him for publication of notice, and for survey, together with all fees and money paid the register and receiver; which sworn statement is required to be transmitted to the General Land Office for the information of the Commissioner, who will take action with the view of correcting any abuses, in cases of excessive or exorbitant charges by any surveyor or publisher.[3]

§ 183. **Special instructions of Commissioner, as ex-officio Surveyor-General.**—The Commissioner of the General

---

[1] Instructions June 10th, 1872, Subdivision 74; February 1st, 1877, 82-92.
[2] Ibid. Subdivisions, 75-77; Ibid.
[3] Ibid. Subdivisions, 78, 79; Ibid.

Land Office, acting as ex-officio Surveyor-General, has issued the following instructions to deputy mineral surveyors in Arkansas, and they will apply to all States where the Commissioner acts as such ex-officio Surveyor-General:

"In the discharge of your duties as deputy mineral surveyor you will be governed by the instructions herein contained, and the circular instructions from this Office. No official survey will be made except on application of the claimant or his duly authorized agent. The claimant must in all cases make satisfactory arrangements with the United States deputy mineral surveyor, for the payment of his services and those of his assistants in making the survey, as the United States will not be responsible for the payment of the same. In making a survey of a claim you will begin at some corner of the public surveys, and run a line either by course and distance, or by triangulation, to a corner of the claim, designating this corner as 'Corner No. 1; beginning.' You will then calculate the true course and distance in a direct line from the corner of the public surveys to said 'Corner No. 1.' From Corner No. 1 you will proceed with the survey of the claim, giving courses and distances of the exterior boundaries, establishing a corner at each angle of the survey. You will describe the corners fully, stating whether a post or stone, the size, depth in the ground, and how marked. The corner monuments will be marked No. 1, No. 2, etc., as you proceed with the survey; also with the number of the survey. You will note all objects crossed by your lines of survey, such as prior surveys, lodes, ditches, ravines, or lines of the public surveys. You will note all shafts and their depths, all adits, cuts, drifts, shaft-houses, mills, etc., and represent the respective locations of the same upon the plats. After describing fully the improvements on the claim, you will give your opinion in regard to the actual value thereof. You will give the names of adjoining claimants, if any, and state the quarter-section, township, and range in which the claim is situated. On the plats the section lines will be represented in *black* ink; the quarter-section lines in *red*. The field-notes will be made upon paper of uniform size. The plats will be prepared upon paper 12x18 inches in size. In each case *four plats* and *one copy* of the original field-notes will be transmitted to this Office for approval.

When the same have been examined and approved, the original field-notes will be retained in this Office; one copy of the plat will be transmitted to the register of the proper land district, to be retained on his files for future reference, and two plats and one copy of the field notes will be returned to you to be handed the applicant, to be disposed of as follows, viz: 1st. One copy of the plat to be posted on the claim; and 2d. One plat and the copy of field-notes to be filed by the applicant with the register and receiver with his application for patent. Accompanying the plat and field-notes transmitted by you to this Office for approval, you will forward the affidavits of at least two responsible parties, that an amount of not less than five hundred dollars has been expended upon the claim in actual labor and improvements. Great care should be exercised to have the courses and distances expressed in the field-notes correspond with those represented on the plats. Your attention is called to circular instructions from this Office dated June 10th, 1872, and November 20th, 1873." [1]

§ 184. Authority of deputies outside the district.—A deputy mineral surveyor is not authorized to make surveys of mineral claims outside of the State or district for which he is appointed.[2]

[1] Instructions of Commissioner, Feb. 19th, 1875; 2 Copp's Land-owner, 34.
[2] Decision of Commissioner, Aug. 6th, 1872; Copp's U. S. Mining Decisions, 131.

CHAPTER XII.

INTERSECTION OF VEINS.

§ 185. Intersection of veins.
§ 186. Conflicts as to surface ground.
§ 187. Identity of lodes.
§ 188. Interference of claims.
§ 189. Abandonment of surface ground.

§ 185. Intersection of veins.—Sec. 2336 of the Revised Statutes reads: "Where two or more veins intersect or cross each other, priority of title shall govern, and such prior location shall be entitled to all ore or mineral contained within the space of intersection; but the subsequent location shall have the right of way through the space of intersection for the purposes of the convenient working of the mine. And where two or more veins unite, the oldest or prior location shall take the vein below the point of union, including all the space of intersection." [1]

The construction which has been given to this part of the law is that a party has a right to a patent for the number of feet along his lode or vein to which he has the local title, upon full compliance with the law and instructions; provided, however, that where another lode crosses, the ore at the space of intersection of the two lodes belongs to the party who owns the prior location of the two, whether patented first or second.

The law clearly refers to cross lodes, and provides that the ore at the crossing of the two lodes shall belong to the first valid location, and hence, where a patent issues for a mining claim which crosses one already patented, the surface ground in conflict is excepted from the second patent, but the subsequent patentee has the right under his patent to the lode for the distance patented, with the proviso hereinbefore referred to, viz: that the ore at the space of intersection of the cross lodes shall belong to the prior location.[2]

---

[1] Rev. Stats. 2336; Sec. 14, Act of 1872; 17 U. S. Stats. 96.
[2] Decision of Acting Commissioner, Feb. 25th, 1876; 2 Copp's Land-owner, 178.

Until two lodes have been developed it cannot be ascertained with certainty that they are one and the same. If the same, the law provides which shall have the better right: if separate, both parties have the right under the law to follow their vein to any depth, although it may enter the land adjoining.

Ordinarily, a few words of explanation will convince the holder under a patent, that a plea or adverse claim is unnecessary where a survey for another lode crosses his own premises, as the ground in conflict is already patented to him, and will be excepted from the patent issued under the subsequent application. Should the patentee persist in filing an adverse claim, the register will receive it, and give him the usual notice in writing that the same is rejected on the grounds above recited, when he may appeal to the General Land Office if he desires to do so.

But where it appeared that the premises conveyed by the patent were incorrectly described therein; that the land conveyed lay considerably east of that claimed, and it became a duty, therefore, to protest against the issuance of a patent on a conflicting survey until a second patent was issued for the lode, correctly describing the claim, the register was directed to receive such plea or protest as the party might desire to file, and transmit the same to the General Land Office with the other papers in the case, after the entry had been perfected as usual.[1]

§ 186. Conflicts as to surface ground.

Where a record showed that the claimants had fully complied with all the requirements of the law, and that no adverse claim was filed in time, and the survey showed a partial conflict as to surface ground with the patented claim of another company, whose lode left the surface ground patented to it, and extended under the surface ground of the other applicant, it was held that the second section of the Act of July 26th, 1866, under which the company made its location, authorized the patentee to follow the vein or lode, "although it may enter the land adjoining, which land adjoining shall be sold subject to this condition." This provision made it proper to recite the "condition" in the patent for the "land adjoining," whether it was absolutely necessary to

---

[1] In re Searle Lode, Decision of Commissioner, Oct. 8th, 1875; 2 Copp's Landowner, 115.

make such recital or not. It may be that the law would sufficiently protect the patentee without any such recital, but it can do no harm to insert it, and the Land Office may properly make the insertion whenever it is shown, by its own records, that there has been a previous patent for a mineral lode on land adjoining that applied for. The Commissioner had directed that the exception should be in these words: " Excepting from this conveyance the surface ground and lode conveyed to the said International Mining and Exchange Company by said patent, dated September 3d, A. D. 1872."

It was objected to this form of expression, that it found that the lode referred to ran under the premises of the other, and that there was no right to find such a fact.

The Secretary ruled, on appeal, that the rights of all parties would be protected by inserting in the patent the following clause, which was directed to be done : " Excepting from this conveyance the surface ground conveyed to the said International Mining and Exchange Company by its patent, dated September 3d, 1872, and also excepting from this conveyance so much of the Hercules lode, if any there be, as was legally conveyed to the said International Mining and Exchange Company by its aforesaid patent,"[1] and the decision of the Commissioner was modified to that extent.

§ 187. **Identity of lodes.**—The bare possibility that two lodes which are separate and distinct on the surface may subsequently converge so as to form, at some indefinite distance under ground, one and the same lode, is not sufficient basis for an adverse claim, nor for a protest against the issuance of a patent. It would not be expedient to carry any such vague and undeterminable question as this into the Courts, for the reason that until sufficient exploration and development have been made to establish the fact that the lodes unite and are identical, the judgment, in view of the developments of one day, might be reversed by the same tribunal, by other and further developments, the next. Such a construction of the law would suspend

---

[1] In re Seven Thirty and Hercules Lodes, Decision of Secretary of Interior, March, 4th, 1875, modifying Decision of Commissioner, August 17th, 1874, 2 Copp's Land-owner, 18; 1 Ibid. 82.

the disposal of the mineral lands until the attempted adjustment of hypothetical controversies, and will not be entertained.

Besides, under the law, if lodes be found to unite, the parties who have the prior location and patent are as fully invested with title to the lode beyond the point of union, including all the space of intersection, as if the other claim had not been patented. Where, therefore, there is no controversy about the possession of the surface of the claims, no sufficient adverse claim is made out.

A protest, therefore, in the following language: "That on the 4th of February, 1870, a patent was issued by the United States to said Chollar-Potosi Mining Company, for their claim on the Comstock Lode; that they are still the owners of the property described in said patent; that said lodes for which said Julia Gold and Silver Mining Company has made application for patents conflict with the claim of said Chollar-Potosi Mining Company; that said lodes have no existence as separate and distinct lodes from said Comstock Lode; but, on the contrary, all lodes of quartz or other rock in place, or otherwise, bearing gold or silver, heretofore found, or that may hereafter be found within the boundaries described in said application of said Julia Gold and Silver Mining Company, are parts and parcels of the said Comstock Lode, and belong and appertain thereto, and there is no lode within said boundaries separate and distinct from said Comstock Lode," was held insufficient to stay proceedings.[1]

§ 188. Interference of claims.—In commenting on Section 2336, it has been remarked that it appears to be plain, but when applied to the facts in mining cases, and when compared with a section giving all veins within his lines to the locator, and especially when involved with different degrees of title, (patented against possessory) or to the case of several overlapping patents, it may be found ambiguous. "The leading idea of the act is, that a lode is a straight vein whose course can be readily ascertained and indicated by a straight line or a series

[1] Case of the Julia Gold and Silver Mining Company's Application, Decision of Commissioner, May 27th, 1872; Decision of Secretary, Feb. 24th, 1873, Copp's U. S. Mining Decisions, 96, 101.

of straight lines, and that occasionally such a vein is crossed by another in a similar straight line, merely requiring the right of way to give each lode its proper claim; but in fact, a lode is scarcely ever a straight line, and is seldom to be traced without confusion for more than a few feet, and in its course other veins are absorbed into, and offshoots, not only spurs, but, perhaps, better developed veins than itself, run from it in all tortuous directions; and in its extension downward, it invariably dips laterally, and often shows a fork, of which both parts approach the surface; and it will divide, and may or may not unite at another point; and it will abut suddenly upon country rock, and so be thrown far to one side; and instead of showing distinct lines, mineral veins are as irregular, as disproportioned in length and width, as much intermingled, as uncertain to segregate from each other, as are the veins of the hand, or the veins on a block of marble.

"It is as the result of these natural facts that the same lode is so often claimed at various openings by as many sets of claimants, by equally honest and valid or invalid locations.

"If this irregularity were once admitted in any case, the remedy might be obvious, but the practical difficulty consists in compelling such admission; with our present superficial mining, and our present knowledge of mineral deposits, the question whether two claims are upon the same or separate veins, or whether there is a junction or crossing, is always a disputed fact upon which parties will stand, and witnesses will disagree.

"But the greatest objection to the Land Office practice, under the mining acts, is to the granting of overlapping patents. A glance at the plat of any late patent in a well-developed district, will introduce the subject to the reader; three or four surveys, partly crossing, partly parallel, and intersecting at all angles, are frequently seen, so that, unless the plat is colored, the eye can scarcely distinguish one from another; only the rigid application of the rule of preference to prior patents can ever relieve this matter from difficulty; for while the words of a patent always except the *surface* of previous surveys, they still proceed upon the supposition that each survey indicates a separate *vein*.

"The theory that each survey covers a distinct vein, or that a survey covers any vein at all, or that its center or discovery-shaft is sunk on a vein, is all bare assumption—these points depend upon underground developments, and not on diagrams or surface surveys."[1]

Where a protest was filed against an application, for the reason that the survey of the claim applied for conflicted with and embraced a portion of the survey of another lode, a clause was ordered to be inserted in the patent, excepting from the conveyance the surface ground of the latter lode which had already been patented.[2]

§ 189. Abandonment of surface ground.—The applicants have the right, of course, under the law, to follow their vein or lode to the intersection of any other lode. The ore at the space of intersection of the two lodes belongs to the prior location. If the applicants have abandoned from their application all the surface ground claimed adversely, and identity of lodes be not alleged, no necessity exists for a further suspension of proceedings upon the application for patent.[3]

---

[1] Morrison's Mining Rights in Colorado, 27, 28.

[2] In re Equator Lode, Decision of Commissioner, October 26th, 1875, 2 Copp's Land-owner, 114.

[3] Decision Equator Mine; Sacramento Mining Co. v. Last Chance Mining Co., Decision of Acting Commissioner, June 17th, 1876, 3 Copp's Land-owner, 66.

CHAPTER XIII.

MILL SITES—PATENTS FOR NON-MINERAL LANDS.

§ 190. Patents for non-mineral lands.
§ 191. Location of mill sites.
§ 192. Procuring patent.
§ 193. A mill site must be non-mineral in character.
§ 194. Improvements.
§ 195. Mill sites in railroad grants.

§ 190. Patents for non-mineral lands.—Section 2337 of the Revised Statutes of the United States provides that: "Where non-mineral land not contiguous to the vein or lode is used or occupied by the proprietor of such vein or lode for mining or milling purposes, such non-adjacent surface ground may be embraced and included in an application for a patent for such vein or lode, and the same may be patented therewith, subject to the same preliminary requirements as to survey and notice as are applicable to veins or lodes; but no location hereafter made of such non-adjacent land shall exceed five acres, and payment for the same must be made at the same rate as fixed by this chapter for the superficies of the lode. The owner of a quartz-mill or reduction works not owning a mine in connection therewith, may also receive a patent for his mill site, as provided in this section."[1]

§ 191. Location of mill sites.—Mill sites may be located under the provisions of the mining act, and if located should be recorded.

Locators of mining claims, their heirs and assigns, have the exclusive right of possession of the surface ground included within the lines of their locations, upon compliance with the laws of the United States, and with the State, Territorial, and

---

[1] Rev. Stats. 2337, Sec. 15, Act of 1872, 17 U. S. Stats. 96. See Rev. Stats. 2320, 2324.

local regulations governing their possessory titles, where no adverse claim thereto existed on the 10th of May, 1872.

The parties having the right of possession to the surface have also the right of possession to the timber growing thereon.[1]

§ 192. **Procuring patent.**—To avail themselves of the provisions of the law in regard to mill sites, when parties hold the possessory right to a vein or lode, and to a piece of non-mineral land not contiguous thereto, for mining or milling purposes, not exceeding the quantity allowed for such purposes by the local rules, regulations, or customs, the proprietors of such vein or lode may file in the proper Land Office their application for a patent, under oath, which application, together with the plat and field-notes, may include, embrace, and describe, in addition to the vein or lode, such non-contiguous mill site; and after due proceedings as to notice, etc., a patent will be issued conveying the same as one claim.[2]

In making a survey in a case of this kind, the lode claim should be described in the plat and field-notes as "Lot No. 37, A," and the mill site as "Lot No. 37, B," or whatever may be its appropriate numerical designation; the course and distance from a corner of the mill site to a corner of the lode claim to be invariably given in such plat and field-notes; and a copy of the plat and notice of application for patent must be conspicuously posted upon the mill site, as well as upon the vein or lode, for the statutory period of sixty days. In making the entry, no separate receipt or certificate need be issued for the mill site; but the whole area of both lode and mill site will be embraced in one entry, the price being five dollars for each acre and fractional part of an acre embraced by such lode and mill site claim.[3]

In case the owner of a quartz mill or reduction works is not the owner or claimant of a vein or lode, the law permits him to make application therefor, in the same manner prescribed for mining claims, and after due notice and proceedings, in the ab-

---

[1] Decision of Commissioner, October 21st, 1875, 2 Copp's Land-owner, 114.
[2] Instructions June 10th, 1872, Subdivision 87; Land Office Report, 1872, 44; Instructions Feb. 1st, 1877, Subdivision 72.
[3] Ibid. Subdivision 88; Instructions Feb. 1st, 1877, Subdivision 73.

sence of a valid adverse filing, to enter and receive a patent for his mill site at a fixed price per acre.[1]

In every case there must be satisfactory proof that the land claimed as a mill site is not mineral in character, which proof may, where the matter is unquestioned, consist of the sworn statement of the claimant, supported by that of one or more disinterested persons capable from acquaintance with the land to testify understandingly. The law expressly limits mill site locations made from and after its passage to five acres, but whether so much as that can be located, depends upon the local customs, rules, or regulations. The registers and receivers must preserve an unbroken, consecutive series of numbers for all mineral entries.[2]

§ 193. **A mill site must be non-mineral** in character, and where application for patent for a mill site is made, satisfactory proof must be furnished that the land claimed is not mineral in character. Where affidavits did not allege the non-mineral character of the mill site, but only alleged that the same "did not to his (the claimant's) knowledge contain any vein or lode of quartz or other rock in place, bearing gold, silver, cinnabar, lead, tin, or copper," before patent was allowed to issue, additional proof was required that there were no valuable deposits, such as placer or gulch mines, embraced within its exterior boundaries.[3]

§ 194. **Improvements.**—The Surveyor-General's certificates attached to plats of either lode, placer, or mill site claims, should contain a clause in regard to the value of improvements upon such claims—that $500 worth of labor has been expended or improvements made upon the mill site.[4]

§ 195. **A mill site passes to a railroad** if located after the land inured to the road. Where the record of the Land Office

---

[1] Instructions June 10th, 1872, Subdivision 89; Land Office Report, 1872, 44; Instructions Feb. 1st, 1877, Subdivision 74.
[2] Ibid. Subdivisions 90, 91, 92; Ibid. 75, 76, 77.
[3] Decision of Acting Commissioner, May 20th, 1873; Copp's U. S. Mining Decisions, 201; Ibid, July 29th, 1872; Ibid, 129.
[4] Decision of Commissioner, April 16th, 1873; Copp's U. S. Mining Decisions, 193.

showed that the rights of the railroad company to a section of land took effect on the 18th day of December, 1866, that being the date upon which the route of the road was definitely located, it was held that subsequently to that time no adverse right thereto could attach where the land was not mineral in character.[1]

[1] In re Golconda Mine, Decision of Commissioner, Oct. 11th, 1872; Copp's U. S. Mining Decisions, 147.

CHAPTER XIV.

WATER AND OTHER VESTED RIGHTS—RIGHT OF WAY FOR CANALS AND DITCHES—EASEMENTS—DRAINAGE—STATE AND TERRITORIAL LEGISLATION—PATENTS SUBJECT TO VESTED RIGHTS—SUTRO TUNNEL ACT.

§ 196. State and Territorial legislation—Easements—Drainage, etc.
§ 197. Conditions inserted in the patent.
§ 198. Vested rights to use of water—Right of way for canals.
§ 199. Patents subject to vested water rights.
§ 200. Possessory water rights confirmed.
§ 201. Local water rights protected.
§ 202. Conditions as to vested water rights inserted in patent.
§ 203. Mining ditch in railroad grant.
§ 204. Conflicting rights of ditch owners and miners.
§ 205. Exercise of eminent domain for a private ditch company's use.
§ 206. Water rights in California under the Codes.
§ 207. Existing water rights obtained by patent, how affected.
§ 208. Effect of the acts upon previous diversion of water upon patented lands.
§ 209. Recognition of the doctrine of prior appropriation.
§ 210. Effect of the statute upon prior appropriation without Government title.
§ 211. Construction of flumes over public lands.
§ 212. Rights of ditch owners on public lands.
§ 213. Sutro tunnel act.
§ 214. Conditions inserted in patents for mines on Comstock Lode, Nevada.
§ 215. Claims rejected.

§ 196. State and Territorial legislation — Easements, drainage, etc.—"As a condition of sale, in the absence of necessary legislation by Congress, the local legislature of any State or Territory may provide rules for working mines, involving easements, drainage, and other necessary means to their complete development; and those conditions shall be fully expressed in the patent."[1]

The local legislatures authorized to make laws for working the mines.—In order to embody such enactments into patents, registers and receivers were ordered to communicate such laws to the General Land Office.[2] The importance of this section

---

[1] Rev. Stats. 2338; Sec. 5, Act 1866; 14 U. S. Stats. 252.
[2] Instructions, Jan, 14th, 1867, Copp's U. S. Mining Decisions, 239.

consists in the conditions to be expressed in the patent, as the tenure under which the title is held. If a patent had issued without such expressed conditions in it, or legally implied from the law, serious questions might arise as to legislative control over a title emanating from the United States, without such conditions.[1]

§ 197. **Conditions inserted in the patent.**—In every patent issued for either a lode or placer claim a condition is inserted to the following effect: "That in absence of necessary legislation by Congress, the legislature of —— may provide rules for working the mine hereby granted, involving easements, drainage, and other necessary means to its complete development."

This condition gives to the legislature of the State or Territory in which a patented claim is situated, ample power and authority for the enactment of all necessary rules and regulations for the proper working and development of the mines, and this as completely in regard to water ditches and flumes as in any other respect, if parties have by virtue of compliance with local laws, customs, or regulations of miners, or by decisions of Courts, acquired the right to construct and maintain ditches or flumes across the mining grounds occupied by others.

It was the intention of the Land Office under the act that acquired rights to construct and maintain ditches or flumes across the mining ground occupied by others, should be protected and not impaired by the issuance of the patent. The aid of the Courts may be invoked as well after as before the issuance of a patent.[2]

§ 198. **Vested rights to use of water—Right of way for canals.**—Sec. 2339 of the Revised Statutes reads: "Whenever, by priority of possession, rights to the use of water for mining, agricultural, manufacturing, or other purposes, have vested and accrued, and the same are recognized and acknowledged by the local customs, laws, and the decisions of Courts, the possessors and owners of such vested rights shall be main-

---

[1] Yale's Mining Claims, 371.
[2] Decision of Commissioner, April 16th, 1871, Copp's U. S. Mining Decisions, 42, Application of Taylor & Smith.

tained and protected in the same; and the right of way for the construction of ditches and canals for the purposes herein specified is acknowledged and confirmed; but whenever any person, in the construction of any ditch or canal, injures or damages the possession of any settler on the public domain, the party committing such injury or damage shall be liable to the party injured for such injury or damage." [1]

§ 199. **Patents subject to vested water rights.**—Sec. 2340 of the Revised Statutes reads: "All patents granted, or pre-emption or homesteads allowed, shall be subject to any vested and accrued water rights, or rights to ditches or reservoirs used in connection with such water rights, as may have been acquired under or recognized by the preceding section." [2]

§ 200. **Possessory water rights confirmed.**—These provisions relate to the appropriation and use of water for agriculture and other purposes, as well as for mining. The State laws and decisions of the Courts and the local customs are recognized and confirmed. The act treats the appropriation of water so recognized as a *vested right*, by which designation something more than a possessory right to mining claims is implied. By the ninth section of the act of Congress for the sale of public lands in the territory northwest of the Ohio and above the mouth of the Kentucky, after the ordinance of 1785, under date of the 18th of May, 1796, it was provided that all navigable rivers within the territory to be disposed of, by virtue of this act, shall be deemed to be and remain public highways;

[1] Rev. Stats. 2339; Sec. 9, Act 1866, 14 U. S. Stats. 253; see Rev. Stats. 2324.
[2] Rev. Stats. 2340. See Rev. Stats. Secs. 2338, 2339, 2344.
Sec. 17, Act of 1870, 16 U. S. Stats. 218, reads: "That none of the rights conferred by Secs. 5, 8, 9, of the act to which this act is amendatory shall be abrogated by this act, and the same are hereby extended to all public lands affected by this act; and all patents granted, or pre-emption or homesteads allowed, shall be subject to any vested and accrued water rights, or rights to ditches and reservoirs used in connection with such water rights as may have been acquired under or recognized by the ninth section of the act of which this is amendatory. But nothing in this act shall be construed to repeal, impair, or in any way affect the provisions of the 'Act granting to A. Sutro the right of way and other privileges to aid in the construction of a draining and exploring tunnel to the Comstock Lode, in the State of Nevada, approved July 25th, 1866.'"

and that in all cases where the opposite banks of any stream not navigable shall belong to different persons, the stream and the bed thereof shall become common to both.[1]

This provision is still regarded as in force, and was by numerous amendatory acts continued in force, and made applicable to other parts of the country containing public lands.

The language of the Act of 1866 confirms the doctrine of appropriation, introduced by the California Courts.

It makes the right a confirmation *in presenti* as to the claims included, without any preliminary proceedings to obtain a title, as in the case of a mining claim. A grant by act of Congress is the highest source of title known to the law.[2]

§ 201. Local water rights protected.—The Act of 1866 was the result of a policy on the part of Congress, seeking to harmonize the right of sovereignty of the soil, inherent in the General Government, with certain possessory rights growing out of the peculiar condition of things found in the mining States and Territories of the West, which had become engrafted upon the public lands through the operation of local customs and legislative enactments. Its object was to furnish a method of dealing with these conflicting interests so as not to impair the validity of either. It recognizes and preserves such possessory claims as are valid and effective under local regulations, but it does not create them. It substantially embodies a stipulation that the General Government, in disposing of the public domain, will proceed in such a manner as to protect such rights of possession to the same as claimants may be entitled to, under such local customs or laws at the time of the sale by the United States. But these rights derive all their vitality from local regulations. The act of Congress imparts none. It respects those existing at the date of the sale of the public lands, but superadds nothing to their efficacy under the local laws.

The United States will, therefore, under the ninth section, maintain and protect such water rights as have vested and accrued by priority of possession, and which at the time of such disposal are recognized and acknowledged by local customs,

---

[1] 1 Stats. 464.   [2] Yale's Mining Claims, 379, 380.

laws, and decisions of Courts, by which those rights are primarily regulated.[1]

§ 202. Conditions as to vested water rights inserted in patent.—To avoid all misapprehension and uncertainty it was determined by the Land Office in all patents granted in mineral regions of the United States, to insert an additional clause or condition expressly reserving and protecting water rights, and making the patent subject thereto the same as before it was granted.[2]

Water privileges are, since May 10th, 1872, located in the same manner as mines, subject to local regulations, *i. e.* by definitely locating the five acres by monuments, and recording with the district or county recorder. If the local rules and decisions of Courts make the privilege forfeitable for non-user, another party may come in and claim the water-right.[3]

§ 203. Mining ditch in railroad grant.—A grantee of a railroad company brought suit to abate a water ditch as a nuisance. The defendant showed that prior to the Act of Congress of July 26th, 1866, it had acquired a right to the use of the water of a mining ditch, "which right was recognized and acknowledged by the local customs, laws, and decisions of Courts." That act operated a grant to it of the right of way, and of the ditch through which the water was running at the date of the passage of the act. The subsequent grantees of the United States of tracts through which the ditch ran, were held to take subject to this easement, and judgment went for defendant.[4]

§ 204. Conflicting rights of ditch-owners and miners.—In an application by a ditch-owner for an injunction to prevent miners from excavating across the plaintiff's ditch, plaintiff claimed under the Act of Congress of 1866 and the Act of 1870. The provisions of these statutes and of the Act of 1872, it was held, should be considered and construed together, and it

---

[1] Decision of Commissioner, November 23d, 1869, Copp's U. S. Mining Decisions, 24.
[2] Ibid. March 21st, 1872, Ibid. 82.
[3] 1 Copp's Land-owner, 31.
[4] Broder *v.* Natoma W. & M. Co. 50 Cal. 621.

was considered apparent that it was the purpose of the legislature, taken as a whole, to recognize in and conform to the respective classes of licenses therein mentioned, the same rights which were accorded to them by the State Courts prior to the passage of the acts of Congress.

It was further said that there was nothing in the ninth section of the Act of 1866 which made the defendant's right to possess and enjoy his mining claim subordinate to the right of plaintiff to construct his ditch. The clause, "and the right of way for the construction of ditches and canals for the purposes aforesaid is hereby acknowledged and confirmed," cannot be construed to enlarge the grant to ditch-owners, so as to include a right not "recognized and acknowledged by the local customs, laws, and the decisions of the Courts." Nor does the proviso authorize the construction of a ditch or canal across the mining claim of another, whatever may be its effect in respect to "settlers" on agricultural lands of the United States.[1]

§ 205. **Exercise of eminent domain for a private ditch company's use.**—In a California case,[2] a plaintiff sought to procure, by condemnation, certain lands to serve as a site for a bedrock flume to carry the dirt and gravel from its mining claims; and also as a place of deposit for the tailings and refuse matter from its claims. A demurrer to the complaint raised the question of the constitutionality of Subdivision five of Section 1238 of the Code of Civil Procedure of that State, authorizing proceedings of this character.

This statute provides that the right of eminent domain may be exercised in behalf of certain enumerated public uses, and in subdivision five, names among other things "tunnels, ditches, flumes, pipes and dumping places for working mines; also, outlets, natural or otherwise, for the flow, deposit or conduct of tailings or refuse matter from the mines."

It was clear from the averments of the complaint that the object sought was the appropriation of the private property of

---

[1] Titcomb v. Kirk, No. 4473, May 5th, 1876, Supreme Court of California, unreported.
[2] Consolidated Channel Company v. Central Pacific Railroad Co., No. 4960, April 3d, 1876. April 3d, 1876, 51 Cal. 269.

the defendants to the private use of the plaintiff. The proposed flume was to be constructed solely for the purpose of advantageously and profitably washing and mining the plaintiff's mining ground. It was not pretended that any person other than the plaintiff would derive any benefit whatever from the structure when completed. Niles, J., delivering the opinion of the Court, said: "No public use can possibly be subserved by it. It is a private enterprise, to be conducted solely for the personal profit of the plaintiff, and in which the community at large have no concern. It is clear that this case does not come within the meaning of that clause of the Constitution which permits the taking of private property for a public use after just compensation made.

"In the case of Loan Association v. The City of Topeka, (20 Wal. 655) the defendant, acting under the authority of an act of the legislature of Kansas, had issued certain bonds to the plaintiff as a donation to encourage that company in its desire of establishing a manufactory of iron bridges in that city. The act gave to the city council 'power to encourage the establishment of manufactories, and such other enterprises as may tend to develop and improve such city, either by direct appropriation from the general fund, or by the issuance of bonds of such city in such amounts as the council may determine.'

"The Court held that the purpose for which the bonds were issued was not of a public character; that the statute authorizing their issue was unconstitutional and void, and that no lawful tax could be levied for their payment. Mr. Justice Miller said, in announcing the opinion of the Court: 'If it be said that a benefit results to the local public of a town by establishing manufactures, the same may be said of any other business or pursuit which employs capital or labor. The merchant, the mechanic, the innkeeper, the banker, the builder, the steamboat owner, are equally promoters of the public good, and equally deserving the aid of the citizens by forced contributions. No line can be drawn in favor of the manufacturer, which would not open the coffers of the public treasury to the importunities of two-thirds of the business men of the city or town.'

"The reasoning of that opinion is applicable to the present

case. It is not competent for the legislature to authorize the levy of a public tax, or the taking of private property, for the encouragement of a purely private industry.

"But it is contended by the counsel for the plaintiff that the statute referred to (Section 1238, Subdivision 5, C. C. P.) is a legislative declaration that the construction of ditches, flumes, and dumping places for working mines are public uses, in behalf of which the right of eminent domain may be exercised; and they invoke the doctrine that the judgment of the legislature upon such questions is conclusive, and not open to review by the judicial department of the Government. Without doubt it is the general rule, that where there is any doubt whether the use to which the property is proposed to be devoted is of a public or private character, it is a matter to be determined by the legislature, and the Courts will not undertake to disturb its judgment in this regard. This question was fully discussed and the doctrine established in the case of S. and V. R. R. Co. v. City of Stockton, 41 Cal. 147. But in the same case an exception to the general rule is recognized. It is said: 'A case might, indeed, be presented, in which it might appear, beyond the possibility of a question, that a tax had been imposed, or the property of a citizen had been taken for a use or purpose in no sense public; or, in the language of Chancellor Walworth, (5 Paige, 159) "where there was no foundation for a pretense that the public was to be benefited thereby"; and in such case it would be our duty to interfere and afford relief.'

"It would be difficult to suppose a case more completely within the exception stated, and in which the absence of all possible public interest in the purposes for which the land is sought to be condemned is more clear and palpable, than in the case at bar."

§ 206. **Water rights in California under the codes.**—The right to the use of running water flowing in a river or stream, or down a cañon or ravine, may be acquired by appropriation.[1]

[1] Civil Code of Cal. Sec. 1410, Eddy v. Simpson, 3 Cal. 249; Irwin v. Phillips, 5 Cal. 140; Kidd v. Laird, 15 Cal. 161; Hoffman v. Stone, 7 Cal. 49; McDonald v. Bear River Co. 13 Cal. 220; Ortman v. Dixon, 13 Cal. 34; Rupley v. Welch, 23 Cal. 452; McDonald v. Askew, 29 Cal. 200; Nevada Water Co. v. Powell 34 Cal. 109;

The appropriation must be for some useful or beneficial purpose, and when the appropriator or his successor in interest ceases to use it for such a purpose, the right ceases.[1] The person entitled to the use may change the place of diversion, if others are not injured by such change, and may extend the ditch, flume, pipe, or aqueduct by which the diversion is made, to places beyond that where the first use was made.[2] The water appropriated may be turned into the channel of another stream and mingled with its water, and then reclaimed; but in reclaiming it, the water already appropriated by another must not be diminished.[3] As between appropriators the one first in time is the first in right.[4] A person desiring to appropriate water must post a notice, in writing, in a conspicuous place at the point of intended diversion, stating therein:

1st. That he claims the water there flowing to the extent of (giving the number) inches, measured under a four-inch pressure.

2d. The purpose for which he claims it, and the place of intended use.

3d. The means by which he intends to divert it, and the size of the flume, ditch, pipe, or aqueduct in which he intends to divert it.

A copy of the notice must, within ten days after it is posted, be recorded in the office of the recorder of the county in which it is posted.[5] Within sixty days after the notice is posted, the

---

Davis v. Gale, 32 Cal. 26. Water flowing in a ditch not the subject of actual partition; sale and distribution, the only mode of disposing of it. McGillivray v. Evans, 27 Cal. 92.

[1] Civil Code Cal. 1411; Weaver v. Eureka Lake Co. 15 Cal. 271; McKinney v. Smith, 21 Cal. 874; Hill v. Smith, 27 Cal. 476; American Co. v. Bradford, 27 Cal. 360; Ortman v. Dixon, 13 Cal. 34; McDonald v. Bear River Co. 13 Cal. 220; Davis v. Gale, 32 Cal. 22; Nevada Water Co. v. Powell, 34 Cal. 109.

[2] Civil Code Cal. Sec. 1412; Kidd v. Laird, 15 Cal. 161; Butte Table Mt. Co. v. Morgan, 19 Cal. 609; Union Water Co. v. Crary, 25 Cal. 504.

[3] Civil Code Cal. 1413; Richardson v. Kier, 34 Cal. 63; Butte Canal and Ditch Co. v. Vaughan, 11 Cal. 143; Hoffman v. Stone, 7 Cal. 46.

[4] Butte Canal and Ditch Company v. Vaughan, 11 Cal. 143; Kidd v. Laird, 51 Cal. 161; Weaver v. Conger, 10 Cal. 233; B. R. & A. W. & M. Co. v. N. Y. Co. 8 Cal. 327; Hill v. King, 8 Cal. 336; Davis v. Gale, 32 Cal. 26; Eddy v. Simpson, 3 Cal. 249; Irwin v. Phillips, 5 Cal. 140; Maeris v. Bicknell, 7 Cal. 261; McDonald v. Askew, 29 Cal. 200; Ortman v. Dickson, 13 Cal. 33; Phœnix Water Co. v. Fletcher, 23 Cal. 481; Civil Code Cal. 3525.

[5] Civil Code Cal. Sec. 1415; Thompson v. Lee, 8 Cal. 275; Weaver v. Eureka

claimant must commence the excavation or construction of the works in which he intends to divert the water, and must prosecute the work diligently and uninterruptedly to completion, unless temporarily interrupted by snow or rain.[1] By "completion" is meant conducting the waters to the place of intended use.[2] By a compliance with the above rules, the claimant's right to the use of water relates back to the time the notice was posted.[3] A failure to comply with such rules deprives the claimant of the right to the use of the water as against a subsequent claimant who complies therewith.[4] Persons who have heretofore claimed the right to water, and who have not constructed works in which to divert it, and who have not diverted nor applied it to some useful purpose, must proceed as in the Title provided, or their right ceases.

The recorder of each county must keep a book in which he must record the notices provided for. The rights of riparian proprietors are not affected by the provisions of the Title.[5]

§ 207. Existing water rights obtained by patent not affected.

The status of water rights obtained through patent was carefully examined in the case of Union M. Co. v. Ferris, 2 Sawyer, U. S. C. C. 176, a case arising in the United States Circuit Court for Nevada. The effect of the Act of 1866 upon existing water rights obtained through a patent from the United States was thoroughly discussed. The action was commenced to enjoin the defendant from an alleged wrongful diversion of water from Carson River, Nevada. Plaintiff's grantors had located as a possessory claim the land upon which a certain mill was constructed. A dam and mill-race for conducting the water to the mill were also made. The mill had been propelled by

---

Lake Co. 15 Cal. 271; Kimball v. Gearheart, 12 Cal. 27; Parke v. Kilham, 8 Cal. 77.

[1] Civil Code Cal. Sec. 1416; Kimball v. Gearhart, 12 Cal. 27; Weaver v. Eureka Lake Co. 15 Cal. 271; Thompson v. Lee, 8 Cal. 275.

[2] Civil Code Cal. Sec. 1417.

[3] Civil Code Cal. Sec. 1418; Kimball v. Gearheart, 12 Cal. 27; Weaver v. Eureka Lake Co. 15 Cal. 271.

[4] Civil Code Cal. Sec. 1419; as to construction of rules as to forfeiture of possessory rights. Coleman v. Clements, 23 Cal. 245; Wiseman v. McNulty, 25 Cal. 230; St. Johns v. Kidd, 26 Cal. 263; Packer v. Heaton, 9 Cal. 568; McGarrity v. Byington, 12 Cal. 426.

[5] Civil Code Cal. Secs. 1420, 1421, 1422.

the water of the river, and been run for the purpose of reducing metalliferous ores. The plaintiff had become the owner in fee of the land, having procured patents from the United States. The waters of the river naturally flowed through the land. It was found that the plaintiff, by virtue of his ownership of the lower premises, had a right to have the water of the river flow through the premises, unaffected by any right arising out of an adverse use as against the upper premises, unless there was something in the Act of July 26th, 1866, qualifying that right in respect to the lower premises. The effect of this act was then considered, and especially the effect of the ninth section.[1] The act was held not to qualify in any manner the patent of either plaintiff or defendant, as the act was general and did not operate retrospectively, and was passed subsequently to the patents. Hillyer, U. S. District Judge for Nevada, in arriving at the above conclusion, said :

"Prior to the passage of this act, the policy of Congress had been, as shown by its legislation, to grant to purchasers of the public land the bed of a non-navigable stream flowing through the land sold, and the lines of sections were run without reference to the meanderings of such stream ; so that the purchaser of land through which a non-navigable stream flowed, took the bed of the stream and such riparian rights to the water of the stream as belong to the owner of the soil. Several attempts had been made to provide by law for the survey and sale of the mineral lands ; the survey to be rectangular, as in case of other lands. These attempts had always been successfully resisted by mining communities, because, among other reasons, such a survey and sale would have been ruinous to the possessors of quartz lodes, which do not descend perpendicularly, but at a greater or less angle. For seventeen years prior to 1866, the mineral land of California and Nevada had been occupied by citizens of the United States, without objection on the part of the Government; canals and ditches were dug during this time, often at great expense, over the public lands, and the water of the streams diverted by these means for mining and other purposes. Local customs grew up in the mining districts by common con-

---

[1] 14 U. S. Stats. 253, same as Rev. Stats. 2339.

sent, and by rules adopted at miners' meetings for governing the location, recording, and working of mining claims, in the particular mining districts. Possessory rights to public lands, mining claims, and water were regulated by State statutes, and enforced in the State Courts. The rules, customs, and regulations of the miners were also recognized by the Courts and enforced in trials of mining rights. The Courts not applying, in all respects, the doctrines of the common law respecting riparian owners in deciding between these possessors, none of whom had title to the soil, recognized a species of property in running water, and held that he who first appropriated the waters of a stream to a beneficial purpose, had, to the extent of his appropriation, the better right as against persons subsequently locating on the stream above or below ; and that the first appropriator might conduct the water in canals, ditches, and flumes wheresoever he pleased, and apply it to whatsoever beneficial purpose he saw fit, without any obligation to return it to the stream whence it was taken, or preserve its purity or quantity. (Kidd *v.* Laird, 15 Cal. 161 ; Weaver *v.* Eureka Lake Co. Id. 271 ; Lobdell *v.* Simpson, 2 Nev. 272 ; Ophir S. M. Co. *v.* C. Carpenter et al. 4 Nev. 534.) In this posture of affairs, the persons who had constructed these canals and ditches, at an expense of hundreds of thousands of dollars, in many instances, over the public land, saw when the question of the sale of those lands was agitated, that should such sale be made, they, as to these possessory rights, would be at the mercy of the buyer of the legal title, without some protective legislation.

"The Act of 1866, Section 9, of which we have quoted in part, was a consequence of this state of things. It gives the possessor of a quartz lode a right of pre-emption, and it declares that the person who has acquired a right to the use of water, by priority of possession, shall be maintained and protected in the same, if such right is recognized and acknowledged by the local customs, laws, and decisions of Courts. The policy of this enactment—so far, at least, as it relates to agricultural districts —may be doubtful ; but it is the law of the land, and the Courts must carry out what appears to be the intention of the legislature as therein expressed. And that, as indicated by the act, appears to be to grant to the owner of possessory rights to the use of

§ 207   VESTED RIGHTS.   269

water, under the local customs, laws, and decisions, the absolute right to such use, which the Government alone could grant. Under this law, when a possessory right to the use of water is claimed, whether or not such right exists, will be determined by reference to the local customs, laws, and decisions, and the question will be determined just as it would have been had it been raised between occupants before the title to the land had passed from the Government. When the right is thus ascertained, the statute has the force of confirming it to the person entitled under the local laws and decisions. But the act is prospective in its operation, and cannot be construed so as to divest a part of an estate granted before its passage. If it be admitted that Congress has the power to divest a vested right by giving a statute a retrospective operation, that interpretation will never be adopted without absolute necessity. (Blanchard v. Sprague, 3 Sum. 535; Vansickle v. Haines, 7 Nev. 249.)

"But if, when the act was passed, the defendant had such a right, by priority of possession, as that act contemplates, upon the construction which must be given, that right is confirmed in him, and he is entitled to protection as against one claiming as riparian proprietor merely, through a patent issued after, and when no right had vested in the patentee, before the act became a law. The statute is, in effect, incorporated into such subsequent patent, and operates as an exception out of the estate granted to the complainant by the patent of October 10th, 1866. If we have rightly interpreted the act of Congress, and the operation of the patents issued before and after the passage of that act is as we have stated, the case stands in this wise: The defendant's claim, by virtue of adverse enjoyment, falls to the ground, because sufficient time has not elapsed since the lower premises were conveyed by the Government. He cannot sustain his claim by force of the act of Congress, because the complainant's patents of September, 1864, were made before the act was passed, and conveyed the upper premises absolutely, and free from any claims by prior possession merely. We have hitherto been considering the questions of prescription and the act of Congress separately, as it was desirable to determine the effect of the act, and of the patents upon these water rights. But the complainant, having taken the lower premises, subject to such right as the defendant had acquired by priority of possession and the Act of Congress of 1866, if he had also acquired by adverse use a right, as against the proprietors of the upper premises, to divert and use the same quantity of water in the same manner that he would have by virtue of his prior appropriation, this would be a complete defense to this action, for the complainant's right would not be infringed by the diversion, either as proprietor of the upper or lower premises. It is, therefore, necessary to ascertain whether there has been, in fact, such adverse use by defendant as affords a presumption of a grant from the proprietors of the upper premises of the complainant."

After a review of the testimony, this claim of defendants to a prescriptive right was denied. The diversion and, to some extent, unreasonable use by defendant was established, but the case was referred to a master for further testimony, in order to ascertain what kind of a decree should be entered.[1]

As the patent to agricultural land, when issued, relates back to the inception of title, *i. e.* the original entry and payment, one who entered and paid for this land prior to the passage of the Act of 1866, has his land and the water upon it unaffected by that act.[2]

§ 208. Effect upon previous diversion of water on patented lands.—In Van Sickle *v.* Haines,[3] the Supreme Court of Nevada found occasion to discuss the effect of the Act of 1866 upon water rights and the prior diversion of water upon the public lands. The action was for damages, and an injunction to restrain further diversion of a portion of a small tributary of the Carson River in the State of Nevada, and called Daggett Creek. In 1857 the plaintiff had diverted by a ditch, for irrigating and domestic purposes, one-fourth of the water of the creek. He made the diversion at a point then on the public land, but which, in 1864, was patented by the United States to the defendant. In 1865 the plaintiff obtained a patent for his own land, where he *used* the water. In the fall of 1867,

---

[1] Union M. and M. Co. *v.* Ferris, 2 Sawyer, U. S. C. C. 176. See, also, Union M. and M. Co. *v.* Dangberg, Id. 450.

[2] Union M. and M. Co. *v.* Dangberg, 2 Sawyer, U. S. C. C. 451.

[3] 7 Nevada, 249.

defendants constructed a wood flume on defendant Haines' land, and turned into it all the water of the stream, thereby depriving the plaintiff of that part of it which he had been using, and which was the subject-matter of the action. The Court, in deciding the case, maintained that a diversion of water on the public lands confers no right as against the Government; that there is no presumption of a grant as against the Government; that a patent to land from the United States passes to the patentee the unincumbered fee of the soil, with all its incidents and appurtenances, among which is the right to the benefit of all streams of water which naturally flow through it; that the Government of the United States has the absolute and perfect title to and is the unqualified proprietor of all public lands to which the Indian title has been extinguished;[1] and that, as running water is an incident to or part of the soil over which it naturally flows, a patent carries it, if naturally flowing, and also carries the right to its use, and the same right to recover for a diversion of it, as the United States or any other absolute owner could have.

That the right of the riparian proprietor does not depend upon the appropriation of the water by him to any special purpose, but that it is a right incident to his ownership in the land to have the water flow in its natural course and condition, subject only to those changes which may be occasioned by such use by the proprietors above him as the law permits them to make of it, and that the common law was the law of the State and must prevail in all cases where the right to water is based upon the absolute ownership of the soil. Lewis, C. J., delivering the opinion of the Court upon petition for rehearing, said: "As the appellant here claims the water of Daggett Creek as an incident to the land patented to him by the United States, and as it is admitted that he could get only such title and right as was vested in the United States itself, it becomes necessary to ascertain what is the nature of the rights of the Federal Government

---

[1] Van Sickle v. Haines, 7 Nev. 249; Irvine v. Marshall, 20 How. U. S. 561; Jourdan v. Barrett, 4 How. U. S. 185; Bagnell v. Brodnill, 13 Pet. 450; U. S. v. Hughes, 11 How. U. S. 568; U. S. v. Gratiot, 14 Pet. 526; U. S. v. Gear, 3 How. U. S. 20; Colton v. U. S. 11 Id. 231; 1 Opinion U. S. Attorney-General, 471; 1 Wood. & M. 82.

to the public land; and we propose to show: 1st. That it has the absolute and perfect title; 2d. That running water is primarily an incident to or part of the soil over which it naturally flows; 3d. That the right of the riparian proprietor does not depend upon the appropriation of the water by him to any special purpose, but that it is a right incident to his ownership in the land to have the water flow in its natural course and condition, subject only to those changes which may be occasioned by such use by the proprietors above him as the law permits them to make of it; 4th. That the government patent conveyed to Haines not only the land, but the stream naturally flowing through it; 5th. That the common law is the law of this State, and must prevail in all cases where the right to water is based upon the absolute ownership of the soil." [1]

The Court strongly repudiated the idea that the General Government had in any way indicated it to be its policy to permit the diversion of streams from their natural channel on the public lands, and further said: "It is clearly manifest, from the pre-emption laws, that no such policy has ever been sanctioned by it. The only rights which can be acquired to the public agricultural lands are provided for by the pre-emption laws, and the manner of obtaining such rights is specifically set out, and no right to or interest in that character of land can generally be acquired from it, except by means of and by pursuing the requirements of those laws. As it has specifically provided the course to be pursued, and designated the rights which will be recognized, it cannot be said that it has sanctioned any policy or means of acquiring such right, except those designated. But the right to divert water from a natural water-course, it must be admitted, creates an interest in the land from which the diversion is made, in favor of him having the right. (An-

---

[1] Vansickle v. Haines, 7 Nevada, 260. See, also, Corning v. Troy Iron and Nail Factory, 40 N. Y. 206; Mason v. Hill, 5 B. & Ad. 22; Embrey v. Owens, 6 Exch. 353; 3 Kent's Com. 411; 10 Ohio, 297; Gardner v. Newburgh, 2 John. Ch. 166; Ex parte Jennings, 6 Cow. 513; Wadsworth v. Tillotson, 15 Conn. 372; Elliott v. Fitchburg R. R. Co. 10 Cush. 193; Johnson v. Jordan, 2 Met. 239; Page v. Williams, 2 Dev. & B. 55; 8 Cal. 140; Davis v. Fuller, 12 Vt. 190; 26 Wend. 413; Bealey v. Shaw, 6 East, 208; Pugh v. Wheeler, 2 Dev. & B. 50; Crocker v. Bragg, 10 Wend.; United States v. Ames, 1 Woodb. & M. 76; Railroad v. Schurmeir, 7 Wall. U. S. 272; Wilcoxon v. McGhee, 12 Ill. 381; Angell on Water-Courses, Sec. 141; 2 Washb. Real. Pr. 68.

gell on Water-Courses, Sec. 314.) Further than this, the right to divert carries with it the right to go upon the land through which the ditch or flume is conducted, and upon which the dam, by means of which the diversion may be effected, is built, to keep them in repair.

"Suppose, for example, that the dam built by Vansickle for diverting this water from the creek was on land purchased by Haines from the United States, and the ditch through which it was conducted ran through such land: now if Vansickle acquired the right to divert the water as against the United States, he has the same right as against Haines; and that right necessarily gives him the privilege, at any and all times, when he may choose, to go upon the land of Haines to keep his ditch and dam in repair—which, in itself, would be an interest in Haines' land. (Angell on Water-Courses, Sec. 141; 2 Washburn on Real Property, 68.) And thus, contrary to all pre-emption laws and the manifest policy of the government as embodied in them, a person may get an interest in public land beyond his one hundred and sixty acres. All the acts of Congress ever adopted up to 1866, clearly show that it has never been the policy of the United States to sanction anything of the kind; but, on the contrary, to ignore all rights to or interest in its land, except such as might be acquired by means of its own pre-emption laws, or other similar acts expressly conferring or confirming them: in other words, to keep the public land in such condition as that it can give to its patentee an absolute and perfect title, free from all easements and incumbrances of all kinds; no purpose of the General Government is more perfectly manifest, from all the legislation of Congress and decisions of its Courts, than this. The diversion here complained of cannot, then, be said to be sanctioned by any policy of the United States. The Act of Congress of July, 1866, if it shows anything, shows that no diversion had previously been authorized; for, if it had, whence the necessity of passing that act, which appears simply to have been adopted to protect those who at that time were diverting water from its natural channel? Doubtless all patents issued or titles acquired from the United States, since July, 1866, are obtained subject to the rights existing at that time; but this is a different case—for if the appellant has any right to the water, he acquired it by the patent issued to him two years before that time, and with which, therefore, Congress could not interfere. But we do not understand it to be claimed that the act does directly affect this case, but that it is only referred to as exhibiting the policy of the General Government. The answer is, that the policy began with that act, was never in any way sanctioned or suggested prior to the time of its passage, and therefore has nothing to do with this case."[1]

The Government, therefore, being the owner of the soil at the time of the diversion, and, as such owner, having an absolute right to the streams, and not having granted away any rights of water to the plaintiff, nor authorized him to make the diversion complained of, it was held that the United States had a right of action against him for making the diversion; that he did not acquire any right to make such diversion under the preemption laws, and that the right of action passed to the United States patentee, the defendant, who was alone entitled to complain, and not the plaintiff.[2]

The case was carefully distinguished from that large class of cases where it had been held, in California and Nevada, that priority of appropriation gave a right to water, as between appropriators none of whom held the absolute title to the soil. That rule was in nowise disputed in Vansickle v. Haines. The case presented the different phase of one involving the question of title to water as an incident to the soil, where the owner had the same right that the Government of the United States would have had, as against any person diverting it from its land. A United States patent conveys a new title, and the time during which a person diverts water from the public land previous to the issuance of the patent, cannot be set up as an adverse user as against the patentee.[3]

---

[1] Vansickle v. Haines, 7 Nevada, 279.

[2] See, also, Cook v. Foster, 2 Gilman, 652; Wilcoxon v. McGhee, 12 Ill. 381; Colvin v. Burnett, 2 Hill, 620.

[3] Vansickle v. Haines, 7 Nev. 249; Irwin v. Phillips, 5 Cal.; Crandall v. Wood, 8 Cal. 141; Lobdell v. Simpson, 2 Nev. 274; Ophir S. M. Co. v. Carpenter, 4 Nev. 534; Covington v. Becker, 5 Nev. 281; Procter v. Jennings, 6 Nev. 83.

§ 209. Recognition of doctrine of prior appropriation—Reasonable use.

In Atchison v. Peterson,[1] the plaintiffs filed a bill for an injunction to restrain the defendants from carrying on certain mining operations on a creek in Montana Territory. The bill alleged that the water diverted by the complainants from the stream for mining purposes was deteriorated in quality and value. The complainants were the owners of two ditches or canals, known respectively as the Helena Water Ditch and the Yaw Yaw Ditch, by which the creek was tapped and the water diverted and conveyed a distance of about eighteen miles, to certain mining districts, and there sold to miners. The complainants' predecessors in interest asserted a claim to the waters of the creek in November, 1864, and during that year commenced the construction of the ditches, and continued work thereon until August, 1866. The work was then suspended for want of means until the following year, when it was resumed, and in 1867 the ditches were completed and put into operation, at a cost of $117,000.

During the progress of this work, and in 1865, there was some mining on the creek above the point of junction with the ditches, but no continued mining until 1867. From that period the defendants worked mining ground about fifteen miles above the point of junction, washing down earth and "tailings" into the creek, and filling the water with mud, sand, and sediment, impairing its value at that point for further mining. It appeared that the volume of water in the creek, which at the point where defendants worked their mining claims was only about 200 inches, according to the measurement of miners, was increased at the point where the ditches of the complainants tapped the creek, by intervening tributary streams of clear water, to about 1,500 inches. Of this water the Helena Ditch diverted about 500 inches, and took it about eighteen miles to the places where it was sold to miners. The water as it entered the ditch was somewhat muddied and affected with sand, but it appeared that the injury in quality from this cause was so slight as not in any material extent to impair the value of the water for mining, nor render it less salable to the miners at the places where it was carried.

[1] 20 Wall. U. S. 507, affirming S. C. 1 Mont. 561.

With respect to the water diverted by the Yaw Yaw Ditch, it was shown that its deterioration, so far as the deterioration exceeded that of the water in the Helena Ditch, was caused by sand and sediment brought by a tributary which entered the creek below the head of the Helena Ditch.

The mining claims of the defendants were shown to be worth from $15,000 to $20,000 each, and it appeared that the defendants were responsible and capable of responding in damages. The injunction was denied in the Territorial Courts, and the Supreme Court of the United States affirmed the decree.

In view of the foregoing facts, the Supreme Court of the United States considered that the deterioration in quality was very slight, and did not render the water to any appreciable extent less useful or salable for mining purposes at the localities to which it was conveyed, and that no additional labor was required on the ditch on account of the muddied condition of the water. A sand-gate at the head of the ditch was necessary in any event, whether there was mining above the stream or not; and the accumulation of sand from all sources, from the hill-sides as well as from the mining of the defendants, only required the additional labor of one person for a few minutes each day. The injury thus sustained was considered hardly appreciable, in comparison with the damage which would result to the defendants from the indefinite suspension of work on their valuable mining claims. The remedy by injunction was therefore refused, and the parties left to their actions at law.

In commenting upon the doctrine of prior appropriation, and its recognition by Congress, Mr. Justice Field, delivering the opinion of the Court, said:

" By the custom which has obtained among miners in the Pacific States and Territories, where mining for the precious metals is had on the public lands of the United States, the first appropriator of mines, whether in placers, veins, or lodes, or of waters in the streams, on such lands for mining purposes, is held to have a better right than others to work the mines or use the waters. The first appropriator who subjects the property to use, or takes the necessary steps for that purpose, is regarded, except as against the Government, as the source of title in all controversies relating to the property. As respects the use of water for mining purposes, the doctrines of the common law, declaratory of the rights of riparian owners, were, at an early day after the discovery of gold, found to be inapplicable, or applicable only in a very limited extent, to the necessities of miners, and inadequate to their protection. By the common law, the riparian owner, on a stream not navigable, takes the land to the center of the stream, and such owner has the right to the use of the water flowing over the land as an incident to his estate. And, as all such owners on the same stream have an equality of right to the use of the water as it naturally flows in quality, and without diminution in quantity, except so far as such diminution may be created by a reasonable use of the water for certain domestic, agricultural, or manufacturing purposes, there could not be, according to that law, any such diversion or use of the water by one owner as would work material detriment to any other owner below him. Nor could the water by one owner be so retarded in its flow as to be thrown back to the injury of another owner above him. 'It is wholly immaterial,' says Mr. Justice Story, in Tyler v. Wilkinson,[1] 'whether the party be a proprietor above or below in the course of the river: the right being common to all the proprietors on the river, no one has a right to diminish the quantity which will, according to the natural current, flow to the proprietor below, or to throw it back upon a proprietor above. This is a necessary result of the perfect equality of right among all the proprietors of that which is common to all.' 'Every proprietor of lands on the banks of a river,' says Kent, 'has naturally an equal right to the use of the water which flows in the stream, adjacent to his lands, as it was wont to run (*currere solebat*) without diminution or alteration. No proprietor has a right to use the water to the prejudice of other proprietors above or below him, unless he has a prior right to divert it, or a title to some exclusive enjoyment. He has no property in the water itself, but a simple usufruct while it passes along. *Aqua currit et debet currere ut currere solebat.* Though he may use the water while it runs over his land, as an incident to the land, he cannot unreasonably detain it or give it

[1] 4 Mason, 379.

another direction, and he must return it to its ordinary channel when it leaves his estate. Without the consent of the adjoining proprietors, he cannot divert or diminish the quantity of the water which would otherwise descend to the proprietors below, nor throw the water back upon the proprietors above, without a grant or an uninterrupted enjoyment of twenty years, which is evidence of it. This is the clear and settled doctrine on the subject, and all the difficulty which arises consists in the application.'[1]

"This equality of right among all the proprietors on the same stream would have been incompatible with any extended diversion of the water by one proprietor, and its conveyance for mining purposes to points from which it could not be restored to the stream. But the Government being the sole proprietor of all the public lands, whether bordering on streams or otherwise, there was no occasion for the application of the common-law doctrine of riparian proprietorship with respect to the waters of those streams.

"The Government, by its silent acquiescence, assented to the general occupation of the public lands for mining; and, to encourage their free and unlimited use for that purpose, reserved such lands as were mineral from sale and the acquisition of title by settlement. And he who first connects his own labor with property thus situated and open to general exploration, does, in natural justice, acquire a better right to its use and enjoyment than others who have not given such labor. So the miners on the public lands throughout the Pacific States and Territories, by their customs, usages, and regulations, everywhere recognize the inherent justice of this principle; and the principle itself was at an early period recognized by legislation, and enforced by the Courts in those States and Territories. In Irwin v. Phillips,[2] a case decided by the Supreme Court of California, in January, 1855, this subject was considered. After stating that a system of rules had been permitted to grow up with respect to mining on the public lands by voluntary action and assent of the population, whose free and unrestrained occupation of the mineral region had been tacitly assented to by the Federal Govern-

[1] 3 Kent's Commentaries, *439. [2] 5 Cal. 140.

ment, and heartily encouraged by the expressed legislative policy of the State, the Court said : 'If there are, as must be admitted, many things connected with this system which are crude and undigested, and subject to fluctuation and dispute, there are still some which a universal sense of necessity and propriety have so firmly fixed as that they have come to be looked upon as having the force and effect of *res adjudicata*. Among these the most important are the rights of miners to be protected in their selected localities, and the rights of those who, by prior appropriation, have taken the waters from their natural beds, and by costly artificial works have conducted them for miles over mountains and ravines to supply the necessities of gold diggers, and without which the most important interests of the mineral region would remain without development. So fully recognized have become these rights, that without any specific legislation conferring or confirming them, they are alluded to and spoken of in various acts of the legislature in the same manner as if they were rights which had been vested by the most distinct expression of the will of the law-makers.'

"This doctrine of right by prior appropriation was recognized by the legislation of Congress, in 1866."[1]

The limitation of the doctrine of prior appropriation and the restrictions as to reasonable use, were touched upon as follows:

"The right to water by prior appropriation, thus recognized and established as the law of miners on the mineral lands of the public domain, is limited in every case, in quantity and quality, by the uses for which the appropriation is made. A different use of the water subsequently does not affect the right; that is subject to the same limitations, whatever the use. The appropriation does not confer such an absolute right to the body of the water diverted, that the owner can allow it, after its diversion, to run to waste, and prevent others from using it for mining or other legitimate purposes; nor does it confer such a right that he can insist upon the flow of the water without deterioration in quality, where such deterioration does not defeat nor impair the uses to which the water is applied.

"Such was the purport of the ruling of the Supreme Court

[1] Atchison v. Peterson, 20 Wall. U. S. 510.

of California, in Butte Canal and Ditch Company *v.* Vaughn,[1] where it was held that the first appropriator had only the right to insist that the water should be subject to his use and enjoyment to the extent of his original appropriation, and that its quality should not be impaired so as to defeat the purpose of that appropriation. To this extent, said the Court, his rights go, and no further; and that in subordination to them, subsequent appropriators may use the channel and waters of the stream, and mingle with its waters other waters, and divert them as often as they choose; that whilst enjoying his original rights, the first appropriator had no cause of complaint. In the subsequent case of Ortman *v.* Dixon,[2] the same Court held, to the same purport, that the measure of the right of the first appropriator of the water as to extent follows the nature of the appropriation or the uses for which it is taken.

"What diminution of quantity or deterioration in quality will constitute an invasion of the rights of the first appropriator will depend upon the special circumstances of each case, considered with reference to the uses to which the water is applied. A slight deterioration in quality might render the water unfit for drink or domestic purposes, whilst it would not sensibly impair its value for mining or irrigation. In all controversies, therefore, between him and parties subsequently claiming the water, the question for determination is necessarily whether his use and enjoyment of the water to the extent of his original appropriation have been impaired by the acts of the defendant.[3]

But whether, upon a petition or bill asserting that his prior rights have been thus invaded, a Court of Equity will interfere to restrain the acts of the party complained of, will depend upon the character and extent of the injury alleged, whether it be irremediable in its nature, whether an action at law would afford adequate remedy, whether the parties are able to respond for the damages resulting from the injury, and other considerations which ordinarily govern a Court of Equity in the exercise of its preventive process of injunction."[4]

---

[1] 11 Cal. 143.
[2] 13 Cal. 33. See, also, Lobdell *v.* Simpson, 2 Nev. 274.
[3] See, to the same effect, Hill *v.* Smith, 27 Cal. 483; Yale's Mining Claims, 194.
[4] Atchison *v.* Peterson, 20 Wall. U. S. 514.

§ 210. Effect of the statute upon prior appropriation without Government title.—This phase of the subject has been considered in a late case in the Supreme Court of the United States. A review of that case will constitute the most authoritative exposition of the law of the subject. The question on the merits in the case was, whether a right to running waters on the public land of the United States, for the purposes of irrigation, can be acquired by prior appropriation, as against parties not having the title of the Government. The Court said: "Neither party has any title from the United States; no question as to the rights of the riparian proprietors can, therefore, arise. It will be time enough to consider those rights when either of the parties has obtained the patent of the Government. At present, both parties stand upon the same footing: neither can allege that the other is a trespasser against the Government without at the same time invalidating his own claim.

"In the late case of Atchison v. Peterson,[1] we had occasion to consider the respective rights of miners to running waters on mineral lands of the public domain; and we there held, that by the custom which had obtained among miners in the Pacific States and Territories, the party who first subjected the water to use, or took the necessary steps for that purpose, was regarded, except as against the Government, as the source of title in all controversies respecting it; that the doctrines of the common law declaratory of the rights of riparian proprietors were inapplicable, or applicable only to a limited extent, to the necessities of miners, and were inadequate to their protection; that the equality of right recognized by that law among all the proprietors upon the same stream, would have been incompatible with any extended diversion of the water by one proprietor, and its conveyance for mining purposes to points from which it could not be restored to the stream; that the Government by its silent acquiescence had assented to, and encouraged, the occupation of the public lands for mining, and that he who first connected his labor with property thus situated and open to general exploration, did in natural justice acquire a better right to its use and enjoyment than others who had not given such la-

---

[1] 20 Wall. U. S. 507.

bor; that the miners on the public lands throughout the Pacific States and Territories, by their customs, usages, and regulations, had recognized the inherent justice of this principle, and the principle itself was, at an early period, recognized by legislation, and enforced by the Courts in those States and Territories, and was finally approved by the legislation of Congress in 1866. The views there expressed, and the rulings made, are equally applicable to the use of water on the public lands for purposes of irrigation. No distinction is made in those States and Territories, by the customs of miners or settlers, or by the Courts, in the rights of the first appropriator from the use made of the water, if the use be a beneficial one."

"In the case of Tartar v. The Spring Creek Water and Mining Company, decided in 1855, the Supreme Court of California said: 'The current of decisions of this Court goes to establish that the policy of this State, as derived from her legislation, is to permit settlers in all capacities to occupy the public lands, and by such occupation to acquire the right of undisturbed enjoyment against all the world but the true owner. In evidence of this, acts have been passed to protect the possession of agricultural lands acquired by mere occupancy; to license miners; to provide for the recovery of mining claims; recognizing canals and ditches which were known to divert the water of streams from their natural channels for mining purposes; and others of like character. This policy has been extended equally to all pursuits, and no partiality for one over another has been evinced, except in the single case where the rights of the agriculturist are made to yield to those of the miner where gold is discovered in his land. The policy of the exception is obvious. Without it the entire gold region might have been inclosed in large tracts, under the pretense of agriculture and grazing, and eventually what would have sufficed as a rich bounty to many thousands would be reduced to the proprietorship of a few. Aside from this, the legislation and decisions have been uniform in awarding the right of peaceable enjoyment to the first occupant, either of the land or of anything incident to the land.'[1]

[1] 5 Cal. 397.

"Ever since that decision, it has been held, generally throughout the Pacific States and Territories, that the right to water by prior appropriation for any beneficial purpose is entitled to protection. Water is diverted to propel machinery in flour-mills and saw-mills, and to irrigate land for cultivation, as well as to enable miners to work their mining claims; and in all such cases the right of the first appropriator, exercised within reasonable limits, is respected and enforced. We say within reasonable limits, for this right to water, like the right by prior occupancy to mining ground or agricultural land, is not unrestricted. It must be exercised with reference to the general condition of the country and the necessities of the people, and not so as to deprive a whole neighborhood or community of its use, and vest an absolute monopoly in a single individual. The Act of Congress of 1866 recognizes the right to water by prior appropriation for agricultural and manufacturing purposes, as well as for mining."[1]

"It is very evident that Congress intended, although the language used is not happy, to recognize as valid the customary law with respect to the use of water, which had grown up among the occupants of the public land under the peculiar necessities of their condition; and that law may be shown by evidence of the local customs, or by the legislation of the State or Territory, or the decision of the Courts. The union of the three conditions in any particular case is not essential to the perfection of the right by priority; and in case of conflict between a local custom and a statutory regulation, the latter, as of superior authority, must necessarily control."[2]

Several decisions of the Supreme Court of Montana have been cited to us, recognizing the right by prior appropriation to water for purposes of mining on the public lands of the United States; and there is no solid reason for upholding the right when the water is thus used, which does not apply with the same force when the water is sought on those lands for any other equally beneficial purpose. In Thorp v. Freed,[3] the

---

[1] Basey v. Gallagher, 20 Wall. U. S. 681; S. C. 1 Montana, 457; Woolman v. Garringer, 1 Montana, 535.
[2] Ibid. 683; Ibid.
[3] 1 Montana, 652, 665.

subject was very ably discussed by two of the justices of that Court, who differed in opinion upon the question in that case, where both parties had acquired the title of the Government. The disagreement would seem to have arisen in the application of the doctrine to a case where title had passed from the Government, and not in its application to a case where neither party had acquired that title. In the course of his opinion, Mr. Justice Knowles stated that ever since the settlement of the Territory it had been the custom of those who had settled themselves upon the public domain and devoted any part thereof to the purposes of agriculture, to dig ditches and turn out the water of some stream to irrigate the same; that this right had been generally recognized by the people of the Territory, *and had been universally conceded as a necessity of agricultural pursuits*. "So universal," added the justice, "has been this usage, that I do not suppose there has been a parcel of land to the extent of one acre, cultivated within the bounds of this Territory, that has not been irrigated by water diverted from some mining stream.

"We are satisfied that the right claimed by the plaintiffs is one which, under the customs, laws, and decisions of the Courts of the Territory, and the act of Congress, should be recognized and protected." [1]

In the case of Basey v. Gallagher, it was, as we have seen, decided, that a right to running waters on the public lands of the United States, for purposes of irrigation, could be acquired by prior appropriation, as against parties not having the title of the Government.[2]

This doctrine was followed in Barnes v. Sabron, 10 Nevada, 230, which was an action to recover damages for the diversion of water. It was there held that the first appropriator of the water of a stream running through the public lands has the right to insist that the water flowing therein shall, during the irrigating season, be subject to his reasonable use and enjoyment, to the full extent of his original appropriation and beneficial use. But his rights go no further; for in subordination to such rights

---

[1] Basey v. Gallagher, 20 Wall. 685; S. C. 1 Montana, 457; Woolman v. Garringer, 1 Montana, 535.

[2] 20 Wall. U. S. S. C. 1 Montana.

subsequent appropriators may take the balance of the water remaining in the stream. The first appropriator is only entitled to as much water as is necessary to irrigate his land, and is bound to make a reasonable use of it, and what is a reasonable use depends upon the circumstances of each case. The Court said: " The doctrine that the first appropriator has the superior right, ' where the right to the use of running water is based upon appropriation, and not upon ownership in the soil,' has been recognized and acknowledged by the decisions of this Court in Lobdell v. Simpson, 2 Nev. 274, and the Ophir S. M. Co. v. Carpenter et al., 4 Nev. 534.

"The facts of this case do not call in question the correctness of the decision in Vansickle v. Haines, 7 Nev. 249, where the title to the land had been obtained from the Government prior to the acts of Congress herein referred to.

"It logically follows, from the legal principles we have announced, that the plaintiff, as the first appropriator of the waters of Currant Creek, has the right to insist that the water flowing therein shall, during the irrigating season, be subject to his reasonable use and enjoyment to the full extent of his original appropriation and beneficial use. To this extent his rights go, but no further; for, in subordination to such rights, the defendants, in the order and to the extent of their original appropriation and use, had the unquestionable right to appropriate the remainder of the water running in said stream. (The Butte Canal and Ditch Co. v. Vaughn, 11 Cal. 143; The Nevada Water Co. v. Powell et al., 34 Cal. 109.)

"In 1870, Congress amended the Act of 1866, and provided: ' That none of the rights conferred by sections five, eight, and nine of the act to which this act is amendatory shall be abrogated by this act, and the same are hereby extended to all public lands affected by this act; and all patents granted, or pre-emption or homesteads allowed, shall be subject to any vested and accrued water rights, or rights to ditches and reservoirs, used in connection with such water rights, as may have been acquired under or recognized by the ninth section of the act of which this act is amendatory.' (16 U. S. Stats. 218, Sec. 17.) The certificate of plaintiff from the State and the patent of Sabron must, under the provisions of this law, be held subject to such vested and accrued water rights as were acquired by the respective parties under the ninth section of the Act of 1866."[1]

"If plaintiff did not require the full amount of his appropriation, he could not hold the defendants responsible in damages for not turning it down to him; he was only entitled to as much water—within his original appropriation—as was necessary to irrigate his land, and was bound, under the law, to make a reasonable use of it. In a dry and arid country, like Nevada, where the rains are insufficient to moisten the earth, and irrigation becomes necessary for the successful raising of crops, the rights of prior appropriators must be confined to a reasonable and necessary use. The agricultural resources of the State cannot be developed, and our valley lands cannot be cultivated without the use of water from the streams to cause the earth to bring forth its precious fruits. No person can, by virtue of a prior appropriation, claim or hold any more water than is necessary for the purpose of the appropriation. Reason is the life of the law, and it would be unreasonable and unjust for any person to appropriate all the waters of a creek when it was not necessary to use the same for the purposes of his appropriation. The law, which recognizes the vested rights of prior appropriators, has always confined such rights within reasonable limits. 'We say within reasonable limits,' with the Court in Basey *v.* Gallagher, 'for this right to water, like the right by prior occupancy to mining ground, * * * is not unrestricted. It must be exercised with reference to the general condition of the country and the necessities of the people, and not so as to deprive a whole neighborhood or community of its use, and vest an absolute monopoly in a single individual.' What is a reasonable use depends upon the peculiar circumstances of each particular case."[2]

§ 211. Construction of flumes over public lands.—In a case in Nevada, a plaintiff in constructing a flume found it necessary to carry it over certain public land in the possession of the defendant. He proceeded under the State law to condemn the right of way, and had appraisers appointed who valued it.

---

[1] Barnes *v.* Sabron 10 Nevada, 233.
[2] Ibid. 243; Basey *v.* Gallagher, 20 Wall. 685; 1 Montana, 457.

He tendered the sum to the defendant, who refused to accept. After such tender he attempted to carry forward his work, but was prevented by defendant. The plaintiff therefore prayed for an injunction restraining the defendant from further interfering with the work. The inferior Court, after hearing the testimony, ordered a preliminary injunction to issue, and defendant took an appeal from the order to the Supreme Court of the State. In commenting upon Sec. 9 of the Act of 1866, (Rev. Stats. 2339) the Court said: " In its adoption there appear to have been three distinct objects in view: 1st. The confirmation of all existing water rights; 2d. To grant the right of way over the public land to persons desiring to construct flumes or canals for mining or manufacturing purposes; and 3d. To authorize the recovery of damage by settlers on such land against persons constructing such ditches or canals, for injuries occasioned thereby. That this section grants the right of way over the public land to all who may desire to construct ditches or canals for mining or agricultural purposes, is about as clear and certain as the objects and purposes of the acts of Congress usually are."

Under the act, it was considered that nothing is necessary to be shown except that the construction of a canal or ditch is desired for some mining or agricultural purpose, and that the land over which it is to be constructed is public. The land being public, it was held that Congress had a perfect right to grant the right of way over it, for the purpose of constructing flumes and for other purposes, and the injunction was allowed to stand.[1]

§ 212. Rights of ditch-owners on public lands.—In a California case the defendant, in 1853 and 1854, had constructed a ditch to convey water for mining purposes in the gold regions of California. The ditch was about thirty miles in length, and of a capacity to carry 14,000 inches of water, and was excavated to carry water for sale to miners and others, and passed over public lands of the United States, which were surveyed prior to 1865. The plaintiff derived title to a portion of the land through which the ditch passed, by a patent from the United States, dated November 1st, 1867, having filed his declaratory statement

---

[1] Hobart v. Ford, 6 Nev. 77.

as a pre-emptor on August 18th, 1866. He derived title to another portion of the land by deed from a United States patentee, the patent being dated December 1st, 1868; and to still another portion by deed from the Central Pacific Railroad Company, who received the land by grant from the United States on the 27th of June, 1867, under the Acts of Congress of 1862 and 1864, granting lands in aid of a railroad and telegraph line. The lands were cultivated by the plaintiff, and he commenced an action on the 19th of October, 1871, to abate the ditch as a nuisance.

The defendant, therefore, showed that prior to the Act of Congress of July 26th, 1866, it had acquired a right to the use of the water which was "recognized and acknowledged by the local customs, laws, and decisions of Courts." "That act," said the Court, "operated a grant to it of the right of way, and of the ditch through which the water was running at the date of the passage of the act. The subsequent grantees of the United States of tracts through which the ditch ran, took subject to defendant's easement." The patents of the plaintiff, and his grantors, were issued after the date of the act. But it was claimed that the railroad company, one of plaintiff's grantors, had a perfect equity at and before the date of the Act of July 26th, 1866, because it had completed "forty consecutive miles" of the railroad, and the land was within that division. The Act of 1862 (Section 4) provides, "that (on completion of forty miles, etc.,) the President shall appoint three commissioners to examine the same, and report to him in relation thereto; that if it shall appear to him that forty consecutive miles of said railroad and telegraph line have been completed and equipped in all respects as required by this act, then, upon certificate of said commissioners to that effect, patents shall issue; and patents shall, in like manner, issue as each forty miles of said railroad and telegraph line are completed, upon certificate of said commissioners." The Court said: "The law places in the President or board of commissioners, or both, the power of determining whether the railroad company has performed the conditions pre-requisite to the issuing of the patents. It is manifest that, until the commissioners made their certificate, the company had no vested equity which can be recognized by the State Courts.

There is no finding that such certificate was made prior to the passage of the Act of July 26th, 1866."

What the effect of such a finding would have been was not decided.[1]

§ 213. The Sutro Tunnel Act—Provisions not to affect existing rights.

—Section 2344 of the Revised Statutes reads: "Nothing contained in this chapter shall be construed to impair, in any way, rights or interests in mining property acquired under existing laws; nor to affect the provisions of the act entitled 'An Act granting to A. Sutro the right of way and other privileges to aid in the construction of a draining and exploring tunnel to the Comstock Lode, in the State of Nevada,' approved July 25th, 1866."[2]

---

[1] Broder v. Natoma Water and Mining Company, 50 Cal. 621. See, generally, as to water rights, *doctrine of prior appropriation:* Blanchard & Weeks' Leading Cases on Mines, Minerals, and Mining Water Rights, 726–757, and numerous cases there cited: Woolman v. Garringer, 1 Montana, 535; Columbia M. Co. v. Holter, Ibid. 296. Diversion of water: Harris v. Shoutz, Ibid. 212; Right of way for ditches, Noteware v. Sterns, Ibid. 311.

[2] Rev. Stats. 2344. See Sec. 2340, Ibid. See Sec. 17 of the Act of 1870, 16 U. S. Stats. 218. Sec. 8 of the Act of 1866, 14 U. S. Stats. 253, read: "Sec. 8. That the right of way for the construction of highways over public lands, not reserved for public uses, is hereby granted." The last clause of Sec. 16 of the Act of 1872, 17 U. S. Stats. 96, read: "*Provided,* That nothing contained in this act shall be construed to impair, in any way, rights or interests in mining property acquired under existing laws."

Following is the text of the so-called Sutro Tunnel Act, approved July 25th, 1866 (14 U. S. Stats. 242):

"An Act granting to A. Sutro the right of way, and granting other privileges to aid in the construction of a draining and exploring tunnel to the Comstock Lode, in the State of Nevada. [*Approved July 25th, 1866.*]

"*Be it enacted by the Senate and House of Representatives of the United States of America, in Congress assembled,* That, for the purpose of the construction of a deep draining and exploring tunnel to and beyond the 'Comstock Lode,' so-called, in the State of Nevada, the right of way is hereby granted to A. Sutro, his heirs and assigns, to run, construct, and excavate a mining, draining, and exploring tunnel: also, to sink mining, working, or air shafts along the line or course of said tunnel, and connecting with the same at any point which may hereafter be selected by the grantee herein, his heirs or assigns. The said tunnel shall be at least eight feet high and eight feet wide, and shall commence at some point to be selected by the grantee herein, his heirs or assigns, at the hills near Carson River, and within the boundaries of Lyon County, and extending from said initial point in a westerly direction seven miles, more or less, to and beyond said Comstock Lode; and the same right of way shall extend northerly and southerly on the course of said lode, either within the same, or east or west of the same: and also on or along any other lode which may be discovered or developed by the said tunnel.

§ 214. Conditions inserted in patents for mines on Comstock Lode, Nevada.—In issuing patents for the Comstock Lode, Nevada, the following clause has been inserted:

"That the claim hereby granted and conveyed shall be subject to the condition specified in the third section of the Act of Congress, approved July 25th, 1866, 'granting the right of way and other privileges to aid in the construction of a draining and exploring tunnel to the Comstock Lode, in the State of Nevada,' and the grantee herein shall contribute and pay to the owners of the tunnel, constructed pursuant to said act, for drainage or other benefits derived from said tunnel or its branches, the same rate of charges as have been or may hereafter be named in agreement between such owners and the companies representing a majority of the estimated value of said Comstock Lode,

"SEC. 2. *And be it further enacted,* That the right is hereby granted to the said A. Sutro, his heirs and assigns, to purchase, at one dollar and twenty-five cents per acre, a sufficient amount of public land near the mouth of said tunnel for the use of the same, not exceeding two sections, and such land shall not be mineral land or in the bona fide possession of other persons who claim under any law of Congress at the time of the passage of this act, and all minerals existing or which shall be discovered therein are excepted from this grant; that upon filing a plat of said land, the Secretary of the Interior shall withdraw the same from sale, and upon payment for the same a patent shall issue. And the said A. Sutro, his heirs and assigns, are hereby granted the right to purchase, at five dollars per acre, such mineral veins and lodes within two thousand feet on each side of said tunnel, as shall be cut, discovered, or developed by running and constructing the same, through its entire extent, with all the dips, spurs, and angles of such lodes, subject, however, to the provisions of this act, and to such legislation as Congress may hereafter provide: *Provided,* That the Comstock Lode, with its dips, spurs, and angles, is excepted from this grant, and all other lodes, with their dips, spurs, and angles, located within the said two thousand feet, and which are or may be, at the passage of this act, in the actual bona fide possession of other persons, are hereby excepted from such grant. And the lodes herein excepted, other than the Comstock Lode, shall be withheld from sale by the United States; and if such lodes shall be abandoned or not worked, possessed, and held in conformity to existing mining rules, or such regulations as have been or may be prescribed by the Legislature of Nevada, they shall become subject to such right of purchase by the grantee herein, his heirs or assigns.

"SEC. 3. *And be it further enacted,* That all persons, companies, or corporations owning claims or mines on said Comstock Lode or any other lode drained, benefited, or developed by said tunnel, shall hold their claims subject to the condition, (which shall be expressed in any grant they may hereafter obtain from the United States,) that they shall contribute and pay to the owners of said tunnel the same rate of charges for drainage or other benefits derived from said tunnel or its branches, as have been, or may hereafter be, named in agreement between such owners and the companies representing a majority of the estimated value of said Comstock Lode at the time of the passage of this act."

at the time of the passage of said act, as provided in said third section."

Both the Acts of 1870 and 1872 contained clauses guarding the rights of the owners of the Sutro Tunnel, and the land embraced by the location of the tunnel was withdrawn from sale.[1]

In May, 1876, protests were filed in the General Land Office, by the Sutro Tunnel Company, against the issuance of any patents for mining claims in certain townships, unless the condition referred to in the third section of the Act of July 25th, 1866, should be inserted therein. The company claimed that under that act, commonly known as the Sutro Tunnel Act, no patents should be issued for mining claims in the townships except to parties holding claims on the Comstock Lode, and to them only subject to that condition.[2] It was demanded:

"1st. That the patents on the Comstock Lode for its whole length be only granted with the restriction made in Sec. 3 of the above act.

"2d. That the mines located in T. 16 and 17 N., 7 R. 21 E., are all within a reasonable distance from the tunnel—probably on lodes cut by the same; and may easily be reached by its branches, and therefore should be withheld from sale."

On the 20th January a decision in the case was made by the Land Office, and on the 1st of February, 1869, Mr. Sutro filed another argument, and requested a re-examination of the matter. In this argument he stated that "all the mines contained in T. 16 and 17 N., R. 21 E., Mount Diablo Meridian, may conveniently be reached by branches from said main tunnel, or may otherwise be benefited by the same: it will be necessary, therefore, that the above clause be inserted in *all* patents issued for mines situated in the above named townships."

He then summed up his claim as follows, viz:

"I claim under the law of July 25th, 1866, as follows:

"1st. All mines embraced within 2,000 feet on each side of said tunnel for seven miles in length, as indicated by blue shading on the map filed with the Commissioner of the General Land Office on the 30th day of July, 1866, excepting the Comstock Lode, are to be withheld from sale by the United States.

---

[1] Decision Commissioner, March 8th, 1873, Copp's U. S. Mining Decisions, 162.
[2] In re Sutro Tunnel Company, Decision Acting Commissioner, May 27th, 1876,
3 Copp's Land-owner, 34.

"2d. Patents may be issued to all mines on the Comstock Lode, including those situated on said lode within 2,000 feet on each side of said tunnel, also to all mines in T. 16 and 17 N., R. 21 E., Mount Diablo Meridian, outside of said 2,000 feet on each side of said tunnel, provided that these patents shall contain the conditions specified in the third section of the Act of July 25th, 1866."

On the 25th of February, 1860, the Office, after a re-examination of the case, decided that:

"1st. The right to purchase mineral veins or lodes granted to A. Sutro, his heirs and assigns, by the second section of the act, is limited to 2,000 feet on each side of the tunnel, constructed from the initial point at the hills near Carson River, to and beyond the Comstock Lode, and applies only to lodes other than the Comstock within said limits, *cut, discovered, or developed* by mining and constructing said tunnel, and not at the passage of the act in the actual bona fide possession of other persons.

"2d. Veins or lodes other than the Comstock, lying within 2,000 feet on either side of said tunnel, at the passage of the act in the actual bona fide possession of other persons, are to be withheld from sale by the United States; but if, after the construction of the tunnel, it shall be found that some of the lodes so withheld from sale are not *cut* by the tunnel, the restriction as to sale will no longer be applied to them; but all of the lands within said 2,000 feet limits will be reserved from sale until after the construction of the tunnel, unless its commencement and prosecution should be procrastinated for such a length of time as to imply its abandonment or the inability of the grantees under the act to accomplish the undertaking.

"3d. That all patents issued to claimants of mines on the Comstock Lode must contain the condition specified in the third section of said act.

"4th. That the like condition is to be inserted in patents issued for mines on any other lode, drained, benefited, or developed by said tunnel.

"5th. That the only patentable lodes other than the Comstock, capable of being drained, benefited, or developed by said tunnel, are such as may be thus affected by means of branches connecting with the tunnel, there being no patentable lodes within 2,000 feet of said tunnel, and that the construction of such branches must be authorized by the act.

"6th. That the act authorizes the construction of branches only along the Comstock Lode, and along any other lode which may be *discovered* or developed by said tunnel.

"7th. That the lodes here referred to are what are called 'blind lodes,' the existence of which it is expected will be disclosed by the construction of the tunnel, and that lodes 'which may be discovered or developed by the construction of the tunnel,' do not include lodes already discovered or which may hereafter be discovered before its construction, and that there is consequently no authority granted by the act for the construction of branches along these, and the condition specified in the third section is not applicable to patents issued to these claims.

"8th. And that consequently the only mines or lodes in any way affected by the Act of July 25th, 1866, are: First—The mines on the Comstock Lode. Second—Those lying within 2,000 feet of the proposed line of said tunnel. Third—Such new lodes as may be discovered or developed by the construction of the tunnel, the existence of which remaining unknown until thus brought to light.

"9th. That the only patents subject to the condition specified in the third section, or that became subject to it prior to the construction or commencement of the tunnel, are those issued for mines on the Comstock Lode."

From this decision Mr. Sutro took an appeal to the Secretary of the Interior, who, on the 6th of July, 1870, affirmed the decision of the General Land Office.

From the foregoing it will be seen that more than five years had elapsed since the very questions presented by the Sutro Tunnel Company were decided by the appellate authority, and the matter was *res judicata*. The Commissioner said: "The request of the attorneys for the Sutro Tunnel Company that the decision of this Office, as affirmed by the Honorable Secretary of the Interior, shall not be adhered to, is denied, as this Office is bound by the construction of a statute as given by the head of the Department.

"It may be proper in this connection to decide what claims shall be considered as on the Comstock Lode, within the meaning of the Act of July 25th, 1866, and therefore coming within the provisions of the third section of said act.

"Said act in the first section declares 'that for the purpose of the construction of a deep draining and exploring tunnel to and beyond the "Comstock Lode," *so called*, in the State of Nevada, the right of way is hereby granted,' etc. In the third section reference is made to mines on said 'Comstock Lode.'

"To determine what patents should contain said condition, it is only necessary to determine what claims had been located on the Comstock Lode, *so called*, at the date of the passage of this act.

"On the 31st July, 1866, five days after the passage of the said act, Mr. Sutro filed in this office a 'map, showing the locations of the Sutro Tunnel and the Comstock Lode, State of Nevada.'

"On this map the Comstock Lode is represented as extending from the Utah claim on the north to the North American on the south; and it is to be presumed, as he was the party in interest, that he represented upon said plat the extent of what was called the Comstock Lode at the time of the passage of said act. In this map, the mines lying in the direction of Silver City from Gold Hill, are not represented as being upon what was called the Comstock Lode, as for instance the Dayton, Pride of the West, Kossuth on the 'Monte Christo Lode,' the Boston, St. Louis, Alpha, Succor, etc.

"The treatise on 'Mining and Metallurgy of Gold and Silver,' by J. Arthur Phillips, published in 1867, 'gives the names of the various mining claims on the Comstock Lode as far as its continuity has been ascertained.' Mr. Phillips refers to the Utah as the northern claim and the Baltimore American as the southern claim on said Comstock Lode.

"The State Surveyor-General of Nevada, (S. H. Marlette) in his official report for the year 1865, gives a list of the mining claims on the Comstock as extending from the Utah to the Baltimore American, both inclusive.

"J. Ross Browne, in his report for 1868, page 341, states that the continuity of the Comstock Lode has been ascertained 'for a length of about three and a half miles,' and gives the same claims as those stated in the State Surveyor-General's report.

"Raymond, in his report for 1868, says that the Comstock Lode has a general north and south course, and has been traced on the surface more than 27,000 feet, and that about 19,000 feet have been actually explored, to wit, all the locations from the Utah mine to the south part of the Overman.

"Raymond, in his report for 1869, again refers to the fact that the Comstock extends from the Utah to the South Overman, and states under the title, 'Mines on the continuation of the Comstock,' that 'so many of the *Comstock mines proper* have been compelled to extract from their old workings ores once cast aside as not worth enough to pay for treatment, that the mills as well as the mines have found their advantage in reducing the prices of custom work, to enable these operations to be continued; and this reduction of prices has in turn caused the resumption of active work on many a mining claim *beyond the limits of the recognized Comstock Ledge*, though on the extensions north and south of its supposed course, south of the Overman, are three or four claims which have lain idle for years; * * * north of the Utah work has been done for some time on small claims.'

"Raymond, in his report for 1870, again refers to the fact that the Comstock has been explored from the Utah to the South Overman, and adds: 'There are further locations both north of the Utah and south of the Overman.'

"Clarence King, in his report, vol. 3, page 37, states that the course of the Comstock is about north 25° east; 'In Seven Mile Cañon, near the base of Cedar Hill, is the most northern known portion of the lode. From that point it continues south in a nearly direct line underneath Virginia City, across the divide, past Gold Hill to American Flat.' On page 40 he states that 'in general, then, the lode has a longitudinal expansion of 22,000 feet.' On page 41 he refers to the '4,300 feet of the southern end of the lode' as extending 'from the furthest workings of the Uncle Sam to the North Alpha line.' Page 98. 'The course of the Comstock Lode is nearly north and south, maintaining a general conformity in direction with the trend of the Washoe Mountains, in which it is contained.' Page 99. The extent to which the vein has been clearly traced, and on which mining claims have been located, is about four miles. At either extremity of this ground, however, and particularly on the north, the vein has been but little explored, and has not been thus far proved to be very valuable. On page 99 Mr. King gives a 'list of the mining claims located on the course of the lode as far as its continuity has been traced with any certainty,' extending from the Utah on the north to the south boundary of the Baltimore American. Page 188. 'There are a number of mines in the Washoe region, that, being located on other veins than the Comstock, are generally classed as "outside." They are on "various ledges."' Among the 'outside' mines, he refers to the Occidental, Monte Christo, Lady Bryan, and the Twin.

"The commission appointed under the authority of the Act of Congress, approved April 4th, 1871, 'to examine and report upon the Sutro Tunnel, in the State of Nevada,' in their report (Ex. Doc. No. 15, Forty-second Congress, Second Session) state on the fourth page thereof, that the lode generally known as the Comstock Lode 'has an extent not yet fully developed, but which reaches certainly from the Ophir mine on the north to the Uncle Sam and Overman on the south, a distance of 12,000 feet. Beyond these points the lode is supposed to extend to the north and south, to the Seven Mile Cañon in the former direction, and to the American Flat in the latter. It may therefore be said that the fissure itself is believed to have been traced from the diggings known as the Utah mines on the north to the locality known as the American Flat or 'American City' on the south, a length of about 22,000 feet.'

"All of the authorities hereinbefore referred to concur in regard to the locality and course of the lode commonly known as the Comstock, and agree in their statements in regard to the extent thereof. They all describe the Comstock Lode, so called, as extending from the Utah and northerly to the Baltimore American, and southerly in the same direction. The mines lying in the direction of Silver City are not referred to by these authorities as being on the so-called Comstock Lode, nor yet are they located in the same general direction as are those mines which they refer to as being on the Comstock.

"The only mining claim which has been entered in T. 16 N., R. 21 E., Mt. Do. Mer., lying in the direction of Silver City and southerly from surveys 49 and 55, which *was located* as being on the Comstock Lode, is that embraced by survey No. 79, located by H. J. F. Scheel, December 30th, 1872, as the South Comstock.

"The Sutro Tunnel Company, in its argument in this case, refers to the recent decision of the Supreme Court in case of the United States *v.* The Union Pacific Railroad Company, as supporting the theory advanced by it in regard to the construction of the said Sutro Tunnel Act.

"In this decision the Supreme Court declare that in construing an act of Congress we are not at liberty to recur to the views of individual members in debate, nor to consider the motives which influenced them to vote for or against its passage. The act itself speaks the will of Congress, and this is to be ascertained from the language used. But Courts may, with propriety, in construing a statute, recur to the history of the times when it was passed; and this is frequently necessary in order to ascertain the reason as well as the meaning of particular provisions in it. * * * 'No argument can be drawn from the wisdom that comes after the fact. Congress acted with reference to a state of things supposed to exist at the time, and no aid can be derived in the interpretation of its legislation from the consideration that the theory on which it proceeded turned out not to be correct.'

"Whatever obligations therefore rest on the company incorporated to accomplish this purpose must depend on the true meaning of the enactment itself, viewed in the light of cotemporaneous history.

"Following these rules in the construction of this statute, to wit, that the true meaning of a statute is to be ascertained from the language used, viewed in the light of cotemporaneous history, but one conclusion can be reached in regard to what claims should be considered as on the Comstock Lode within the meaning of the Act of July 26th, 1866, and therefore subject to the provisions of the third section of said act.

"The question is not what is *now known* as the Comstock Lode, what nor *in the future may prove to be* the Comstock Lode, but what *was known* as and called the Comstock Lode at the date of the passage of said act. The language used in the first section of the act is 'that for the purpose of the construction of a deep draining and exploring tunnel to and beyond the "Comstock Lode," *so called*, in the State of Nevada,' etc. All authorities upon the subject which I have been able to examine, agree in regard to what was known as the Comstock Lode at the date of the passage of the act, and for several years thereafter. These authorities, including Mr. Sutro himself, represent and describe the Comstock Lode as extending from the Utah claim and northerly to the Baltimore American, and southerly in the same direction.

"From a careful and thorough examination of this case, I am clearly of the opinion that the only patents which should contain the condition specified in the third section of the Act of July 25th, 1866, are such as may be issued for mining claims on the Comstock lode as hereinbefore defined and described, to wit, on the lode extending from the Utah and northerly to the Baltimore American, and southerly in the same direction."

§ 215. **Claims rejected.**—A claim within the Sutro Tunnel Grant was rejected under the Act of 1866. It was not within the Comstock Lode, which was excepted from the grant, and therefore could not be patented.[1]

[1] In re McKibben Lode, Decision of Commissioner March 29th, 1873, Copp's U. S. Mining Decisions, 179.

CHAPTER XV.

HOMESTEADS AND TOWN SITES—HOMESTEAD RIGHTS ON NON-MINERAL LANDS—TOWN-SITE ENTRIES.

§ 216. Non-mineral lands open to homesteads.
§ 217. Pre-emption of homesteads on agricultural lands formerly designated as mineral.
§ 218. Homestead entries including mineral deposits.
§ 219. Rights of pre-emptioners and homestead claimants.
§ 220. Conflicts between homestead and mill-site claimants.
§ 221. Title to town lots subject to mineral rights.
§ 222. Conflicts between mineral and town-site claimants.

§ 216. Non-mineral lands open to homesteads.—Section 2341 of the Revised Statutes is as follows: "Wherever, upon the lands heretofore designated as mineral lands, which have been excluded from survey and sale, there have been homesteads made by citizens of the United States, or persons who have declared their intention to become citizens, which homesteads have been made, improved, and used for agricultural purposes, and upon which there have been no valuable mines of gold, silver, cinnabar, or copper discovered, and which are properly agricultural lands, the settlers or owners of such homesteads shall have a right of pre-emption thereto, and shall be entitled to purchase the same at the price of one dollar and twenty-five cents per acre, and in quantity not to exceed one hundred and sixty acres;[1] or they may avail themselves of the provisions of chapter five of this Title, relating to Homesteads."

In carrying out the provisions of the tenth section of the Act of 1866, (which was substantially the same as Rev. Stat. 2341) it was held, that if the deputy surveyor returns the land as agricultural, and there is no data to the contrary, and no one

---

[1] NOTE.—Sec. 10 of the Act of 1866, 14 U. S. Stats. 253, was same as above, with the following after the words "160 acres": "Or, said parties may avail themselves of the provisions of the Act of Congress, approved May 20th, 1862, entitled 'An Act to secure homesteads to actual settlers on the public domain,' and acts amendatory thereof."

files an affidavit of its being more valuable for mineral than for agricultural purposes, the settler will be allowed to enter it under the provisions of the tenth section. If an affidavit is filed alleging the land to be mineral, a trial must be had to determine whether it is more valuable for mining purposes than for agricultural. In such cases, a day will be fixed for the hearing, giving to the claimants and to the party filing the affidavit sufficient notice to enable them to be present with their witnesses; and when the tract had been occupied for agricultural purposes and improved as such before the date of the Act of July 26th, 1866, the burden of proof was upon the party seeking to establish its mineral character, and the testimony should be of a nature clearly proving the truth of such allegations before a decision is rendered against the right of the settler to enter the land. Should the deputy surveyor return the land as mineral, the settler will be required to furnish satisfactory proof of the error of such return prior to entering the land. The return of a deputy surveyor is not conclusive in these cases when disputed, but the matter must be investigated by the examination of witnesses, capable from experience and observation, and from previous examination, to testify understandingly in reference to the existence of minerals upon any particular tract, and whether the deposit is of sufficient extent to render it more valuable for mining than for agriculture. When lands had not been filed upon, the officers were required to satisfy themselves as to which class they belonged, before taking steps looking to their disposal.[1]

§ 217. Pre-emption of homestead on agricultural lands formerly designated as mineral.—From the statute it is probable that the right to enter lands as agricultural, wh'ch as mineral lands were previously excluded from survey and sale, is confined entirely to actual settlers coming within the requirements of the pre-emption laws, who, upon making competent proof that the tracts actually settled upon, occupied, and improved by them as homesteads, contain no mines of gold, silver, cinnabar, or copper, are entitled to the execution in their favor of the pre-emption or homestead laws.

[1] Instructions May 16th, 1868, Zabriskie's Land Laws, 208–211, Copp's U. S. Mining Decisions, 218.

Where an attempt was made to prove the agricultural character of a certain tract previously reserved as mineral land, to the end that it might inure to a railroad company and not by a party coming within the purview of the tenth section, it was held that the whole proceeding was without legal sanction, and the land was treated as mineral.[1]

§ 218. Homestead entries including mineral deposits.

—Where lands containing valuable mineral deposits have been included in a homestead entry, the entry will be canceled at any time prior to the issuance of the patent, upon satisfactory evidence of the existence of such valuable deposits. Lands containing gold, silver, etc., cannot be taken under the homestead or other laws for the disposal of agricultural lands. If a party undertakes to homestead mineral land, an affidavit to that effect, setting forth the facts, should be filed in the local land office, and a hearing will be ordered and a decision rendered by the General Land Office in accordance with the facts proven. But when land has passed by patent to agricultural claimants, and the mineral deposits are not discovered until afterward, they have been held to pass by the patent, (in the absence of words of reservation) and the Land Office to have no further jurisdiction in the matter, the lands ceasing to become a part of the public domain.[2]

Where a party obtains a patent to a tract of land under the pre-emption or homestead laws, which at the date of the patent embraced *a known mine*, he does not obtain title to such mine by virtue of such patent. The only way to obtain Government title to mines and mineral lands is by compliance with the mining acts. The only restrictions specified in the pre-emption or homestead laws are in regard to *known* mines. When the mineral character of a specified tract first became known subsequent to the issuance of a patent therefor as agricultural land, the Land Office has not heretofore pursued the inquiry respecting it.[3] Owners of known mines may make applications for

---

[1] Decision of Commissioner, Oct. 21st, 1871, Copp's U. S. Mining Decisions, 60.
[2] Decision of Commissioner, Nov. 11th, 1873, Copp's U. S. Mining Decisions, 233; 1 Copp's Land-owner, 77.
[3] Decision of Commissioner, June 21st, 1876, 3 Copp's Land-owner, 50

patents, the same as though no homestead entries had ever been allowed covering their claims.[1]

§ 219. Rights of pre-emptioners and homestead claimants.—Congress did not intend to abolish the long-established distinction between mineral and agricultural lands, or to allow mineral lands to be classed and disposed of as agricultural, but provided that the public surveys might be extended over a region that was so clearly mineral in character, that it had previous to the passage of the mining acts been reserved, and that such tracts as should appear to be "properly" and "clearly agricultural" might be disposed of under the laws applicable to agricultural lands. The act gives no rights to agricultural claimants except to such lands as are clearly and properly agricultural. Where land is returned by the Surveyor-General as mineral, and is in a well-known mineral district, the burden of proof is upon the party who seeks to establish its agricultural character. Where testimony failed to establish the incorrectness of the Surveyor-General's return, or to affirmatively establish that the land in question was *clearly* and *properly* agricultural in character, and the land did not appear to have been thoroughly prospected, but the evidence showed that nearly the whole of it had been located by different persons as mineral land, the tracts were held reserved as such.[2]

The object of the act was to give to persons who have in good faith made agricultural settlements on public lands theretofore designated as mineral, but subsequently determined to be agricultural, a preference in pre-empting or entering the land as homesteads.[3]

After a consideration of testimony, certain land was held mineral in character, and subject to be disposed of under the mining statutes. It was claimed, on behalf of the pre-emption claimant, that the tenth section gave to qualified persons who had, prior to the passage of the act, made homesteads on lands theretofore

---

[1] Rev. Stats. 2258, 2289. In re Champion Mine, Decision Commissioner, March 26th, 1877, 4 Copp's L. O. 17.

[2] Carron v. Curtis, Decision of Commissioner, Oct. 24th, 1876, 3 Copp's Landowner, 130.

[3] Smith v. Stewart, Decision of Acting Secretary, Dec. 14th, 1872, Copp's U. S. Mining Decisions, 133.

designated as mineral, and excluded from survey and sale, a right of pre-emption or homestead therein, unless, before the passage of the act, valuable mines of gold, silver, cinnabar, or copper had been discovered thereon; and further, that the subsequent discovery of such mines did not affect the right of pre-emption or homestead thus acquired. This was held by the Secretary not to be the proper construction.[1]

§ 220. Conflicts between homestead claimants and mill-site owners.—The question of bad faith and insufficiency of cultivation on the part of a homestead claimant cannot enter into a controversy during the period allowed by law. It has been uniformly held that a homestead entry prior to the expiration of the time allowed by law, can only be canceled for abandonment, or when in conflict with a properly asserted prior right. The homestead entry of a party was ordered to remain suspended until a company claiming the land as a mill-site, presented the following evidence:

1st. Duly certified copy of the local laws in force at the date of locating their mill-site.

2d. Copies of abstracts of the deed from their grantors.

3d. Evidence of full compliance with the local laws and customs relating to mill-sites.

In case such evidence was found satisfactory, it was said that the entry of the homestead claimant would be canceled for so much of the land embraced therein as properly belonged to the company claiming the mill site.[2]

§ 221. Title to town lots subject to mineral rights.—Section 2386 of the Revised Statutes reads: "Where mineral veins are possessed, which possession is recognized by local authority, and to the extent so possessed and recognized, the title to town lots to be acquired shall be subject to such recognized

---

[1] Smith v. Stewart, Decision of Acting Secretary, Dec. 14th, 1872, Copp's U. S. Mining Decisions, 133.

[2] Newark Mill and Mining Co. v. Meinke, Decision of Commissioner, August 13th, 1875, affirmed by the Secretary, April 29th, 1876, 3 Copp's Land-owner, 67. General provisions as to homesteads, see Revised Statutes of the United States, Secs. 2289-2317; Instructions thereunder, June 17th, 1875, Copp's Public Laws, 182; Forms, Ibid. 195.

possession and the necessary use thereof; but nothing contained in this section shall be so construed as to recognize any color of title in possessors for mining purposes as against the United States." [1]

The Town-site Acts of March 2d, 1867, and June 8th, 1868, declared that no title should be acquired under their provisions to any mine of gold, silver, cinnabar, or copper, or to any valid mining claim or possession held under existing laws. A bona fide mining claim, therefore, held in compliance with the local laws and regulations, and the Congressional enactments, situate within the exterior boundaries of the premises embraced by the town-site applications, might be entered in accordance with the law and the instructions.[2]

Section 2392 Rev. Stats. provides "that no title shall be acquired under the foregoing provisions of this chapter, to any mine of gold, silver, cinnabar, or copper; or to any valid mining claim or possession held under existing laws"; to which, therefore, no title can be acquired by a town-site entry or patent. Lands which embrace *lode* claims may be included within a town-site entry or patent; and in such cases, when patents issue for the town sites, the following clause is inserted: "*Provided*, that no title shall be hereby acquired to any mine of gold, silver, cinnabar, or copper, or to any valid mining claim or possession, held under existing laws; *and provided further*, that the grant hereby made is held and declared to be subject to all the conditions, limitations, and restrictions contained in Section 2386 of the Revised Statutes of the United States, so far as the same are applicable thereto." Mining claims may, therefore, be patented when within town sites; and applications for mining claims are allowed to be filed, even though the same may conflict with or be embraced by the exterior boundaries of a town-site application.[3]

The fact, therefore, that a given tract of land has been entered or patented as a town site, in no way prevents the owner of a lode claim from securing a patent to his mine upon compliance with

---

[1] Rev. Stats. 2386; Sec. 2, Act of March 3d, 1865, 13 U. S. Stats. 530.

[2] In re Application of Nagler, Decision of Commissioner, Jan. 21st, 1873, Copp's U. S. Mining Decisions, 156.

[3] Decision of Commissioner, Aug. 19th, 1872, Copp's U. S. Mining Decisions, 135.

the terms of the mining act, as the same is excluded from the operation of such town-site patent.

The rule had been laid down that placer mining ground could not be included within the exterior boundaries of a town-site patent, but this rule was subsequently declared erroneous, and it was held that the law clearly contemplated that towns would exist in mineral localities, and that entries might be made of such town sites.

As the Government, in issuing patents for town sites, conveys the premises within the exterior boundaries of the town-site entry only in accordance with the provisions of law, the title to all mines of gold, silver, cinnabar, or copper, and to all valid mining claims or possessions held under existing laws, which are situated within such exterior boundaries, still remains in the United States after patent has issued for such town site. Title to these mining claims or possessions can only be acquired under the provisions of law regulating the disposal of mineral lands by parties who show compliance with the terms of the mining acts.

Patents issued for town sites in mining regions contain a clause in accordance with the terms of the law, (Rev. Stat. Sec. 2392) under these provisions, the patents for town sites containtaining such clause are held subject to "*any valid mining claim or possession*," and a purchaser of a lot from the town-site authorities holds the same subject to the same conditions.

In accordance with this view, previous instructions were recalled, and it was decided that patent might issue for a town-site for the premises embraced by a former survey, mine-owners within the town-site entry being allowed to make application for patents for their claims.[1]

The Government, reserving from all township patents all *valid mining claims or possessions*, as well as all mines of gold, silver, cinnabar, or copper, has the right to dispose of such reserved mines and possessions to parties who show compliance with the terms of the law relating to the disposal of mineral lands.

---

[1] Decision of Commissioner, Nov. 23d, 1876, 3 Copp's Land-owner, 131. See In re Township of Butte, Montana, Decision of Commissioner, October 27th, 1876, 3 Copp's L. O. 114; Ibid. Decision of Commissioner, August 19th, 1872, Copp's U. S. M. D. 135.

Where a company had shown compliance with the terms of the mining act from the date of the location of their claims; had presented proof of occupation and possession of the premises described in their patent; had given public notice, by publication and posting, in the manner and for the length of time required by law, of their intention to apply for a patent; and, in short, had filed such proofs of compliance with the law and instructions as are required previous to the issuance of a patent, it appeared that a portion of this mining claim was situated within the exterior boundaries of the town site of Silver City, Nevada; but as no title can be acquired by virtue of a town-site patent " to any mine of gold, silver, cinnabar, or copper, or to any valid claim or possession held under existing laws,"[1] and no error being found in the patent, and it being legally issued, the Office refused to recall it.[2]

§ 222. Conflicts between mineral and town-site claimants.—In making and approving town-site entries, patents will issue therefor in due course, but with the proviso above mentioned. The Land Office is not vested with a discretionary authority in the matter of the disposal of the public lands. It can neither grant without express authority of law, nor can it limit or qualify, by form of conveyance, the substance, conditions, or extent of the subject-matter granted, save as the same may be authorized to be done by express legislation. The proviso above quoted embraces by recitation and reference all that Congress has seen fit to enact by way of qualification in the matter under consideration, contains all of that to which appeal can be had, should the Courts be applied to for the settlement of conflicting claims, and must therefore be held to be the limit of executive authority.

The town-site laws clearly contemplate that towns will exist in mining localities; by clear implication, town-site entries are to be permitted on mineral lands. This is indicated by the clause excepting title to mines from the title acquired by the town. It is inevitable that where the surface is suitable, it will,

---

[1] Rev. Stat. U. S. 2392.
[2] In re South Comstock G. & S. M. Co., Decision of Commissioner, December 29th, 1875, 2 Copp's Land-owner, 147.

in a mining vicinity, be populated, and attain the character of a town or city. Where any branch of business flourishes, there capital and population will concentrate. The various trades and callings will center there. Hotels will be a necessity. Dwellings will be built, and permanent homes established; all the various interests which constitute valuable property rights as connected with the soil will be created. And this is not necessarily antagonistic to the miners. The protection of municipal government is in the miner's interest, as it is in the interest of any other class of business men.

In the case of Theodore H. Becker *v.* Citizens of Central City, Colorado, Becker was a mineral claimant to 3,000 linear feet of the Gunnel Extension, or White Lode, under Act of July 26th, 1866. He claimed compliance with law, and was opposed by certain citizens of the town, who represented that the lode extended to a considerable distance under town lots and improvements, owned and occupied by them in said city. In this case the Secretary of the Interior decided, August 7th, 1871, that "in the present case the application for a patent includes the surface and soil as well as the mineral. I am of the opinion that the persons in possession of this surface are adverse claimants within the meaning of this law, and are entitled to be heard in the local Courts before a patent is issued." The exception in the mining patents, for claims within the exterior limits of a town, having in view the legality of the possession of the surface ground by the inhabitants, is as follows, to wit: "Excepting and excluding, however, from these presents, all town property rights upon the surface, and there are hereby expressly excepted and excluded from the same all houses, buildings, and structures, lots, blocks, streets, alleys, or other municipal improvements on the surface of the above described premises, not belonging to the grantees herein, and all rights necessary or proper to the occupation, possession, and enjoyment of the same." The Commissioner said: "By this exception, the surface in the actual possession and occupation of the mine-owner, or covered by his improvements, is as distinctly assured and conveyed to him as is that surface to which town property rights have attached, or on which improvements by other parties have been placed excepted from his patent. These correlative exceptions, inserted in the town-site and mineral patents, secure the objects contemplated in the town site and mineral laws. They assure to all parties just what, under the law, they are respectively entitled to claim. To grant to the miner the entire surface ground, along the whole line of the lode, with the prescribed width, without regard to the acquired surface rights of others, would be to ignore the principle announced. The two laws must be so construed that both may stand. Under the system established of inserting exceptions in the patents to towns and mine-owners, there were no occupants in Central City presenting their claims adversely in the manner provided in the mining statutes, and for the reason that, by said exceptions, the rights of all parties are respected and so defined that they are easily susceptible of definite ascertainment. To except from the town patent definite surveys of mineral claims, initiated or extended after surface occupation by other parties, would obviously be ignoring, to an unjustifiable extent, adverse *rights* which have not been presented for adjustment by the Courts prior to the mineral entry, simply for the reason that under the practice of the Office, indicated by said exceptions, it was wholly unnecessary. This non-action was based on the practice of the Office, on which they had the right to rely, and no power to control; and this practice itself was based on the reasonable and essentially necessary construction of the town-site and mineral laws, whereby both might be executed and claimants under them secured in such rights as they had respectively acquired. It should also be remembered, in this connection, that the Government does not act upon the individual claims of town occupants, but does adjust and patent mineral claims directly to the mine-owners.[1]

"The request of the mineral claimants, however, as presented in their protest and claim now under consideration, constitutes a proposition never hitherto before this Office for decision. It is, in brief, that every mine discovered, or *hereafter* to be discovered, throughout its entire length, with a width of one hundred feet, surface ground included, be excepted from the town patent.

"Where and when will these mines be discovered and opened?

---

[1] Decision of Commissioner, Dec. 23d, 1875, 2 Copp's Land-owner, 150.

What and whose property will they then embrace and practically confiscate? What foot of surface ground will ever be held by a town occupant under a clear title where the same is not purchased from the mine-owner? Was the mineral law designed by Congress as a repeal of the town-site statutes? These points would assume vital significance were the present claim conceded.

"The town of Central City was incorporated in 1864. The first patents were issued to mine-owners in 1869. Precisely *when* mining claims attached to any particular piece of ground I cannot determine. Precisely *when* a legal surface claim by a town occupant attached to any particular lot I have no means of ascertaining. How, then, with deference to those laws under which these claims have attached, can I defer the one absolutely to the other?

"The necessity of so construing both laws as not to defeat either—that respect for rights under each, which, of itself, seems, must control my action—the fact that the exceptions in mineral patents secure a shield of protection to town occupants and mineral claimants alike, and on which town occupants have relied, and that the exception in the town-site patent is as broad as the law suggests, and almost in its exact language—render it improper for me to grant the present claim.

"The mine-owner is protected by the local rules and customs, and these are recognized by the United States. The town patent is executed to a trustee, who is controlled by the legislature of his State or Territory. The local Courts are open for those particular adjustments which this Office cannot reach; and I conclude that the present demand is entirely outside what, in the proper execution of said laws, can be legitimately claimed or conceded. I therefore decline to grant the application; and in conformity to the views herein set forth, I hereby revoke my letters to the register and receiver, of August 26th, 1874, and to John A. Dix, Esq., of April 24th, 1875, so far as they conflict with this decision, and decline to except by name and survey any mine whatever in said town." [1]

---

[1] See, generally, Provisions as to Town Sites, Rev. Stats. 2380, ——, 2394; See Instructions, Copp's Pub. L. L. 678, 679, 680, 700; Effect of Deed under Congressional Town-site Act: Treadway v. Wilder, 8 Nev. 92; S. C. 9 Nev. 67.

CHAPTER XVI.

SEGREGATION OF MINERAL AND AGRICULTURAL LANDS—WITHDRAWAL FROM AGRICULTURAL ENTRY.

§ 223. Manner of setting apart mineral lands as agricultural.
§ 224. Segregation of agricultural from mineral lands.
§ 225. Mineral affidavits.
§ 226. Mineral affidavits on timber land.
§ 227. Segregation under Acts of 1866 and 1870.
§ 228. Withdrawal of certain lands from agricultural entry.
§ 229. Surveyors' returns.
§ 230. Their prima facie accuracy.
§ 231. Hearings to determine the character of land—Publication.
§ 232. What is mineral land.
§ 233. Burden of proof.
§ 234. Evidence as to agricultural character of land.
§ 235. The testimony.
§ 236. Proof as to mineral character of land.
§ 237. Discovery of mines on agricultural lands.
§ 238. Agricultural patent covering mines already worked.
§ 239. Fraud in pre-emption entry.
§ 240. Compromises between miners and settlers.
§ 241. Attempt by railroad to disprove mineral character of lands.
§ 242. Non-mineral proof by settlers on lands within railroad limits.

§ 223. Manner of setting apart mineral lands as agricultural.—Section 2342 of the U. S. Revised Statutes reads: "Upon the survey of the lands described in the preceding section, the Secretary of the Interior may designate and set apart such portions of the same as are clearly agricultural lands, which lands shall thereafter be subject to pre-emption and sale as other public lands, and be subject to all the laws and regulations applicable to the same."[1]

§ 224. Segregation of agricultural from mineral lands, for sale and pre-emption.—It was apparently the intention of this section to throw a quantity of land open to pre-emp-

---

[1] Rev. Stats. 2342, Sec. 11, Act 1866; 14 U. S. Stats. 253. See Rev. Stats. 2341, 2258, 2406.

tion which had been under the restrictive system reserving mineral lands considered as mineral.[1]

§ 225. **Mineral affidavits.**—Under the early instructions issued in 1868, the Surveyors-General in the mining States and Territories were instructed to require their deputies to describe in their field-notes, and designate on township plats, such lands as were agricultural. After the filing of the plats in the district land offices, if no counter affidavits were presented, the tracts designated " agricultural lands " might have been filed upon under the pre-emption, or taken under the homestead laws; but pre-emptors were not permitted to prove up and enter until after such a period of actual settlement and cultivation as showed good faith—say not less than six months from the date of settlement embraced in the filing of the declaratory statement; and if before the expiration of such time, an affidavit was filed alleging the mineral character of the particular tract claimed, a trial was to be had before entry made to determine the question.[2]

Affidavits alleging particular lands to be more valuable as mineral than as arable were required to apply to each of the smallest legal subdivisions. Of a quarter-section, 120 acres may be mineral and the remaining forty acres arable, and the mineral character of the former is no reason why the latter should not be entered as agricultural. Nor was it sufficient that such affidavits were based upon opinion or belief. They were required to contain a statement of facts within the knowledge of deponents, derived from actual observation or examination, furnishing a strong presumption that the particular subdivision was mineral land, and more valuable as such than as agricultural. Mere speculation, based upon no positive knowledge and disclosing no material facts, was not received, and claimants were not to be put to the trouble and expense of meeting affidavits of this character. The utmost good faith was required from both sides. Lands more valuable for mineral than for other purposes were not to be taken under the pre-emption or

---

[1] Yale's Mining Claims, 282.
[2] Instructions May 16th, 1868, Zabriskie's Land Laws, 208; Copp's U. S. Mining Decisions, 248.

homestead laws; nor on the other hand were parties to be intimidated from making settlements upon, and ultimately entering under these laws, such lands as were most useful for agricultural purposes, simply because valuable mineral lands might lie in the immediate vicinity. Both classes of land were to be disposed of under their appropriate laws, and neither interest to be subordinated to the other, to the detriment of the public welfare.[1]

§ 226. **Mineral affidavits on timber land.**—If the land officers in the various districts had reason to believe that mineral affidavits were being placed on timber land on which no mineral had been found, for the purpose of keeping the same from settlement until after the timber had been removed, they were ordered to inform the parties filing such papers that affidavits in which it was not alleged that either gold, silver, cinnabar, or copper had been found were insufficient; and however specific and full such affidavits might be, they were not to be deemed conclusive as to the matters alleged. Any settler filing an affidavit of a contrary character, and desiring to enter the particular tract as agricultural land, was entitled to have the question examined, and on proving it to be more valuable as arable than as mineral, could file upon it under the pre-emption or enter it under the homestead law, and in due time consummate the title.

It mattered not whether the previous metals were found in quartz ledges, or in placer or hydraulic mines: if the particular subdivisions of the public lands containing them were more valuable for mining than for agriculture, they could not be entered under the pre-emption or homestead law. If placers, once valuable, had become exhausted, so as no longer to be valuable as mines, the land might then be entered as arable.[2]

These affidavits were to be filed with the register, and copies transmitted to the General Land Office. They were required to be registered and carefully filed, but it was not necessary to record them. All witnesses were to be carefully cross-examined,

---

[1] Instructions May 16th, 1868; Zabriskie's Land Laws, 208; Copp's U. S. Mining Decisions, 248.

[2] Instructions May 16th, 1868, Zabriskie's Land Laws, 208; Copp's U. S. Mining Decisions, 248.

in order to elicit the truth, and the testimony, with the papers in each case, together with the joint opinions of the district land officers, transmitted to the General Land Office for examination and review.[1]

In order to enable the Department properly to give effect to this section of the law, the Surveyors-General were ordered to describe in their field-notes of surveys, in addition to the data required to be noted in the printed Manual of Surveying Instructions, on pages 17 and 18, the agricultural lands, and represent the same on township plats by the designation of "Agricultural Lands."[2]

§ 227. Segregation of agricultural from mineral land under Act of 1866-1870.—It was at first thought that under the operation of this law, recognizing ten-acre lots as legal subdivisions of the public lands in mining regions, much of the difficulty theretofore experienced in proving the mineral or non-mineral character of lands might be obviated. It had been necessary to file mineral affidavits on each forty-acre tract, that being the smallest legal subdivision of public lands; and to disprove the mineral character of lands so filed upon, it had been necessary to establish the fact that such forty-acre tract was as a whole more valuable for agricultural than for mining purposes. This it was often found impossible to do, for the reason that although parties could be readily produced to testify to the fact that one-half or perhaps three-fourths of a given tract was only fit for farming, yet inasmuch as a small fraction of the land was intersected by a gulch, ravine, or quartz lode yielding mineral, the value of which deposit there was no definite means of ascertaining, the deponents would be unable to testify that the entire forty-acre subdivision was of greater value for farming than for mineral purposes. In this way, although thirty and even thirty-five acres in a forty-acre tract might be shown to contain no mineral whatever, or none in quantities sufficiently abundant to be remunerative to the miner, yet on account of the known mineral character of a small fragment of the land the

---

[1] Instructions May 16th, 1868, Zabriskie's Land Laws, 208; Copp's U. S. Mining Decisions, 248. As to fees authorized in pre-emption and homestead cases, see Instructions Sept. 17th, 1867.

[2] Instructions Jan. 14th, 1867, Copp's U. S. Mining Decisions, 239.

bona fide agriculturist had been debarred from securing a title to his land, at least to the extent of that forty-acre tract. For these reasons, the local officers were ordered May 6th, 1871, when an application was filed to enter land as agricultural which had been returned by the United States deputy surveyor as mineral, or upon which mineral affidavits had been filed, to publish at the expense of the applicant a notice of such application for thirty consecutive days, in a newspaper of general circulation, published nearest to the land in question, or if in a weekly paper, for five consecutive weeks, giving the name and address of the applicant; the designation of each forty-acre tract covered by the application, the names of any miners or mining companies whose claims were upon the land, the names of the parties who filed the mineral affidavits, and when such filing was made, and finally the notice named a day, after the thirty days had expired, upon which a hearing was had before the register and receiver to determine the facts as to the mineral or non-mineral character of the land, when such witnesses as might be brought by the parties in interest were to be examined and their testimony reduced to writing; and the depositions of such witnesses as were unable to be present, whether from distance, infirmity, or other good cause, were received and examined; after which, the proceedings were submitted to the Commissioner of the General Land Office for review, prior to a final award of the land. A copy of this notice was posted in a conspicuous place upon each forty-acre tract embraced in the application, for the period of thirty consecutive days; proof of which was required on the day of the hearing by the sworn statements of at least two witnesses, one of whom might be the applicant, the deponents to state where the notice was posted, the date of the posting, and how long continued, and a copy of the printed notice was to be also filed, with the publisher's affidavit attached, stating when the notice was first published, and for how long.

In every case where practicable, in addition to this publication and posting, personal notice was required to be served in the usual manner upon the parties who filed the mineral affidavits, and upon those who were actually engaged in mining upon the land; on the day of hearing, the witnesses were to be examined by the register and receiver, with the view of eliciting the truth as to the mineral or non-mineral character of the land; and in cases where it was established that a portion of the land in a forty-acre tract was mineral, and the remainder agricultural, the testimony was required to be of a nature clearly showing what particular portion or portions of the land were actually covered by placer or quartz claims, or used in connection therewith, and fixed by local customs or rules of miners; and it was suggested that if, prior to such hearing, the respective parties could come to an agreement as to the proper boundaries of the mineral and agricultural lands, in the same forty-acre tract, they file, on the day of hearing, a diagram and description, showing in what portions of the tract such mines and grounds used in connection therewith existed, stating whether the same were placers, or vein or lode claims, by way of assisting the officers in the discharge of their duties.[1]

Where the applicant claimed the pre-emption right to the land filed upon, at the hearing all the customary proof was exacted, usual in cases of pre-emption contests, as required by the law and the instructions. The same rule applied to homestead applicants.

After the hearing, all the papers and testimony were to be transmitted, together with the joint opinion of the register and receiver, to the Commissioner of the General Land Office, for review. In cases where a survey was necessary, to set apart the mineral from the agricultural land, in any forty-acre tract, the necessary instructions were issued to enable the agricultural claimant, at his own expense, to have the work done, at his option, either by United States deputy, county, or other local surveyor, under the 16th Section, Act of 1870.[2]

§ 228. **Segregation of mineral and agricultural lands, June 17th, 1872.**—*Large quantities withdrawn from agricultural entry till proof given of their non-mineral character.*— Prior to November, 1871, the practice had prevailed of allowing pre-emption and railroad rights to attach to lands in the mineral region, when the same were returned by the surveyors as

---

[1] Instructions May 6th, 1871, Copp's U. S. Mining Decisions, 231.
[2] Ibid.

agricultural land, without making an investigation as to the correctness of such return, unless affidavits were filed alleging the tracts to be mineral. By reason of erroneous and false returns by the surveyors, patents were issued conveying valuable mining premises as agricultural land, which never would have been done if the land had been properly returned by the surveyors, the grants generally expressly excepting and excluding all mineral lands except those containing coal or iron.

In consequence of numerous frauds practiced upon the Department, the Commissioner of the General Land Office at length became impressed with the conviction that it was neither in harmony with the spirit or intent of the laws of Congress, nor with true public policy, to sanction the indiscriminate absorption of the lands, in what had theretofore been known as the reserved mineral belt in the public domain, under laws only applicable to lands clearly non-mineral, simply because the deputy surveyors failed to return the same as mineral in character, especially as the majority of mineral patents were found, upon consulting the official township plats, to be within subdivisions not reported as mineral in character. In many of the hearings had before the local officers, to determine the true character of those tracts, the testimony showed conclusively that, of a whole quarter or half quarter-section sought to be entered under the pre-emption law, the only portion really agricultural in character was confined to two or three acres upon which the pre-emption party had a house and garden, the agricultural utility of the remainder consisting in its adaptability to grazing cattle. The mere fact that an individual used one of these isolated garden spots, situate in an imperfectly developed mineral region, as a homestead or ranch, was not finally considered by the Commissioner as investing the settler with an equitable right to a Government title to an entire quarter-section of land, the real mineral character of which had not been tested, or at all events not sufficiently to enable parties to tell with any certainty whether the land contained valuable mines or not.

When a bona fide agricultural claimant desired the segregation of the ground containing his improvements from the adjoining mineral land, he could have the same effected under existing circular instructions. From the fact that but few of these pre-emption claimants appeared disposed to avail themselves of the privilege of this segregation, the inference was, that it was not so much on account of the agricultural value of the tract as of its probable mineral deposits, that title was desired.

To illustrate the correctness of these views, it was shown that in Nevada County, California, the length of the county from east to west was about sixty-five miles, having an average breadth of twenty, and containing about 1,300 square miles. It is near the middle of the great gold region that stretches along the westerly slope of the mountain chain, extends entirely across the auriferous belt, and in the preceding nineteen years had produced more gold than any tract of country of equal extent in the world.

The deep placer or hill diggings, in the channels of ancient streams, in many places underlying hundreds of feet of alluvial deposits and volcanic material, were almost inexhaustible. No estimate approaching to accuracy could be made of the amount of gold contained in the placer mines of that county, and which might still be extracted. In some of the deep placers, deposits of gold, it was thought, might be found in such quantities as would materially diminish the value of the metal. The extent and value of the gold-bearing quartz ledges were scarcely of less magnitude.[1]

Notwithstanding all this, Township 16 North, Range 8 East, Mount Diablo Meridian, embracing the towns of Grass Valley, Gold Hill, and part of Nevada City, all in the heart of this rich mining district, was not returned on the official plat of the township, as to any portion of it, as mineral in character, nor was any portion segregated from the agricultural portion as required by the law and the instructions.

After the survey and return of the township, numerous contests arose between the miners and the pre-emption claimants, as regarded the character of the land, and several applications were made and titles issued for mines therein, no intimation of the existence of which was given, however, upon the official

---

[1] Official Report of J. Ross Browne to the Secretary of the Treasury, March 5th, 1868.

plat of the township. A gold quartz claim, yielding $49,000 per month, was situated in a quarter-section of the township,[1] and yet the tract was not shown by the plat to contain any mineral land or claim whatever. The widely-known Ophir Hill, or Empire Mine, near the town of Grass Valley, was in Section 35 of said township. This mine had been worked since 1852, had had $250,000 expended in improvements thereon, and was reported to have yielded nearly $2,000,000 in gold. Yet the plat gave not the slightest indication that there was any such mine in existence, or that the subdivision in which it lay was other than agricultural land. Numerous other instances of a like character existed.

In other townships another condition of things existed. In some of the township plats, certain of the subdivisions were shaded yellow, and designated "Mineral Land," the remainder being designated "Agricultural Land." On inspection of some of these plats, it was found that within those agricultural tracts were marked "Quartz Ledges," "Placer Mines," "Hydraulic Mines," etc., the plat thus contradicting itself. Prior to the Act of 1866, this mineral region had been excluded from survey and sale by the laws of Congress. From the indefinite nature of the returns made by the deputy surveyors, the impracticability of carrying into effect the eleventh section of the Act of 1866[2] became at once apparent, experience having shown that little reliance could be placed upon those plats in determining the true character of the land. To set apart the lands clearly agricultural from such data, partook for the most part of the nature of guess-work.

Public considerations, therefore, induced the Commissioner of the General Land Office to ask authority for withdrawing from disposal as agricultural lands, such townships or parts of townships in that region as might reasonably be presumed, from common report, from official and other data, to be properly classed as mineral land, and that no entries thereof be permitted except by legally qualified citizens holding mineral claims, in accordance with the mining statute, except in cases where the

---

[1] Official Report of J. Ross Browne to Secretary of the Treasury, March 5th, 1868.
[2] Rev. Stats. 2342.

agricultural character should first be established by competent testimony, in accordance with existing regulations applicable to the subject.[1]

Accordingly, authority was given by the Secretary of the Interior to the Commissioner to make the necessary withdrawal, and to instruct the local officers not to permit any of the tracts which might be withdrawn to be entered as agricultural land, unless the non-mineral character of the same should have been first fully and clearly established by competent testimony.[2]

Acting under this authority, and stating further that experience had shown that the Office could not, with any degree of safety, judge of the character of the lands from the data furnished by the returns, and there being no authority of law for the employment of a competent geologist to investigate the matter, the head of the Department had, in consideration of the public interests, and to prevent the indiscriminate absorption of the mineral lands of the public domain through the instrumentality of insufficient returns, found it imperatively necessary to adopt the course announced, both for the protection of the parties who had already expended time, capital, and labor in opening and developing mines, and those of the citizens of the United States who might thereafter desire to exercise their legal right to do so, the Commissioner designated a number of townships as within the order.[3]

---

[1] Letter of Commissioner, November 24th, 1871, Copp's U. S. Mining Decisions, 297.

[2] Letter of Secretary of the Interior, November 24th, 1871, Copp's U. S. Mining Decisions, 301.

[3] Decisions of Commissioner, December 2d, 1871; December 7th, 1871; January 22d, 1872; March 11th, 1872; March 20th, 1872; April 20th, 1872, Copp's U. S. Mining Decisions, 301, 302, 304, 311, 314.

NOTE.—*Mount Diablo base and meridian.*—The following Townships were included in the order: Township 1 North, Ranges 11, 12, 13, 14, and 15 East; Township 2 North, Ranges 11, 12, 13, 14, and 15 East; Township 3 North, Ranges 10, 11, 12, and 13 East; Township 4 North, Ranges 10, 11, 12, and 13 East; Township 5 North, Ranges 10, 11, and 12 East; Township 6 North, Ranges 10, 11, 12, and 13 East; Township 7 North, Ranges 9, 10, 11, and 12 East; Township 8 North, Ranges 9, 10, 11, and 12 East; Township 9 North, Ranges 9, 10, 11, and 12 East; Township 10 North, Ranges 8, 9, 10, 11, and 12 East; Township 11 North, Ranges 6, 7, 8, 9, 10, and 11 East; Township 12 North, Ranges 6, 7, 8, 9, 10, and 11 East; Township 13 North, Ranges 6, 7, 8, 9, 10, and 11 East; Township 14 North, Ranges 6, 7, 8, 9, 10, and 11 East; Township 15 North, Ranges 6, 7, 8, 9, 10, and 11 East; Township 16 North, Ranges 5, 6, 7, 8, 9, 10, 11, 12, 15, 16, and 17 East; Township 17 North, Ranges 5, 6, 7, 16, and 17 East; Township 18 North, Ranges

The order directing the Commissioner to suspend from disposal, as agricultural lands, certain townships therein designated, until the non-mineral character thereof should first be established by competent proof, taken at a hearing, to be had after due notice, was subsequently modified with respect to entries which had already been made and reported to the Office before the instructions of December 2d, 1871, were issued, but it was ordered to be strictly enforced with regard to all subsequent applications.

The cases which had been reported prior to the date of the instructions were each ordered to be carefully examined in its turn, and if, from the facts in any case, further hearing or additional proof should be necessary, the proper rulings were to be made in each individual case.[1]

It being objected to this action of the Department that it necessitated additional expense and delay, and that the real meaning and object intended to be conveyed and effected by the order was a suspension of the lands in question from settlement, and a denial in toto of the right of any settler to secure title to any tract whatever within the suspended townships, the persons objecting omitting, however, to consider that upon making proof

---

4, 5, 6, 7, 16, and 17 East; Township 19 North, Ranges 3, 4, 5, 6, 7, 16, and 17 East; Township 20 North, Ranges 3 and 4 East; Township 21 North, Ranges 3 and 4 East; Township 22 North, Range 3 East; Township 23 North, Range 3 East; Township 24 North, Range 4 East; Township 1 South, Ranges 12 and 14 East; Township 4 South, Range 16 East; Township 5 South, Ranges 16 and 17 East; Township 6 South, Range 18 East; Township 7 South, Ranges 17 and 18 East; Township 8 South, Range 18 East; Township 9 South, Range 18 East; Township 10 South, Ranges 20 and 22 East; Township 11 South, Ranges 22 and 23 East.

Withdrawals from agricultural entry in Colorado.—Under instructions of November 15th, 1875, all land lying within the Del Norte Land District, Colorado, (except Townships relieved from suspension by letter of June 15th, 1874) were withheld from sale as agricultural lands, until the non-mineral character should be established.

Township 40 North, Ranges 4, 5, and 6 East, N. M. M., were also reported as mineral, and ordered withheld in the same manner, and so much of the letter of June 15th, 1874, as relieved those Townships from suspension was revoked. The register was ordered, in case he became satisfied that any Township or part of a Township included was *clearly agricultural*, to report the fact to the General Land Office, accompanying his report with the sworn statement of parties who, from their knowledge of the land, could testify understandingly as to the true character of the land. (Decision of Commissioner, November 15th, 1875, 2 Copp's Land-owner, 130.)

[1] Decision Commissioner, April 20th, 1872, Copp's U. S. Mining Decisions, 314.

of the non-mineral character of any tract so suspended, the settler's rights would be fully recognized, the Department sustained its position by further argument and proofs.

The Commissioner called attention to the fact that Congress had, from its earliest legislation in reference to public lands, made a distinction between lands which are mineral and those which are not, and this distinction had invariably been enforced in every public land law enacted by that body, and so long as the legislative branch of the Government saw fit and proper to specially make such distinction, the executive had before it the plain duty of enforcing the same, and was without power under the law to waive it.[1]

As recited in a previous communication, under the Act of 1866, it was not the intention of the latter statute to abolish or do away with the distinction between mineral and agricultural lands, or to allow mineral lands to be classed and disposed of as agricultural; but it simply provided that the public surveys might be extended over a region that was so clearly mineral in character, that before that time it had been all reserved for mineral purposes, and the tract that should appear to be clearly agricultural set apart for disposition under the laws relating to such lands, while the mineral lands should be still reserved for disposition under the laws relating to lands of that class, the Department having no more right to dispose of mineral lands in large tracts than it had before the enactment of this law.[2]

Owing to the fact that the two classes of land in the mineral belt are so interlaced as to prevent, in most cases, their segregation by the rectangular system of surveys, the proper execution of the requirement of setting apart the "clearly agricultural" portions, is one of the greatest difficulty, and in many cases it is almost impossible. During the lapse of ages the melting of snows and washing of rains have had the effect of disintegrating the quartz lodes or other auriferous deposits in the mountains, which are washed down into the valleys, flats, and ravines, the gold, from its greater specific gravity, settling to the bottom or

[1] Letter of Commissioner, March 11th, 1872, Copp's U. S. Mining Decisions, 304.
[2] Ibid.

bed rock; these deposits forming the "placers," or "diggings," some of which are quite shallow and soon exhausted; others again being very deep, and overlaid with good soil, the surface in the latter case being "clearly agricultural," while the deposit underlying the same is of such a character as to render the land of great value for mineral purposes. Again, there exists in the State of California what are called "Blue Leads," "Cement" or "Gravel" claims, supposed to be the beds of ancient river channels, very deep, rich in gold, and practically inexhaustible. These immense deposits are frequently covered to a depth of from fifty to one hundred feet, the surface of the overlying mass being perhaps tillable land, and presenting no indication whatever of the valuable underlying deposit. Arable land is also sometimes found to overlie quartz lodes, the existence of which may not have been known at the time the settler began his improvements, but were afterward discovered by prospecting shafts or otherwise, by miners, who thereupon claimed such lodes, under local rules and customs, together with a sufficient area of surface ground for the convenient working of such mines.[2]

But aside from the obstacles growing out of the peculiar character of the lands, which rendered it very difficult, even with the utmost circumspection and care, to carry out the intention of Congress and prevent the disposition of mineral lands as agricultural, it had been found that, owing to the grossly careless, not to say fraudulent, manner in which deputy surveyors executed their work in the field, and made their returns, the distinction which Congress had drawn between agricultural and mineral lands was not observed, and whole townships containing the richest mineral land in the world, including well-known mines which had been worked successfully for years, and which were still being worked successfully and profitably, were returned to the General Land Office and to the local land offices as agricultural land, and so posted on the tract-books, and became from the date of such return subject to sale and to selection by railroads, etc., as agricultural lands, in direct violation of the plain intent of Congress, as expressed not only in previous legislation, but in the very act under which these lands were surveyed and brought into market.

[2] Letter of Commissioner, March 11th, 1872, Copp's U. S. Mining Decisions, 304.

The action, therefore, taken by the Office in requiring agricultural claimants to submit satisfactory proofs as to the non-mineral character of the lands sought to be entered by them, was considered imperatively necessary to carry out the will of Congress, clearly and repeatedly expressed in regard to the reservation of mineral lands for mineral purposes.

To illustrate the unreliability of the surveyors' returns as to the character of the lands, and the absolute necessity for the rule adopted, the names and locations of fifteen gold quartz claims, six other quartz claims, and thirteen placer claims, all within one district, were given, the lands embracing which were returned on the official township plats as agricultural in character, the existence of mines therein not becoming known to the Office until after the receipt of the applications for mining titles.[1]

The rule had prevailed that upon the survey of these reserved lands, homesteads, pre-emptions, and railroad grant rights took effect upon all such lands as were returned by surveyors as "agricultural," except in cases where, before such lands were patented, affidavits were filed alleging their true character to be mineral, in which case a hearing would be had before the register and receiver to determine whether the tract was of more value for mineral than for agricultural purposes, mineral lands being expressly excluded from land grants to railroads, and from the operation of the pre-emption or homestead laws.

But the Commissioner became convinced that this rule failed to afford adequate protection to the miners as a class, or prevent the disposition of mineral lands as agricultural. It was admitted that parties engaged in the real estate business or in land speculations, and who were therefore well informed as to the regulations governing the land offices, and also those miners who had acquired a knowledge of the reckless manner in which returns had been made by deputy surveyors, could protect themselves from the consequences of such erroneous and false returns by making affidavits as to the mineral character of the lands in which they were directly interested, and filing the same with the register and receiver, thus necessitating a hearing be-

---

[1] Letter of Commissioner, March 11th, 1872, Copp's U. S. Mining Decisions, 304.

fore the land so filed on could be disposed of as agricultural; but inasmuch as the law does not provide for or require such affidavits to be filed, but does authorize the Secretary of the Interior, when the surveys are made, to segregate the agricultural from the mineral lands before they can be classed or disposed of under the law relating to agricultural lands, the miners and owners of mining claims had a right to suppose, and the great mass of them did undoubtedly suppose, that they were protected by the law without action on their part; but if this had not been so, no reason was perceived why, in a region confessedly mineral, and in which Congress had seen proper to hold all lands as mineral except those specially designated as agricultural by the Secretary of the Interior, the burden of filing proof as to the character of the land should be imposed on the mineral instead of the agricultural claimant. If a mine should be discovered in a region where agricultural lands predominate, such a rule would be reasonable and proper; but in a mineral region this burden of proof should be on the agricultural claimant, and Congress has so provided in effect by considering and treating all the lands as mineral which have not been specially designated by the Secretary of the Interior as agricultural. The work of designating and setting apart agricultural lands in the mineral region is not left to the Surveyor-General, nor even to the Commissioner of the General Land Office, but is by the statute thrown upon the Secretary of the Interior; and therefore, according to the most liberal construction, the disposition of these lands as agricultural, under the rule previously prescribed, was by the Commissioner declared unauthorized and illegal, March 11th, 1872.[1]

But, admitting the propriety and legality of the proceedings under these mineral affidavits, they did not prevent the mineral lands from being disposed of as agricultural, except in special cases where contests arose between parties claiming adversely. Where the mineral and agricultural interests are both vested in the same person or persons, the lands will be taken as agricultural, unless the parties in interest are required to give notice and submit proof under oath as to the non-mineral character of the land.

---

[1] Letter of Commissioner, March 11th, 1872, Copp's U. S. Mining Decisions, 304.

Again, in many localities the mineral-bearing lands had not been occupied or worked because of the lack of water, or other necessary facilities; but it did not follow because they were not then occupied or worked by some one who was ready to contest the right of the agricultural claimant, that they were not mineral lands, or that they might not under a changed condition of things become productive mines.

A ditch of a few miles in length frequently rendered mines very profitable, that could not be successfully worked without water. But, as there was no law which authorized or required these mineral affidavits, so there was nothing to prevent their being withheld or withdrawn for fraudulent purposes. In some cases, where the matter came on for hearing before the register and receiver, the mineral affiants failed to appear, and instances came to the knowledge of the Office of private arrangements being entered into between the respective mineral and agricultural claimants, by which the latter were not to be opposed in obtaining titles to the land which, upon being patented as agricultural, was to be held by the several parties in pursuance of such previous agreement, and thus the only obstacle to the disposition of mineral lands as agricultural—the mineral affidavit—was withheld or removed. The order of withdrawal was, therefore, directed to be adhered to and strictly enforced.[1] In pursuance of this order, and to save as much as possible the expense, trouble, and delay incident to the existing manner of taking proofs as to the mineral or agricultural character of lands, directions were given in March, 1872, that testimony upon this point might be taken before a clerk of a Court of Record in and for the county in which the land is situate, after due notice.

At the hearing, the claimants and witnesses will be thoroughly examined with regard to the character of the land; whether the same has been thoroughly prospected; whether or not there exists, within the tract or tracts claimed, any lode or vein of quartz or other rock in place, or other valuable mineral deposits, which has ever been claimed, located, recorded, or worked; whether such work is entirely abandoned, or whether occasion-

---

[1] Letter of Commissioner, March 11th, 1872, Copp's U. S. Mining Decisions, 304.

ally resumed; if such lode does exist, by whom claimed, under what designation, and in what subdivision of land it lies; whether any placer mine or mines exist upon the land; if so, the character thereof—whether of the shallow surface description, or of the deep cement, blue lead, or gravel deposits; to what extent mining is carried on when water can be obtained, and what the facilities are for obtaining water for mining purposes; upon what particular ten-acre subdivisions mining has been done, and at what time the land was abandoned for mining purposes, if abandoned at all.[1]

§ 229. **Surveyor's returns.**—Some of the acts passed prior to 1866 prohibited the extension of subdivisional surveys over large districts which were held and reserved as mineral lands, and subject to exploration as such. The policy of restricting subdivisional surveys to agricultural lands was abandoned by the Act of 1866, which provided, among other things, for the extension of the subdivisional surveys over mineral lands, and recognized homestead and pre-emption rights to lands therein which are "properly agricultural," and authorized the Secretary of the Interior to designate and set apart such portions as are "clearly agricultural," to be thereafter subject to disposal as other lands of that class. Congress did not intend to abolish or do away with the distinction between mineral and agricultural land, or to allow mineral lands to be classed and disposed of as agricultural; but it simply provided that the public surveys might be extended over regions that were so clearly not agricultural in character that they had, previously to that time, been regarded and treated as exclusively mineral, and that such tracts as should appear to be clearly agricultural might be set apart for disposition as agricultural lands, while the mineral lands, which under that act included all lands in mineral regions not *properly and clearly* agricultural, should be still reserved for disposition under the laws relating to lands of that class; or, in other words, the Act of 1866 did not abolish the previous rule and policy of the Government, but merely modified them so far as to extend subdivisional surveys over mineral lands, and

---

[1] Instructions March 20th, 1872, Copp's U. S. Mining Decisions, 311; Instructions Feb. 1st, 1877.

authorized the Secretary of the Interior to designate and set apart for agricultural entry such tracts as should be found to be *clearly agricultural* in character.

Under this law, the Secretary of the Interior is not authorized to set apart and designate lands as agricultural simply because there is no proof to show that they are not mineral, for this would defeat the object and purposes of the laws which hold mineral lands open to exploration and development, and enable parties to appropriate undeveloped mineral land as agricultural. The power of the Secretary is confined to designating and setting apart such tracts as are *clearly and properly agricultural*, and no others.

Where the lands are returned as mineral by the Surveyor-General, the burden of proof is on the parties who question the correctness of the return, and unless they establish the fact that they are *clearly and properly agricultural lands*, they must fail in the assertion of any claim or right to them under laws relating to the disposition of agricultural lands. Proof of the fact that no paying mines have ever been discovered is not sufficient.

A case arose wherein the Surveyor-General caused to be indorsed upon the township plat the following words: "The above township is a rough, barren, volcanic country, the hills generally broken and rocky. The south half of the township contains gold and silver-bearing quartz veins, the great Comstock Lode passing through Sections 20, 29, 31, and 32, in which are situate Virginia City and Gold Hill" (Nevada). The proof sustained the return of the Surveyor-General. To disprove the mineral character of the tracts, a hearing was held, after due notice, and in the prescribed manner. The testimony established the fact that no mines had yet been discovered upon the tracts, but that they lay in the mineral belt, and in the immediate vicinity of the Comstock Lode; and notwithstanding no mines had been developed, the character of the land, its location, and the testimony indicated the existence of mineral therein, while the fact was clearly established that the lands were not of any value whatever for agricultural purposes, and that if they were of any value at all, it was for undeveloped mines.

In pursuance of the doctrine above enunciated, the claims of the State of Nevada and the Central Pacific Railroad Company of California, who disputed the mineral character of the land, were denied, and the tracts reserved as mineral.[1]

§ 230. Prima facie accuracy of the surveyor's return.— The surveyor's return of the character of land is prima facie correct. The burden of proof is upon the party seeking to disprove such return. If the testimony does not show that the land is more valuable for agricultural than for mining purposes, the application to enter as a pre-emptor will be denied.

In such contests to determine the character of land any person who has a knowledge thereof, whether interested or not, is permitted to testify in behalf of the surveyor's return.[2]

But a failure of a Government surveyor to segregate mineral from agricultural lands cannot operate to defeat the rights of occupant miners. In the face of an open and notorious possession of the miner, the fact that the claims were not segregated and listed as mineral lands was held not available for the settler. Segregation, when required, must be made by the surveyor; and to hold that the failure of the surveyor to fully discharge his duty could operate to defeat the rights of the miner, was considered violative of the plainest principles of justice. The returns of the surveyor are not conclusive as to the character of the lands; for the Land Office allows affidavits as to the character of the lands to be made in impeachment of the returns.[3] The open, notorious possession of the miner was considered sufficient to charge the settler with notice of the character of the lode, and also to bring the lode within the description of known mineral deposits. Nor were the miner's rights regarded as forfeited, nor in the least abridged by failure to procure a patent for the claims. There was nothing obligatory on the miner to proceed under the Act of 1866, and where they failed to do so, there being no adverse interest, they held the same relations to the premises they worked as they did before the passage of the

---

[1] In re Claims of the State of Nevada and C. P. R. R. Co. of Cal.; Decision of Commissioner, April 3d, 1874, 1 Copp's Land-owner, 18, 114.

[2] Decision of Commissioner, Dec. 22d, 1875, 2 Copp's Land-owner, 146; Circulars May 6th, 1871; March 20th, 1872; June 10th, 1872; Rules of Practice, Nov. 29th, 1875; Post.

[3] Gold Hill Quartz Mining Co. v. Ish, 5 Oregon, 104.

act, with the additional guarantee that they possessed the right of occupancy under the statute. As the settler was never the owner of the lode and never obtained title, he could not be declared trustee for the miner nor could he execute a deed conveying to the plaintiff the legal title. The proper relief was considered to be an injunction preventing the settler from asserting title to the lode and from interference with the claim.[1]

§ 231. **Hearings to determine the character of land—Publication.**—Published notices of hearings to disprove the mineral character of land should not be signed by the applicants themselves, nor should they make their own arrangements for hearing testimony and publishing notices.

The notice of the hearing should be prepared by the local officers and signed by them, in order to secure a correct description of the land and to insert the names of mineral affiants, should any mineral affidavits covering the land applied for be on file in the local office.

The register should designate the paper of general circulation near the land in which to publish the notice; and, in all cases where practicable, the hearings should be held before him. Where distance or other good cause renders it advisable, he should designate an officer using a seal, or other person authorized to administer oaths, whose character is known to him, and residing near the land, as the proper person before whom the hearing should be held.

The testimony submitted should be as far as possible by questions and answers, and the officer by whom the testimony is taken should endeavor to elicit full information as to the mineral and agricultural qualities of each ten-acre tract of the claim.[2]

§ 232. **What is mineral land.**—This question was considered in a case in California,[3] prior to the passage of the Act of 1866.

Section eight of an act of the legislature of California of April 16th, 1859, provided that the act was not to be construed

---

[1] Gold Hill Quartz Mining Co. v. Ish, 5 Oregon, 104.
[2] Decision of Commissioner, August 14th, 1875, 2 Copp's Land-owner, 98.
[3] Ah Yew v. Choate, 24 Cal. 562.

so as to authorize or confirm the location or purchase of any of the mineral, swamp, or overflowed lands in the State as school lands (Stat. 1859, p. 340). All lands containing gold, it was held, were not necessarily mineral lands within the meaning of the act. The Court said: "It is often a matter of difficulty to determine whether any given piece of land should be classed as mineral land, or otherwise. The question may depend upon many circumstances: such as whether it is located in those regions generally recognized as mineral lands, or in a locality ordinarily regarded as agricultural in its character. Lands may contain the precious metals, but not in sufficient quantities to justify working them as mines, or make the locality generally valuable for mining purposes, while they are well adapted to agricultural or grazing pursuits; or they may be but poorly adapted to agricultural purposes, but rich in minerals; and there may be every gradation between the two extremes. There is, however, no certain, well-defined, obvious boundary between the mineral lands and those that cannot be classed in that category. It is to be considered whether, upon the whole, the lands appear to be better adapted to mining or other purposes. It is necessary to know the condition and circumstances of the land itself, and of the surrounding locality. It is the duty of the officers of the Government, before making a grant, to ascertain these facts, and to determine the problem whether the lands are mineral or not."

Where, therefore, the lands appeared to have been surveyed with a view of bringing them into the market, and they were described by range, township, and section, according to the official survey of the United States, and the location was approved by the Government, the purchase-money paid to the State, the notice of intention to apply for a patent published as required by law, and the patent duly issued by the State, the lands were considered agricultural, and will be so declared whenever the regular proceedings prescribed by law have been taken, and the officers of the Government have ascertained these facts, and adjudged the lands subject to be granted. The Court further said: "The patent is the record of the State that the land was subject to location under the grant of the United States, and has been located through its officers in pursuance of the terms of the donation, (Doll *v.* Meador, 16 Cal. 324) and in this case it is a record of the judgment of the State, by its officers duly appointed for that purpose, that the conditions and characteristics of the land are not such as to constitute it mineral land within the meaning of the provisions of the statute, and the verity of this record is not overthrown by the mere fact appearing that the land patented has been ascertained to contain a sufficient amount of gold to induce a party to mine it for that metal."[1]

§ 233. **Burden of proof.**—In the absence of proof as to the mineral or non-mineral character of tracts upon which mineral affidavits have been filed, they will be considered as mineral until proved otherwise; but where land has been returned as agricultural, has been entered under the pre-emption laws, and proof and payment have been made, the burden of proof is upon the mineral claimants.[2]

Land having been returned as mineral, the burden of proof is upon the agricultural claimants. If the testimony shows that portions of the surface are susceptible of cultivation, and the balance is used for grazing, while it is not shown that valuable deposits have been found in the particular location, yet, if discoveries have been made in immediate proximity, the entry may be suspended.

If the land has been returned as mineral, an agricultural entry must not be allowed without a hearing.

Whenever land is returned on the township plat as mineral, or upon which mineral affidavits have been filed, or which is suspended for non-mineral proof by order from the Land Office, a hearing, after due notice, must be held in accordance with circular instructions. At this hearing, if no adverse claimant appears, there is required full non-mineral proof of every legal subdivision; and if such proof and the evidence required of agricultural applicants are satisfactory, the entry is allowed; but if an adverse claimant appears, the register must await the

---

[1] Ah Yew *v.* Choate, 24 Cal. 562.
[2] Decision of Commissioner, Jan. 24th, 1872, and Decision of Acting Secretary, Feb. 12th, 1872, Copp's U. S. Mining Decisions, 77.

action of the Land Office on the testimony presented before permitting entry of the land.[1]

When lands are withdrawn as mineral in character, such withdrawal shifts the burden of proof on the agricultural claimant, and having the affirmative, he must produce conclusive evidence of the non-mineral character of the land he applies to enter. But if the testimony establishes the fact that though a small portion of the land at one time contained gold, that portion has been exhausted and abandoned for many years, and that no part of that tract now contains mineral in sufficient quantities to pay for working, and that nearly all the land is valuable for agricultural purposes, the patent will go to the qualified preemptor under the agricultural entry upon full compliance with the laws.[2]

§ 234. **What is satisfactory evidence of the agricultural character of land.**—Where it was shown that there were on the land agricultural improvements, to the value of $1,000 or $1,200; that the greater portion was inclosed with fencing, and seven or eight acres of it cultivated in fruits, vines, vegetables, and grain; that the only active mining done on the tract was at a quartz mine and some placer diggings; and that the owners of the quartz mine had abandoned it, declaring that it was exhausted and worthless, and the diggings were abandoned for the same reasons; and that there were some quartz veins on adjoining lands that might run into the land in question one quarter, but that they were all either exhausted or unprofitable; that all the witnesses testified that in their opinion the land was more valuable for agricultural than for mining purposes, and the mineral affiants, though present at the trial, in person and by attorney, offered no testimony in support of their affidavits, but contented themselves with cross-examining the opposing witnesses, the agricultural character of the land was considered established.[3]

---

[1] Ewing v. Hartman, Decision of Commissioner, Feb. 18th, 1875, 1 Copp's Land-owner, 180.

[2] Mulls v. Rolls & Ross, Decision Acting Secretary, April 5th, 1877, 4 Copp's Land-owner, 19.

[3] Clark v. Ellis, Decision Secretary Interior, July 10th, 1872, Copp's Mining Decisions, 128.

§ 235. The testimony.—The testimony should show the agricultural capacities of the land, what kinds of crops are raised thereon, and the value thereof; the number of acres actually cultivated for crops of cereals or vegetables, and within which particular ten-acre subdivisions such crops are raised; also, which of these subdivisions embraces the improvements, giving in detail the extent and value of the improvements, such as house, barn, vineyard, orchard, fencing, etc.

It is thought that bona fide settlers upon lands really agricultural will be able to show by a clear, logical, and succinct chain of evidence that their claims are founded upon law and justice, while parties who have made little or no permanent agricultural improvements, and who only seek title for speculative purposes, on account of the mineral deposits known to themselves to be contained in the land, will be defeated in their intentions. The testimony should be as full and complete as possible, and, in addition to the leading points indicated above, everything of importance bearing upon the question of the character of the land should be elicited at the hearing. If, upon a review of the testimony at the General Land Office, a forty-acre tract should prove to be properly mineral in character, that fact will be no bar to the execution of the settler's legal right to the remaining non-mineral portion of his claim, if contiguous.[1]

Notice of hearing and verification of affidavits.—Sec. 2335 of the Revised Statutes reads as follows: "All affidavits required to be made under this chapter may be verified before an officer authorized to administer oaths within the land district where the claims may be situated, and all testimony and proofs may be taken before any such officer, and, when duly certified by the officer taking the same, shall have the same force and effect as if taken before the register and receiver of the Land Office. In cases of contest as to the mineral or agricultural character of land, the testimony and proofs may be taken as herein provided on personal notice of at least ten days to the opposing party; or if such party cannot be found, then by publication of at least once a week for thirty days in a newspaper, to be designated by the register of the land office as published

---

[1] Instructions March 20th, 1872, Copp's U. S. Mining Decisions, 311; Instructions Feb. 1st, 1877.

nearest to the location of such land, and the register shall require proof that such notice has been given."[1]

§ 236. **Special cases.**—Where it appeared that the tract was in a mineral belt, and in the center of a deposit that had been worked on either side, and found to yield gold in paying quantities; that the agricultural claimant had boasted of his land being valuable for mining, and had offered to dispose of it at a large price for such purpose; that a tunnel to reach the land had been almost continuously in process of construction for a considerable time, and large sums of money had been and were being expended to bring water to that and neighboring tracts in order to afford the necessary facilities for hydraulic mining, and prospecting also yielded evidences of mineral character, the land was considered more valuable for mining than for agriculture, and the land held to be mineral.[2]

And where, although land was shown to be in the mineral belt, and in the immediate vicinity of valuable placer and lode claims, yet it was worthless for mining purposes, and if it was ever paying ground had been worked out, and it was established that the land was of very great value for agricultural purposes, that the applicant had been in possession for twenty years, had cultivated it nearly all of the time, and had very valuable and lasting improvements on it, the land was held agricultural.[3]

Mines only become valuable when they can be developed and the precious metals extracted; and in cases where the land is of little, if any, value for agricultural purposes, and is essential to the proper working of deep gravel mines, it should be withheld from sale under the laws regulating the disposal of agricultural lands, and disposed of only to such parties as may be entitled to the same under the mining acts of Congress.

Where the testimony submitted at the hearing established the fact that the land was of little value for agricultural purposes, and that it was bounded on the south by valuable gold-bearing gravel mines or deep hydraulic diggings which could be success-

---

[1] Act May 10th, 1872, Sec. 13, 17 U. S. Stats. 95.
[2] Pulliam v. Hunter, Decision of Secretary, Feb. 5th, 1876, affirming Decision of Commissioner April 21st, 1875, 2 Copp's Land-owner, 180.
[3] Decision of Secretary, July 10th, 1872, August 6th, 1872, Copp's U. S. Mining Decisions, 128, 130.

fully worked and developed only by means of tunnels passing through this land to a creek, which was the only natural and practicable outlet for these mines; and it was shown that portions of the land in dispute were claimed and held by mine-owners, and that several tunnels were being run through this land for the purpose of developing and working said gravel mines, and it also appeared that the agricultural applicant had conveyed by deed to a mining company certain mining rights upon the land in dispute, and that he acknowledged the mineral character of the S. ½ of S. ½ of the N. W. ¼ of the N. W. ¼ of the section, the land was held to be only valuable on account of its location with reference to the mining claims, and of far greater value for mining purposes than for agricultural purposes.[1]

Where a small portion of the land at one time contained gold in paying quantities, yet that portion has been exhausted and abandoned, and nearly all the land is valuable for agriculture, the agricultural character of the land may be considered established. While the mining interests are entitled to and must receive protection against the encroachments of persons who, under the guise of agricultural claimants, seek to secure title to large tracts of mining land; the rights of bona fide pre-emption and homestead claimants to lands proven to be agricultural are also entitled to the same protection against adverse combinations of miners.[2]

In any case where there is a contest, or where the non-mineral character of the land and the bona fides of the claimant are not entirely clear, the entry will not be permitted until the testimony has been reviewed at the General Land Office, whose power to review, revise, or reverse the action of the register is not taken away or impaired by the act of paying for the land, but the claimant under such purchase only acquires a vested right on the condition that the officers of the Department shall concur with local officers.[3]

Lands reserved as mineral are only subject to the pre-emption laws after their segregation by the Secretary of the Interior.[4]

---

[1] Decision of Commissioner, Jan. 3d, 1876, 2 Copp's Land-owner, 146.
[2] McKenna v. Dillon, Decision of Acting Secretary, May 6th, 1872, Copp's U. S. Mining Decisions, 93.
[3] Decision of Commissioner, May 10th, 1872, Copp's U. S. Mining Decisions, 94.
[4] Tong v. Hall, Decision of Secretary of Interior, Feb. 5th, 1876, 3 Copp's Land-owner, 2.

Land adjudged agricultural by the Secretary of the Interior, after a hearing on appeal and upon evidence taken in the Land Office, cannot be entered under the mining acts, unless such discoveries or developments have been made since the date of the hearing, as will show that the tract described is of more value for purposes of mining than of agriculture.[1]

§ 237. Discovery of mines on agricultural lands.—In case valuable deposits of mineral are discovered upon a legal subdivision of the public lands, after the same has been entered as agricultural, but before patent has issued therefor, the parties owning the possessory right to said mine may make application for patent for the same, and the agricultural entry will be canceled to that portion of the tract embraced by the mining claim.[2]

§ 238. Agricultural patent covering mines already worked.—Where parties have a valid mining claim under the local laws, and were engaged in mining on the land embraced in an agricultural claim at the time entry thereof was made, and that fact is established to the satisfaction of the Land Office, the latter will afford all the aid in its power to set aside the patent so as to enable the miner to acquire title to the mine.[3]

§ 239. Fraud in pre-emption entry.—Where one made a pre-emption cash entry, charges of fraud in the entry were made and supported by affidavits. The entry was suspended, and an investigation ordered. The Commissioner canceled the entry, the land being considered mineral. No appeal was taken. A rehearing was applied for, denied, and the decision affirmed by the Secretary.

The rehearing was denied, but it was said that if the applicant could bring himself within the law and show that the mineral in the land had been exhausted, or that later developments demonstrated its non-mineral character, he could make a new claim and initiate a new contest, after proper notice before the local officers.[4]

---

[1] Decision of Commissioner, Dec. 2d, 1872, Copp's U. S. Mining Decisions, 150.
[2] Ibid. March 12th, 1873, Ibid. 163.
[3] Ibid. July 17th, 1873, Ibid. 212.
[4] Decision of the Secretary, March 24th, 1876, 3 Copp's Land-owner, 2.

§ 240. Compromises between miners and settlers.—
In a case where, upon a forty-acre tract, there appeared to be both agricultural and mineral land, it was suggested that an amicable arrangement be made between the claimants in order that the agricultural might be segregated from the mineral portion in accordance with circular instructions, otherwise it was said it would be reserved as mineral.[1]

§ 241. Attempt by railroad to disprove mineral character of land.—
It is probable that the right to enter lands as agricultural, which, as mineral lands, were previously excluded from survey and sale, is confined entirely to actual settlers, coming within the requirements of the pre-emption laws, who, upon making competent proof that the tracts actually settled upon, occupied, and improved by them as homesteads, contain no known mines, are entitled to the execution in their favor of the pre-emption or homestead laws.

Where an attempt was made to prove the agricultural character of a certain tract previously reserved as mineral land, to the end that it might inure to a railroad company, and not by a party coming within the purview of the 10th section, it was held that, even if the proof were satisfactory, the whole proceeding would fail for want of legal sanction to support it. The land was therefore treated as mineral, the same as if no proceedings had been had.[2]

Title to known mines does not pass to railroad companies—Conditions in patent.—All patents issued to the California and Oregon Railroad Company contain a clause in accordance with the requirements of law, as follows:

" Excluding and excepting from the transfer by these presents all mineral lands, should any such be found to exist, in the tracts described in the foregoing; but this exclusion and exception, according to the terms of the statute, shall not be construed to include coal and iron lands."

The patent, therefore, does not pass title to mineral lands other

---

[1] Tremaine v. Brydon, Decision Acting Commissioner, Nov. 14th, 1872, Copp's U. S. Mining Decisions, 148; Instructions of Commissioner, May 6th, 1871, Copp's U. S. Mining Decisions, 261. See 309, 313.

[2] Decision of Commissioner, Oct. 21st, 1871, Copp's U. S. Mining Decisions, 60.

than coal and iron. If, therefore, mining claims exist upon such a tract, no title to the same is acquired by the railroad company, but on the contrary the title remains in the Government.

Should parties who have the possession and the right of possession to mining claims upon such a tract desire to secure titles thereto, their applications for patents will be received.[1]

Railroad selections.—In every case reported from the district land offices of selections made under the Acts of 1862 and 1864, for the Pacific Railroad, the agent of the company in the first place is required to state in his affidavit that the selections are not interdicted, mineral, nor reserved lands, and are of the character contemplated by the grant. Upon the filing of lists with such affidavits attached, it is made the duty of registers and receivers to certify to the correctness of the selections in the particulars mentioned, and in other respects. They subsequently undergo scrutiny in the land office, are tested by the plats, and by all the data on the files, sufficient time elapsing after the selections are made for the presentation of any objections to the Department before final action is taken; and to more effectually guard the matter, there is inserted in *all* patents issued to said railroad company a clause to the following effect : " Yet excluding and excepting from the transfer by these presents *all mineral lands*, should any such be found to exist in the tracts described in this patent, this exception, as required by statute, ' not extending to coal and iron land.' "

A person in the occupancy of mineral lands under the local customs and rules of miners, is protected by the license granted in the first section of the Act of July 26th, 1866. He cannot be ejected by a railroad company having no title to the land at all. The Government license, it is reasonable to suppose, would constitute a sufficient defense against any one not able to show a better title. The grantee of such license is no trespasser upon the public lands, and the license cannot be considered as revoked by a patent to a railroad company, when such instrument expressly excepts and excludes from the grant all interdicted mineral land. Claimants authorized to apply for and to obtain patents under the Mining Act have an efficient remedy in its pro-

---

[1] In re California and Oregon Railroad Company, Decision Commissioner, March 21st, 1877, 4 Copp's Land-owner, 2.

visions, and by taking the proper steps may obtain patents for their claims, even should they happen to be embraced within tracts patented to railroad companies, as the exceptions in such patents enable the United States to segregate the mineral lands included, by distinct and separate conveyance to mining claimants.

Placers more valuable for mining than for agriculture cannot be entered as pre-emption or homestead lands, nor selected by railroad companies.

§ 242. Non-mineral proof by settlers on lands within railroad limits.—The non-mineral proof required of parties who enter land under the provisions of the Act of March 3d, 1875, "for the relief of settlers on lands within railroad limits," in cases where the applicants are not personally acquainted with the character of the land they desire to enter, is as follows:

1st. The affidavit of the applicant that, to the best of his knowledge and belief, the land sought to be entered is non-mineral in character; the usual non-mineral affidavit being modified by omitting the words, "that he is well acquainted with the character of said described land, and with each and every legal subdivision thereof, having frequently passed over the same; that his knowledge of the same is such as to enable him to testify understandingly with regard thereto," and substituting for the words "to his knowledge" the words, "to the best of his knowledge and belief"; otherwise the usual affidavit to remain the same.

2d. An agent's non-mineral affidavit in the usual form, with changes to indicate agency.

3d. The applicant's affidavit that the person so acting is his authorized agent.[1]

[1] Decision of Commissioner, August 4th, 1875, 2 Copp's Land-owner, 84.

CHAPTER XVII.

COAL LANDS—RIGHT OF ENTRY AND OF PRE-EMPTION—PRESENTATION OF CLAIMS—LIMITATION OF ENTRY—CONFLICTING CLAIMS—EXISTING RIGHTS.

§ 243. Entry of coal lands.
§ 244. Pre-emption of coal lands.
§ 245. When claims are to be presented.
§ 246. Only one entry allowed.
§ 247. Conflicting claims.
§ 248. Existing rights.
§ 249. Departmental regulations and instructions.
§ 250. Restrictions as to purchase.
§ 251. School sections containing coal.
§ 252. Coal lands and town sites.
§ 253. Actual possession of coal mines upon railroad sections.
§ 254. Coal lands in Minnesota, Wisconsin, and Michigan.

§ 243. Entry of coal lands.—Section 2347 of the Revised Statutes is as follows: "Every person above the age of twenty-one years, who is a citizen of the United States, or who has declared his intention to become such, or any association of persons severally qualified as above, shall, upon application to the register of the proper land office, have the right to enter, by legal subdivisions, any quantity of vacant coal lands of the United States not otherwise appropriated or reserved by competent authority, not exceeding one hundred and sixty acres to such individual person, or three hundred and twenty acres to such association, upon payment to the receiver of not less than ten dollars per acre, for such lands where the same shall be situated more than fifteen miles from any completed railroad, and not less than twenty dollars per acre, for such lands as shall be within fifteen miles of such road."[1]

§ 244. Pre-emption of coal lands.—Section 2348 of the Revised Statutes reads: "Any person or association of persons

[1] Rev. Stats. 2347; Act 1873, Sec. 1; 17 U. S. Stats. 607.

severally qualified, as above provided, who have opened and improved, or shall hereafter open and improve, any coal mine or mines upon the public lands, and shall be in actual possession of the same, shall be entitled to a preference right of entry, under the preceding section, of the mines so opened and improved; *provided*, That when any association of not less than four persons, severally qualified as above provided, shall have expended not less than five thousand dollars in working and improving any such mine or mines, such association may enter not exceeding six hundred and forty acres, including such mining improvements."[1]

§ 245. **Pre-emption of coal lands—When claims are to be presented, etc.**—Section 2349 of the Revised Statutes is in the following words: "All claims under the preceding section must be presented to the register of the proper land district within sixty days after the date of actual possession and the commencement of improvements on the land, by the filing of a declaratory statement therefor; but when the township plat is not on file at the date of such improvement, filing must be made within sixty days from the receipt of such plat at the district office; and where the improvements shall have been made prior to the expiration of three months from the third day of March, eighteen hundred and seventy-three, sixty days from the expiration of such three months shall be allowed for the filing of a declaratory statement, and no sale under the provisions of this section shall be allowed until the expiration of six months from the third day of March, eighteen hundred and seventy-three." [2]

§ 246. **Only one entry allowed.**—Section 2350 of the Revised Statutes reads: "The three preceding sections shall be held to authorize only one entry by the same person or association of persons; and no association of persons any member of which shall have taken the benefit of such sections, either as an individual or as a member of any other association, shall enter or hold any other lands under the provisions thereof; and no member of any association which shall have taken the benefit

[1] Rev. Stats. 2348; Act 1873, Sec. 2; 17 U. S. Stats. 607.
[2] Ibid. 2349; Ibid. Sec. 3; Ibid. 607.

of such sections shall enter or hold any other lands under their provisions; and all persons claiming under Sec. 2348 shall be required to prove their respective rights and pay for the lands filed upon within one year from the time prescribed for filing their respective claims; and upon failure to file the proper notice, or to pay for the land within the required period, the same shall be subject to entry by any other qualified applicant." [1]

§ 247. **Conflicting claims to coal lands.**—Section 2351 of the Revised Statutes reads: " In case of conflicting claims upon coal lands where the improvements shall be commenced after the third day of March, 1873, priority of possession and improvement, followed by proper filing and continued good faith, shall determine the preference right to purchase. And also where improvements have already been made prior to the third day of March, 1873, division of the land claimed may be made by legal subdivisions, to include as near as may be, the valuable improvements of the respective parties. The Commissioner of the General Land Office is authorized to issue all needful rules and regulations for carrying into effect the provisions of this and the four preceding sections." [2]

§ 248. **Existing rights.**—Section 2352 of the Revised Statutes reads: " Nothing in the five preceding sections shall be construed to destroy or impair any rights which may have attached prior to the 3d day of March, 1873, or to authorize the sale of lands valuable for mines of gold, silver, or copper." [3]

§ 249. **Regulations and instructions of the Department concerning coal lands.**—The sale of coal lands is provided for:
1st. By ordinary private entry under Sec. 1. (Rev. Stats. 2347. Act of March 3d, 1873, 17 U. S. Stats. 607.)
2d. By granting a preference right of purchase based on priority of possession and improvement under Sec. 2.[4] (Rev. Stats. 2348.)
The land entered under either section must be by legal sub-

---

[1] Rev. Stats. 2350; Act 1873, Sec. 4; 17 U. S. Stats. 607.
[2] Ibid. 2351; Ibid. Sec. 5; Ibid.
[3] Ibid. 2352; Ibid. Sec. 6; Ibid.
[4] Instructions April 15th, 1873, Subdivision 1.

divisions, as made by the regular United States survey; entry is confined to surveyed lands; to such as are vacant, not otherwise appropriated, reserved by competent authority, or containing valuable minerals other than coal.[1] (Rev. Stats. 2347.) A person is not disqualified by the ownership of any quantity of other land, nor by having removed from his own land in the same State or Territory.[2]

The price per acre is $10, where the land is situated *more than fifteen miles* from any completed railroad, and $20 per acre where the land is *within* fifteen miles of such road. Where the land lies *partly within* fifteen miles of such road, and in *part outside* such limit, the *maximum* price must be paid for all legal subdivisions the greater parts of which lie within fifteen miles of such road.[3] The term "completed railroad" is held to mean one which is actually constructed on the face of the earth; and lands within fifteen miles of any point of a railroad so constructed will be held and disposed of at $20 per acre.[4] Possession by agent is recognized as the possession of the principal. The clearest proof on the point of agency must, however, be required in every case, and a clearly defined possession must be established.[5]

The opening and improving of a coal mine, in order to confer a preference right of the purchase, must not be considered as a mere matter of form; the labor expended and improvements made must be such as to clearly indicate the good faith of the claimant.[6]

These lands are intended to be sold, where there are adverse claimants therefor, to the party who, by substantial improvements, actual possession, and a reasonable industry shows an intention to continue his development of the mines, in preference to those who would purchase for speculative purposes only. With this view, there is requisite such proof of compliance with the law, when lands are applied for under Sec. 2, Rev. Stats. 2348, by adverse claimants, as the circumstances of each case may justify.[7]

In conflicting claims, where improvement has been made prior to March 3d, 1873, if each party make subsequent compliance

---

[1] Instructions April 15th, 1873, Subdivision 2.   [2] Ibid. Sub. 5
[3] Ibid. Subs. 9, 10.   [4] Ibid. Sub. 11.   [5] Ibid. Sub. 13.   [6] Ibid. Sub. 14.
[7] Ibid. Sub. 15.

with the law, the land is to be awarded by legal subdivisions, so as to secure to each, as far as possible, his valuable improvements; there being no provision in the act allowing a joint entry by parties claiming separate portions of the same legal subdivision.[1]

In conflicts when improvements, etc., have been commenced subsequent to March 3d, 1873, or shall be commenced, priority of possession and improvement governs the award, when the law has been fully complied with by each party. A mere possession, however, without satisfactory improvements, will not secure the tract to the first occupant when a subsequent claimant shows his full compliance with the law.[2]

Contests and investigations.—After an entry has been allowed to one party, no investigation concerning it is to be made by the register, at the instance of any person, except on instructions from the Land Office. All affidavits, however, are to be received concerning such case, and forwarded to the General Land Office, accompanied by a statement of the facts as shown by the records of the local land office.[3]

Prior to entry, it is competent for the register to order an investigation, on sufficient grounds set forth under oath of a party in interest, and substantiated by the affidavits of disinterested and credible witnesses.[4]

Notice of contest, in every case where the same is practicable, must be made by reading it to the party to be cited and by leaving a copy with him. This notice must proceed from the local office, and be signed by the register or receiver. Where such personal service cannot be made by reason of the absence of the party, and because his whereabouts are unknown, a copy may be left at his residence, or, if this is unknown, by posting a copy in a conspicuous place on the tract in controversy, and by publication in a weekly newspaper having the largest general circulation in the vicinity of the land, (where no newspaper is specified by the General Land Office) for five consecutive insertions, covering a period of four weeks next prior to the trial; and in each case requiring such notice, a copy must be forwarded with the returns to the General Land Office, accompanied

---

[1] Instructions April 15th, 1873, Subdivision 16; Rev. Stats. 2351.
[2] Ibid. Sub. 17; Ibid.   [3] Ibid. Sub. 18; Ibid.   [4] Ibid. Sub. 19; **Ibid.**

with proof of service by affidavit indorsed thereon.[1] In every case of contest, all papers in the same must be forwarded to the General Land Office for review before an entry is allowed to either party.[2] Thirty days from the decision of the register is allowed to enable any party to take an appeal, or file argument to be forwarded to the General Land Office.[3] And no appeal is entertained unless the same be forwarded through the district land office.[4] A party may still further appeal from the decision of the Commissioner of the General Land Office to the Secretary of the Interior. This appeal must be taken within sixty days after service of notice on the party. This may be filed with the district land officers, and by them forwarded, or it may be filed with the Commissioner, and must recite the points of exception.[5] If not appealed, the decision is by law made final.[6] After appeal, thirty days are usually allowed for filing arguments, and the case is then sent to the Secretary, whose decision is final and conclusive.[7]

Manner of obtaining title by private entry—Application.— The form of application for coal lands is prescribed by subdivision 26 of the Instructions.[8]

Thereupon, the register, if the tract is vacant, will so certify to

---

[1] Instructions April 15th, 1873, Subdivision 20; Rev. Stats. 2351.
[2] Ibid. Sub. 21; Ibid.   [3] Ibid. Sub. 22; Ibid.   [4] Ibid. Sub. 23; Ibid.
[5] Ibid. Sub. 24; Ibid.   [6] Sec. 10, Act June 12th, 1858, U. S. Stats. Vol. 11, p. 317.
[7] Instructions April 15th, 1873, Subdivision 25; Rev. Stats. 2351.
[8] The following form is prescribed by the Department:

APPLICATION FOR COAL LAND.

I, ——, hereby apply, under the provisions of the Revised Statutes of the United States, (Secs. 2347 to 2352 inclusive) providing for the sale of the lands of the United States containing coal, to purchase the —— quarter of section —— in township —— of range ——, in the district of land subject to sale at the land office at ——, containing —— acres, and I solemnly swear that no portion of said tract is in the possession of any other party; that I am twenty-one years of age, a citizen of the United States, [or have declared my intention to become a citizen of the United States] and have never held nor purchased lands under said act, either as an individual or as a member of an association; and I do further swear that I am well acquainted with the character of said described land, and with each and every legal subdivision thereof, having frequently passed over the same; that my knowledge of said land is such as to enable me to testify understandingly with regard thereto; that there is not, to my knowledge, within the limits thereof, any vein or lode of quartz or other rock in place bearing gold, silver, or copper; and that there is not, within the limits of said land, to my knowledge, any valuable mineral deposits, other than coal, so help me God.

(To this application the register will append the usual jurat.)

the receiver, stating the price, and the applicant must then pay the amount of the purchase-money.[1] The receiver will then issue to the purchaser a duplicate receipt, and at the close of the month the register and receiver will make returns of the sale to the General Land Office, from whence, when the proceedings are found regular, a patent or complete title will be issued; and on surrender of the duplicate receipt such patent will be delivered at the option of the patentee, either by the Commissioner at Washington or by the register at the district land office.[2] This disposition by private entry will be subject to any valid prior adverse right which may have attached to the same land, and which is protected by section six, Rev. Stats. Sec. 2352.[3]

When the application to purchase is based on a priority of possession, as provided for in section two, the claimant must, when the township plat is on file in the local office, file his declaratory statement for the tract claimed within sixty days from and after the first day of his actual possession and improvement. Sixty days, exclusive of the first day of possession, etc., must be allowed.[4]

Subdivision 31 prescribes substantially the form of the declaratory statement.[5]

[1] Instructions April 15th, 1873, Subdivision 27.
[2] Ibid. Sub. 28.   [3] Ibid. Sub. 29.   [4] Ibid. Sub. 30.
[5] The following form is prescribed as the

DECLARATORY STATEMENT ON APPLICATION FOR COAL LAND.

I, ——, being —— years of age, and a citizen of the United States, [or having declared my intention to become a citizen of the United States] and never having, either as an individual or as a member of an association, held or purchased any coal lands under the Act approved March 3d, 1873, entitled, "An Act to provide for the sale of the land of the United States containing coal," or under chapter six of title thirty-two of the Revised Statutes of the United States, do hereby declare my intention to purchase, under the provisions of said chapter six of title thirty-two of the Revised Statutes of the United States, the —— quarter of section ——, in township ——, of range ——, of lands subject to sale at the district land office at ——, and that I came into possession of said tract on the —— day of ——, A. D. 18—, and have ever since remained in actual possession continuously, and have expended in labor and improvements on said mine the sum of —— dollars, the labor and improvements being as follows:
[Here describe the nature and character of the improvements.]

And I do furthermore solemnly swear that I am well acquainted with the character of said described land, and with each and every legal subdivision thereof, having frequently passed over the same; that my knowledge of said land is such as to enable me to testify understandingly with regard thereto; that there is not, to my knowledge, within the limits thereof, any vein or lode

When the township plat is not on file at the date of claimant's first possession, the declaratory statement must be filed within sixty days from the filing of such plat in the local office.[1] When improvements shall have been made prior to June 4th, 1873, the declaratory statement must be filed within sixty days from that date.[2] No sale under the act was allowed prior to September 4th, 1873. One year from and after the expiration of the period allowed for filing the declaratory statement is given, within which to make proof and payment; but no party is allowed to make final proof and payment except on notice as aforesaid to all others who appear on the local records as claimants to the same tracts.[3]

A party who otherwise complies with the law may enter *after* the expiration of said year, *provided* no valid adverse right shall have intervened. He postpones his entry beyond said year at his own risk, and the Government cannot thereafter protect him against another who complies with the law, and the value of his improvements can have no weight in his favor.[4]

One person can have the benefit of one entry or filing only. He is disqualified by having made such entry or filing alone, or as a member of an association. No entry can be allowed an association which has in it a single person thus disqualified, as the law prohibits the entry or holding of more than one claim, either by an individual or an association. No entry is allowed, under the sections relative to coal lands, of lands containing other valuable minerals. The character of the land is to be determined under the rules relative to agricultural and mineral lands. Those that are sufficiently valuable for other minerals to prevent their entry as agricultural lands, cannot be entered under the coal sections.[5]

Assignments of the right to purchase under this act will be recognized when properly executed. Proof and payment must be made, however, within the prescribed period, which dates from the first day of the possession of the assignor who initiated the claim.[6]

---

of quartz or other rock in place, bearing gold, silver, or copper; and that there is not, within the limits of said land, to my knowledge, any valuable mineral deposit other than coal.

[1] Instructions April 15th, 1873, Subdivision 32.   [2] Ibid. Sub. 33.
[3] Ibid. Sub. 34.   [4] Ibid. Sub. 35.   [5] Ibid. Sub. 36.   [6] Ibid. Sub. 37.

The act is construed, so as in its application not to destroy or impair any rights which may have attached prior to March 3d, 1873. Those persons who may have initiated a valid claim under any prior law relative to coal lands are permitted to complete their entries under the same.[1]

The local officers are required to report at the close of each month as "sales of coal lands" all filings and entries under the act in separate abstracts, commencing with number one, and thereafter proceeding consecutively in the order of their reception. Where a series of numbers has already been commenced by sale of coal lands, the same is continued without change.[2]

The affidavit required from each claimant under Sec. 2 at the time of active purchase is prescribed by Subdivision 39 of the Coal Land Instructions.[3]

In case the purchaser shows by an affidavit that he is not personally acquainted with the character of the land, his duly authorized agent who possesses such knowledge may make the required affidavit as to its character; but whether this affidavit is made by principal or agent, it must be corroborated by the

---

[1] Instructions April 15th, 1873, Subdivision 38. [2] Ibid. Idem.
[3] The following form is prescribed by the Department:

AFFIDAVIT ON APPLICATION FOR COAL LAND.

I, ——, claiming the right of purchase under chapter six of title thirty-two of the Revised Statutes of the United States, (Secs. 2347 to 2352 inclusive) providing for the sale of the lands of the United States containing coal, to the —— quarter of section ——, in township —— of range ——, subject to sale at ——, do solemnly swear that I have never had the right of purchase under this act, either as an individual or a member of an association, and that I have never held any other lands under its provisions; I further swear that I have expended in developing coal mines on said tract, in labor and improvements, the sum of —— dollars, the nature of such improvements being as follows:

[Here describe the nature and character of the improvements.]

That I am now in the actual possession of said mines, and make the entry for my own use and benefit, and not directly or indirectly for the use and benefit of any other party; and I do furthermore swear that I am well acquainted with the character of said described land, and with each and every legal subdivision thereof, having frequently passed over the same; that my knowledge of said land is such as to enable me to testify understandingly with regard thereto; that there is not, to my knowledge, within the limits thereof, any vein or lode of quartz or other rock in place bearing gold, silver, or copper; and that there is not, within the limits of said land, to my knowledge, any valuable mineral deposit other than coal, so help me God.

I, ——, of the land office at ——, do hereby certify that the above affidavit was sworn and subscribed to before me this —— day of ——, A. D. 18——.

affidavits of two disinterested and credible witnesses having knowledge of its character.[1]

§ 250. Restrictions as to purchase—Coal Land Act.— No person who has, in his individual capacity or as a member of an association, taken the benefit of the coal land acts, can enter or hold any other lands thereunder. If an association of persons enters a less number of acres of coal land than they might have done under said act, they will not be entitled to make a second entry.[2]

Where parties have located or filed upon coal land, they may transfer their rights in the premises to persons duly qualified under the act to enter and hold coal land; but no assignment can be recognized to a party who is not qualified under the act to hold and enter such lands.[3]

§ 251. School sections containing coal.—Lands which are found upon survey to be designated as Secs. 16 or 36 did not pass to the State of California under the Act of March 3d, 1853, when the same are mineral lands. No mineral lands were granted by that act. This applies equally to coal lands and to claimants under the Coal Land Act of March 3d, 1873.[4]

That lands containing valuable deposits of coal have been considered and treated as mineral lands is evident from the text of the Act of July 1st, 1864, entitled "An Act for the disposal of coal lands and town property in the public domain," viz., "That where any tracts embracing coal beds or coal fields, constituting portions of the public domain, and *which as mines* are excluded from the Pre-emption Act of 1841, and which under past legislation are not liable to ordinary private entry," etc.[5]

A case arose wherein it was debated whether Secs. 16 and 36 in each township within the limits of Wyoming Territory which were found to contain valuable deposits of coal, were reserved

---

[1] Instructions April 15, 1873, Subdivision 40.
[2] Thirty-seventh paragraph of Circular Instructions, dated April 15th, 1873.
[3] Decision of Commissioner, June 14th, 1876, 3 Copp's Land-owner, 50.
[4] Decision of Commissioner, Nov. 3d, 1874, 1 Copp's Land-owner, 135; Act March 3d, 1873; 17 U. S. Stats. 607; Rev. Stats. 2347-2352; Act of March 3d, 1853.
[5] Decision in the Keystone Case, April 28th, 1873, Copp's U. S. Mining Decisions, 105; Sherman v. Buick, Oct. Term, 1876, Supreme Court U. S. reversing S. C. 45 Cal. 656.

for school purposes, or could be sold as other coal lands. But it was held that Sec. 14 of the Act of July 25th, 1868, providing for a temporary government for the Territory of Wyoming, made no exception in reserving Secs. 16 and 36 for school purposes in each township, and the General Land Office therefore was without authority of law for disposing of school sections within the Territory, except in cases where, after the passage of the Act of March 3d, 1873, the parties were found in actual occupancy of the lands at the date of survey.[1]

§ 252. **Coal lands and town sites.**—The Town-site Acts provide, among other things, that no title "shall be acquired to any valid mining claim or possession held under the existing laws of Congress," by virtue of the provisions of said Town-site Acts. Where land has been returned as "coal lands" by the Surveyor-General, it cannot be entered as a town site until a hearing has been had to determine the character of land, viz., whether it is mineral or agricultural in character. The coal-land statutes provide for the sale of land *by legal subdivisions only*, and it is necessary to present evidence in regard to *each forty-acre tract* in controversy.[2]

§ 253. **Actual possession of coal mines upon railroad sections.**—In Crismon v. Union Pacific Railroad Company, on January 9th, 1874, the township plat was filed, and on March 6th, 1874, Crismon filed his declaratory statement claiming certain lots as coal land. The land was within the limits of the withdrawal for the Union Pacific Railroad.

The evidence submitted at the hearing showed that one Wild discovered coal upon this tract in June, 1864, and that Wild and one Redden went into possession thereof, developing the coal bed and extracting coal therefrom; that they remained continuously in the possession of the land until the year 1869, when Redden conveyed his interest therein to Charles Crismon and sons, and 1870, when Wild conveyed to Charles Crismon, Sr., and George Crismon his interest therein. It also appeared

---

[1] Decision of Commissioner, July 30th, 1873, 1 Copp's Land-owner, 19.

[2] Decision Acting Commissioner, April 21st, 1874, 1 Copp's Land-owner, 19. See, as to the manner of conducting hearings to determine the true character of lands, Circular Instructions May 6th, 1871; March 20th, 1872; June 10th, 1872.

that Charles Crismon, Sr., and one Groesbeck conveyed their interest in the tract to George Crismon, November 21st, 1874.

The evidence showed that the land had been in the actual possession and occupation of Wild, Redden, and their grantees since the date of the original discovery of coal therein, and that they had during that time expended more than $20,000 in developing the tract and in extracting coal therefrom.

The Union Pacific Railroad was definitely located past this land in June, 1868.

The question was presented whether the tracts inured to the railroad company by virtue of their grant. Mineral lands are excluded from the grant to said company, but the 4th Section of the Act of July 2d, 1864, provides that the term mineral land "shall not be construed to include coal and iron lands." The section also provides that "any lands granted by this Act or the act to which this is an amendment shall not defeat or impair any pre-emption, homestead, swamp land, or other lawful claim."

The Act of 3d March, 1865, (13 Stats. 529) provided "that in the case of any citizen of the United States who, at the passage of this act, may be in the business of bona fide actual coal mining on the public lands * * for purposes of commerce, such citizen, upon making proof satisfactory to the register and receiver to that effect, shall have the right to enter, according to legal subdivisions, a quantity of land, not exceeding one hundred and sixty acres, to embrace his improvements and mining premises." This act also specified when the declaratory statement should be filed, to wit: "In case of lands unsurveyed at the date of the act, such declaratory statement shall be filed within three months from the return to the district office of the official township plat."

In the case under consideration, rights had attached to the coal lands under the Act of 3d March, 1865, before the line of the road was definitely fixed past the land, and hence, by the terms of the Act of July 2d, 1864, the tracts were excluded from the grant to the company.

The right which Wild and Redden had acquired was subject to assignment, and, as already stated, was assigned to Crismon.[1]

[1] Crismon v. U. P. R. R. Co. Decis. of Com. July 26th, 1875, 2 Copp's L. O. 67.

§ 254. Coal lands in Minnesota, Wisconsin, and Michigan.—It was a question not settled May, 1874, whether coal lands in Minnesota, Wisconsin, and Michigan could be purchased at private entry the same as agricultural lands, or must be bought under the Coal Act of March 3d, 1873. It is clear that from February 18th, 1873, the date of the act excepting coal and iron from the operation of the General Mining Act of May 10th, 1872, to March 3d, 1873, coal lands in those States were purchasable at ordinary private entry.[1]

---

[1] 1 Copp's Land-owner, 31.

Registers are required to forward to the General Land Office, with coal-land entries, under Sec. 2348 of the Revised Statutes, the original declaratory statements, retaining copies on file. (Decision of Acting Commissioner, Jan. 21st, 1876, 2 Copp's Land-owner, 162.) See, generally: Instructions of Aug. 20th, 1864, under Coal-land Act of July 1st, 1864, and supplemental Act of March 3d, 1865, Copp's Pub. L. L. 661–664; Instructions of April 15th, 1873, under Coal-land Act of March 3d, 1873, Id. 667; Instructions August 11th, 1873, Id. 672; Instructions August 14th, 1873, Id. 673–677.

CHAPTER XVIII.

MISCELLANEOUS PROVISIONS.

§ 255. Power of the President as to appointments.
§ 256. Pending applications—Existing rights.
§ 257. Possessory actions relative to mines.
§ 258. Practice before the Land Department—Hearings, contests, and appeals—Witnesses and testimony.
§ 259. Appeals, exceptions, evidence.
§ 260. Fees of registers and receivers.
§ 261. Payment pending contest.
§ 262. Decisions of the Land Department—Their authority.
§ 263. Right of inspection of mine.
§ 264. Mining claims in river beds.
§ 265. Timber on mineral lands—Railroad companies.
§ 266. Claims not within any mining district.
§ 267. Removal of machinery.
§ 268. Criminal offenses.
§ 269. Various provisions.

§ 255. Power of the President to provide districts and officers.—Section 2343 of the Revised Statutes provides: "The President is authorized to establish additional land districts, and to appoint the necessary officers under existing laws, whenever he may deem the same necessary for the public convenience in executing the provisions of this chapter." [1]

§ 256. Pending applications—Existing rights.—Section 2328 of the Revised Statutes reads as follows: "Applications for patents for mining claims under former laws now pending may be prosecuted to a final decision in the General Land Office; but in such cases where adverse rights are not affected thereby, patents may issue in pursuance of the provisions of this chapter; and all patents for mining claims upon veins or lodes heretofore issued, shall convey all the rights and privileges conferred

---

[1] Rev. Stats. 2343, Sec. 7, Act of 1866; 14 U. S. Stats. 252.

by this chapter, where no adverse rights existed on the tenth day of May, 1872." [1]

§ 257. Possessory actions relative to mines.—Section 910 of the Revised Statutes provides that "no possessory action between persons, in any Court of the United States, for the recovery of any mining title, or for damages to any such title, shall be affected by the fact that the paramount title to the land in which such mines lie is in the United States; but each case shall be adjudged by the law of possession." [2]

§ 258. Practice before the Land Department—Hearings, contests, and appeals.—With a view to the promotion of greater uniformity in the practice in cases before the United States District Land Offices, the General Land Office, and the Department of the Interior, certain rules were adopted and approved under date of November 29th, 1875.[3]

Hearings and contests.—In the adjustment of conflicting claims to lands under the various statutes, it very frequently becomes necessary to institute regular proceedings in the nature of a formal hearing, with notice to all parties to the record, in order to reach the facts and legal conditions upon which an award of the tract may be based. (Rev. Stats. Sections 2263, 2273, 2297, 2326, 2335, 2351, 2467, and 2488.)

To conduct such investigations in an orderly manner, and with due regard to the rights of all parties, requires the exercise of sound judgment and discretion; and although, from the great variety of particular conditions and circumstances, details must necessarily be governed, in great measure, by the incidents of each case, yet the observance of certain fixed general

---

[1] Sec. 9 of the Act of 1872 reads: "That Secs. 1, 2, 3, 4, and 6 [of the Act of 1866] are hereby repealed, but such repeal shall not affect existing rights. Applications for patents for mining claims now pending, may be prosecuted to a final decision in the General Land Office; but in such cases where adverse rights are not affected thereby, patents may issue in pursuance of the provisions of this act; and all patents for mining claims heretofore issued under the Act of July 26th, 1866, shall convey all the rights and privileges conferred by this act, where no adverse rights exist at the time of the passage of this act."

[2] Rev. Stats. 910; Sec. 9, Act of Feb. 27th, 1865; 13 U. S. Stats. 441.

[3] Rules approved by Secretary of the Interior Nov. 29th, 1875, printed, published, and promulgated by Commissioner, Dec. 1st, 1875, 2 Copp's Land-owner, 153.

rules tends to promote uniformity and dispatch in proceedings, to secure fairness to parties concerned, and materially aids registers and receivers in the performance of this delicate and highly important duty. The following provisions were made generally applicable, leaving exceptional cases to be particularly considered under special instructions, according to the circumstances arising in each :

Investigations are usually commenced upon the application of one or more of the respective parties to make due proof of his or their claims, or to clear the record of an abandoned or defective claim, so as to leave undisturbed and undisputed the right of the party so proceeding. In such case the register and receiver are authorized to issue the proper notices to all parties to the record, or claiming an interest, of a day upon which they will receive testimony touching the legal right to the land.

If, however, an entry of the tract has previously been permitted, and remains of record upon the books of the Office, all applications to attack such record or to impeach the entry must be forthwith submitted to the Commissioner of the General Land Office, with a special report and recommendation from the register and receiver; and the question of ordering such hearing will be determined by the Commissioner upon the matters set forth in the application and report, in connection with the records in his office affecting the status of the land. In no case of this nature should jurisdiction to try the questions involved be assumed by the district officers without a special order to proceed, received from the Commissioner of the General Land Office.

The notice to the parties should give sufficient time for a fair and full trial; and as the Department has no compulsory process for the production of testimony, but must rely on such as parties may be able to procure, at least thirty days should be allowed, after notice, before trial, unless by consent of parties an earlier day can be named, in which case it is always desirable to proceed with the least practicable delay.

Parties making application for contest should provide sufficient security for costs, so that the register and receiver may not be compelled to advance from their own funds the expenses of the hearing, nor incur individual responsibility therefor.

When brought to trial the costs of the case may be equitably apportioned by those officers in the exercise of a sound discretion, and any sum deposited as security, over the proper apportionment to the party, should be returned to him upon the final disposition of the case. Only the actual costs of notice, and the legal fees for reducing testimony to writing, or for acting on applications for mineral lands, can be charged to the parties.[1] Costs of notice will, of course, include the notice of further proceedings by way of appeal, including notices to file argument, etc., up to the final notice of award, should the case be prosecuted before the Secretary of the Interior.

Where hearings are ordered by the office upon discovery of reasons for suspension in the usual course of examination of entries, the preliminary costs will necessarily be provided from the contingent fund for registers and receivers; but when the parties are actually brought before the register and receiver in obedience to the order, such costs should be collected, and provision required for such further notification as may become necessary in the usual progress of the case to final decision.

As before stated, the Department has no means of enforcing by compulsory process the production of testimony. It is, therefore, the more essential that all the material facts bearing upon a case be developed and brought to the knowledge of the officers conducting the investigation, even at the risk of accepting matter not strictly relevant under the rules of the Courts having power to compel the attendance of witnesses, and to reach by reference to text-books and approved precedents the more nice and just discriminations of the judicial tribunals.

The register and receiver will be particularly careful to reach, if possible, the exact condition and status of the tract involved; the nature, extent, and value of alleged improvements; by whom made and at what date; the true date of the settlement of persons claiming as pre-emptors; the steps taken to mark and secure the claim; and the status of the land at that date upon the records of their office. In like manner, under the homestead and other laws, the conditions affecting the inception of the alleged right, as well as the subsequent acts of the re-

[1] Rev. Stats. Sec. 2238.

spective claimants, should be fully and specifically examined, and such testimony as may be offered should be considered— due regard being had in every instance to the necessity of giving opposing claimants the opportunity to confront and cross-examine the witnesses.

When, through ignorance, or for the purpose of avoiding the production of material facts, the parties or their attorneys fail to draw from the witness the facts apparently within his knowledge necessary to lead the judgment of the officers to a correct conclusion respecting any point connected with the case, the register and receiver should by judicious questioning personally direct the examination, and thus obtain the desired information.

If, for good reason, parties are prevented from bringing their witnesses in person to the district office, upon affidavit of that fact continuance may be granted, and the depositions of such witnesses may be received at the adjourned hearing: *Provided*, they have been properly taken, with due notice to the opposite party, by any officer having the powers of a magistrate or commissioner; due regard being had to proximity to the land, residence of witnesses and parties, and the convenience of both parties, so that the opportunity of cross-examination has not been prejudiced. Such depositions should not be received as a matter of course at a first hearing, unless offered by consent of parties for the purpose of avoiding expense and delays in the proceedings, or where it may be shown that due notice has been given beforehand that they will be received, and all the opportunity reasonably required by the opposing claimant has been afforded.

Upon objection being made to the competency or admissibility of evidence, the matter should not be excluded, but should be noted as excepted to, and come up with the case for the consideration of the Commissioner.

Witnesses and testimony—Production of witnesses.—As the law provides no compulsory process to secure the attendance of witnesses before registers and receivers, parties serve their own subpœnas, and as a general thing no difficulty has been experienced. The local officers must, on hearings, afford the right of cross-examination of witnesses, or their testimony will not be received by the General Land Office.¹ Testimony and affidavits taken on a hearing without notice to the opposite party and without opportunity of cross-examination cannot be considered.²

Affidavits sworn to before a Justice of the Peace must be accompanied by a certificate of the proper officer, that the person is a Justice of the Peace, and when before a Notary Public the notarial seal must be attached.³

The officers before whom the affidavit is made should usually be able to certify to the credibility of the witness; yet, where such is not the case, and the deponent is a stranger, it is proper to require his character for truth to be established to the satisfaction of the officer, before giving credit to his affidavit, in all cases where the question deposed to is not merely technical, but goes to the merits of the claim.

The law simply requires that the officers should be satisfied of the truth of the testimony offered, in whatever form presented.⁴

Parties interested.—By the Act of Congress of July 2d, 1864, Sec. 3, Vol. 13, p. 351, Stats. at Large, witnesses are not excluded from testifying in the Courts of the United States on account of being parties to or interested in the issue tried. The same rule should be observed in proceedings before the Executive Department. The weight of the testimony is, as in other cases, a matter about which the officer is to exercise his judgment and discretion.⁵

Having carefully taken and examined the evidence, the register and receiver will render thereon their joint report and opinion, with full and specific reference to the posting and annotations upon their records, subject to the appeal hereinafter provided in these regulations, and will forward the entire record to the General Land Office, with a brief letter of transmittal,

---

¹ In re Brunswick Mine, Decision of the Commissioner, Oct. 19th, 1876, 3 Copp's Land-owner, 111.
² Holland v. Gulichmi, Decision of Commissioner, March 14th, 1873, Copp's U. S. Mining Decisions, 164.
³ In re Cerro Bonito Quicksilver Mine, Decision of Secretary, March 10th, 1872, 4 Copp's Land-owner, 3.
⁴ Decision of the Commissioner, Aug. 15th, 1869, Copp's U. S. Mining Decisions, 16, 17.
⁵ Ibid. Aug. 15th, 1868, Ibid.

describing the case by its title, the nature of the contest, and the tract involved; and thereafter take no further action affecting the disposal of the land until instructed by the Commissioner.

Decisions of registers and receivers, appeals from.—Any person making application to file upon or enter a tract of public land, having complied with the law and regulations touching the presentation of such applications, and feeling aggrieved by the refusal of the register and receiver to recognize his claim, or by any order, direction, or condition affecting the same, may appeal from the action of those officers to the Commissioner of the General Land Office, who is by law invested with the supervision and control of all matters relating to the disposal of the public lands, subject to the direction of the Secretary of the Interior.[1]

For the purpose of enabling such appeal to be taken and perfected, the register and receiver will indorse upon the written application the date when presented and their reasons for refusing it, promptly advising the parties in interest of the facts, and note upon their record a memorandum of the transaction. The party aggrieved will then be allowed thirty days from the receipt of notice of such action within which to file his appeal to the Commissioner.

The appeal should be in writing, and should set forth in brief and clear terms the specific points of exception to the ruling appealed from. It must in all cases be filed with the district officers, to be forwarded by them with a full report of the case to the office.

This report should recite the proceedings had, to wit: The application and rejection, with the reasons therefor, and also the status of the tract involved as shown by the records of the office; together with a reference to all entries, filings, annotations, memoranda, and correspondence shown by such record relating thereto; so as to direct the attention of the Commissioner to all the material facts and issues necessary to a proper determination of the questions presented. No appeal from the decision of the register and receiver will be received at the General Land Office unless forwarded through the local officers in the manner herein prescribed.

---

[1] Rev. Stats. Secs. 453 and 2478.

The report should be forwarded at once upon the filing of the appeal, except in contested cases after regular hearing; when, unless all parties request its earlier transmission, it should not be made until the expiration of the thirty days included in the notice, in order that all parties may have full opportunity to examine the record and prepare their argument upon the question at issue. All documents once received must be kept on file with the cases, and no papers will be allowed under any circumstances to be removed from such files or taken from the custody of the register and receiver; but access to the same under proper rules, so as not to interfere with necessary public business, should be permitted to the parties in interest under the supervision of those officers.

Decisions of the Commissioner of the General Land Office, appeals from.—Upon any question relating to the disposal of the public lands, appeal from the decision of the Commissioner of the General Land Office will lie to the Secretary of the Interior, (Rev. Stats. Secs. 441, 2273) except in cases of interlocutory orders and decisions, and orders for hearing or other matters resting in the sound discretion of the Commissioner. Such latter cases constitute matters of exception which should be noted, and they will be considered by the Secretary on review.

The appeal is required to be made in writing, fairly and specifically stating the points of exception to the decision appealed from, and must be filed either with the register and receiver for transmission, or with the Commissioner, within sixty days from receipt, by the party or his attorney, of the notice of the decision.

After appeal is filed, the fact of its receipt and pendency will be promptly communicated to the district office and to the parties, and thirty days from the service of such notice will be allowed for the filing of argument on the points involved in the controversy. At the expiration of the time prescribed, the papers and record will be forwarded to the Secretary of the Interior. All arguments shall be filed with the Commissioner within the time specified in the notice, in order that they may be referred to and considered in transmitting the case to the Secretary, if deemed expedient by the Commissioner. Examination of cases on appeal to the Secretary will be facilitated by filing in printed form such argument as it is desired to have considered.

Decisions of the Commissioner not appealed from within the period prescribed become final, and the case will be regularly closed. (Rev. Stats. Sec. 2273.)

The decision of the Secretary is necessarily final so far as respects the action of the Executive.

Service of notice under the foregoing regulations.—Notice, to be properly served, must be brought home to the knowledge, actual or presumptive, of the party, so as to bar any future pursuit of his application before the officers of the Department.

It should, therefore, be personally served either upon the claimant or his attorney, in every case where such service is practicable, and the acknowledgment of such service should be taken when obtainable. When not acknowledged the service may be proved by the affidavit of the person, or statement of the officer serving the same, duly indorsed upon the copy of such notice returned into the district office.

Service by mail, when in regular correspondence with the party or his attorney, will be held sufficient. In other cases, where mail service is resorted to, it may be presumed to have reached the party when, after proper time, no question respecting the receipt of the letter may appear to have arisen, and its delivery to the person addressed may have become reasonably certain. Should there be probable cause for doubt respecting the delivery of notice by mail, such publication should be resorted to as will satisfy the Courts and laws of the State or Territory in legal proceedings affecting land titles—proof of such publication to be furnished in all cases to the Commissioner of the General Land Office when report is made by the register and receiver.

In making report of the expiration of time under these rules, or under any special order, sufficient additional time should be allowed for mail delivery, where service has been made by letter addressed by post. Fifteen days will, in ordinary cases, be sufficient. In case of infrequent mails and difficult access, longer time will be found necessary; and careful attention and discretion will be expected on the part of registers and receivers with respect to this matter, so that in all cases actual service of the notice required may be reasonably presumed.

In case of parties known or presumed, upon good information, to be absent from the State or Territory, publication, according to the law of such State or Territory in like cases, must be had, and should be paid for by the opposing claimant, in order that his claim may be the more speedily relieved from embarrassment.

Summary of reference respecting appeals, notices, etc., for convenience of parties having business before the Department.— For appeal from decision of register and receiver, thirty days.

For examination of papers by parties, after receipt by Commissioner, thirty days.

For appeal to Secretary of the Interior, sixty days.

For argument on appeal to Secretary before transmission, thirty days.

For transmission of notice and receipt of answer, when served by mail, fifteen days.

Should anything in the nature of the circumstances of the case require instant decision, or other action, the usual rule, allowing thirty days from receipt of papers before taking them up for decision, will be suspended. Time may be extended, upon proper request, at the discretion of the Commissioner. Continuances will be allowed by the district officers, in proper cases, subject to the exercise of a sound discretion.

Hearings to establish the character of lands.—Beside the general instructions given above, the following are specially applicable to mineral lands: Sec. 2335 provides that "all affidavits required under this chapter may be verified before *any* officer authorized to administer oaths within the land district where the claims may be situated, and all testimony and proofs may be taken before any such officer, and, when duly certified by the officer taking the same, shall have the same force and effect as if taken before the register and receiver of the Land Office. In cases of contest as to the mineral or agricultural character of land, the testimony and proofs may be taken, as herein provided, on personal notice of at least ten days to the opposing party; or, if such party cannot be found, then by publication of at least once a week for at least thirty days in a newspaper to be designated by the register of the land office as published nearest to the location of such land, and the register shall require proof that such notice has been given." Testimony for the purpose of disproving the mineral character of lands may be taken before any officer authorized to administer oaths within the land district, and where the residence of the parties who claim the land to be mineral is known, such evidence may be taken without publication, ten days after the mineral claimants or affiants shall have been personally notified of the time and place of such hearing; but in cases where such affiants or claimants cannot be served with personal notice, or where the land applied for is returned as mineral upon the township plat, or where the same is now or may hereafter be suspended for non-mineral proof, by order of the office, then the party who claims the right to enter the land as agricultural will be required, at his own expense, to publish a notice once each week, for five consecutive weeks, in the newspaper of largest circulation published in the county within which said land is situated; or, if no newspaper is published within such county, then in a newspaper published in an adjoining county; the newspaper in either case to be designated by the register; which notice must be clear and specific, giving the name and address of the claimant, the designation of the subdivision embraced by his filing, the names of any miners or mining companies whose claims or improvements are upon the land, or in the immediate vicinity thereof, the names of the parties who filed the affidavits that the land is mineral, and finally the notice should name a day, which shall not be not less than thirty days from the date of the first insertion of said notice in such newspaper, upon which testimony will be taken to determine the facts as to the mineral or non-mineral character of the land. The notice must also state before what officer such hearing will be held, and the place of such hearing. A copy of this notice must be posted in a conspicuous place upon each forty-acre subdivision claimed, during the publication of the notice, proof of which must be made under oath by at least two persons, who will state when the notice was posted and where posted. At the hearing there must be filed the affidavit of the publisher of the paper that the said notice was published for the required time, stating when and for how long such publication was made, a printed copy thereof to be attached and made a part of the affidavit. In every case, where practicable, in addition to the foregoing, *personal* notice must be served upon the mineral affiants, and upon any parties who may be mining upon or claiming the land.

At the hearing the claimants and witnesses will be thoroughly examined with regard to the character of the land; whether the same has been thoroughly prospected; whether or not there exists within the tract or tracts claimed any lode, or vein of quartz, or other rock in place, bearing gold, silver, cinnabar, lead, tin, or copper, or other valuable deposit, which has ever been claimed, located, recorded, or worked; whether such work is entirely abandoned, or whether occasionally resumed; if such lode does exist, by whom claimed, under what designation, and in which subdivision of the land it lies; whether any placer mine or mines exist upon the land; if so, the character thereof, whether of the shallow surface description, or of the deep cement, blue lead, or gravel deposits; to what extent mining is carried on when water can be obtained, and what the facilities are for obtaining water for mining purposes; upon what particular ten-acre subdivisions mining has been done, and at what time the land was abandoned for mining purposes, if abandoned at all. The testimony should also show the agricultural capacities of the land, what kind of crops are raised thereon, and the value thereof; the number of acres actually cultivated for crops of cereals, or vegetables, and within which particular ten-acre subdivisions such crops are raised; also, which of these subdivisions embrace his improvements, giving in detail the extent and value of his improvements, such as house, barn, vineyard, orchard, fencing, etc.

It is thought by the Department that bona fide settlers upon lands really agricultural will be able to show, by a clear, logical, and succinct chain of evidence, that their claims are founded upon law and justice; while parties who have made little or no permanent agricultural improvements, and who only seek title for speculative purposes, on account of the mineral deposits known to themselves to be contained in the land, will be defeated in their intentions.

The testimony should be as full and complete as possible, and in addition to the leading points indicated above, everything of importance bearing upon the question of the character of the land should be elicited at the hearing.

Where the testimony is taken before an officer who does not use a seal, other than the register and receiver, the official character of such officer must be attested by a clerk of a Court of Record, and the testimony transmitted to the register and receiver, who will thereupon examine and forward the same to the General Land Office, with their joint opinion as to the character of the land as shown by the testimony.

When the case comes before that office such an award of the land will be made as the law and the facts may justify; and in cases where a survey is necessary to set apart the mineral from the agricultural land in any forty-acre tract, the necessary instructions will be issued to enable the agricultural claimant, *at his own expense*, to have the work done at his option, either by United States deputy, county, or other local surveyor; the survey in such case may be executed in such manner as will segregate the portion of land actually containing the mine, and used as surface ground for the convenient working thereof, from the remainder of the tract, which remainder will be patented to the agriculturist to whom the same may have been awarded, subject, however, to the condition that the land may be entered upon by the proprietor of any vein or lode, for which a patent has been issued by the United States for the purpose of extracting and removing the ore from the same, where found to penetrate or intersect the land so patented as agricultural, as stipulated by the Mining Act.

Such survey, when executed, must be properly sworn to by the surveyor, either before a notary public, officer of a Court of Record, or before the register or receiver, the deponent's character and credibility to be properly certified to by the officer administering the oath.

Upon the filing of the plat and field-notes of such survey, duly sworn to as aforesaid, the same must be transmitted to the Surveyor-General for his verification and approval; who, if he finds the work correctly performed, will properly mark out the same upon the original township plat in his office, and furnish authenticated copies of such plat and description both to the proper local land office and the General Land Office, to be affixed to the duplicate and triplicate township plats respectively.

In cases where a portion of a forty-acre tract is awarded to an agricultural claimant, and he causes the segregation thereof from the mineral portion, as aforesaid, such agricultural portion will not be given a numerical designation, as in the case of surveyed mineral claims, but will simply be described as the "Fractional —— quarter of the —— quarter of section ——, in township ——, of range —— meridian, containing —— acres, the same being exclusive of the land adjudged to be mineral in said forty-acre tract."

The surveyor must correctly compute the area of such agricultural portion, which computation will be verified by the Surveyor-General. After the authenticated plat and field-notes of the survey have been received from the Surveyor-General, the office will issue the necessary order for the entry of the land, and in issuing the receiver's receipt and register's patent certificate, the register will be governed by the description of the land given in the order from that office.

The fees for taking testimony and reducing the same to writing in these cases will have to be defrayed by the parties in interest. Where such testimony is taken before any other officer than the register and receiver, the register and receiver will be entitled to no fees.

If, upon a review of the testimony at the General Land Office, a ten-acre tract should be found to be properly mineral in character, that fact will be no bar to the execution of the settler's legal right to the remaining *non-mineral* portion of his claim, if contiguous. The fact that a certain tract of land is decided upon testimony to be mineral in character, is by no means equivalent to an award of the land to a miner.

A miner is compelled by law to give sixty days publication of notice, and posting of diagrams and notices as a preliminary step; and then before he can enter the land, he must show that the land yields mineral; that he is entitled to the possessory right thereto in virtue of compliance with local customs or rules of miners, or by virtue of the Statute of Limitations; that he or his grantors have expended, in actual labor and improvements, an amount of not less than five hundred dollars thereon, and that the claim is one in regard to which there is no controversy or opposing claim. After all these proofs are met he is entitled to have a survey, made at his own cost, where a survey is required, after which he can enter and pay for the land embraced by his claim.[1]

§ 259. **Appeals, exceptions, evidence.**—An appeal brings before the Commissioner the entire proceedings which have taken place prior to the date of the order appealed from, and all exceptions to any of the proceedings must be presented and insisted upon before the Commissioner on the hearing of such appeal, and, unless so presented, are to be deemed waived. After an appeal is dismissed, a party cannot go back and appeal from a former order.[2]

Appellants, taking appeal from the General Land Office to the Secretary of the Interior, must notify the office in writing of the points of exception to its action, within the time allowed for appeal. A notice that "you will please direct my appearance to be registered and cause an appeal from said decision to be entered in due form, the points and argument thereupon to be submitted thereafter," will not be considered nor recognized as an appeal.[3]

No new or additional evidence can be submitted to the Secretary of the Interior on appeal, otherwise his decision would be an original decision, and not a review of that of the Commissioner. New affidavits cannot be considered on the argument on appeal.[4]

In the case of the Overman Silver Mining Co. v. the Dardanelles Mining Company, the question was presented whether the latter company, having once appealed from a decision made by the local officers, and that appeal having been dismissed, could go back of the decision appealed from and appeal from another decision made prior to that one, or whether the first appeal

---

[1] Instructions Feb. 1st, 1877.

[2] Decision of Acting Secretary, July 19th, 1873, reversing decision of Acting Commissioner, and Decision of Commissioner, April 11th, and June 3d, 1873, Copp's U. S. Mining Decisions, 181, 182, 186.

[3] In re Zella Mine, Mountain Tiger Mine, Rockwell Mine, Decision of Commissioner, August 18th, 1873, Copp's U. S. Mining Decisions, 217.

[4] Decision of Commissioner, Aug. 21st, 1872, Copp's U. S. Mining Decisions, 136.

brought the whole case before the Commissioner and gave him jurisdiction thereof, and required that all objections to the proceedings up to that date should be presented to the Commissioner, or, if not presented, be regarded as waived. The latter view was adopted as the true one, any other practice being condemned as contrary to the analogies derived from legal proceedings, inconvenient in practice, and productive of a multiplicity of appeals in a single case.[1]

§ 260. Fees of registers and receivers.—The fees payable to the register and receiver for filing and acting upon applications for mineral land patents, made under Act of 1872, are five dollars to each officer, to be paid by the applicant for patent at the time of filing, and the like sum of five dollars is payable to each officer by an adverse claimant at the time of filing his adverse claim. All fees or charges may be paid in United States currency.

The register and receiver is required at the close of each month to forward to the General Land Office an abstract of mining applications filed, and a register of receipts, accompanied with an abstract of mineral lands sold, and an abstract of adverse claims filed. The fees and purchase-money received by registers and receivers must be placed to the credit of the United States in the receiver's monthly and quarterly account, charging up in the disbursing account the sums to which the register and receiver may be respectively entitled as fees and commissions, with limitations in regard to the legal maximum.[2]

Registers and receivers, in addition to their salaries, are to be allowed each the following fees and commissions:

1st. For each declaratory statement filed, and for services in acting on pre-emption claims, one dollar.

2d. On all moneys received at each receiver's office, a commission of one per cent.

3d. A commission, to be paid by the homestead applicant at the time of entry, of one per centum on the cash price, as fixed

---

[1] Overman Silver Mining Co. v. Dardanelles Silver Mining Co., Decision of Acting Secretary, July 19th, 1873, reversing on this point Decision of Acting Commissioner, June 3d, 1873, Copp's U. S. Mining Decisions, 181, 182, 186.

[2] Instructions February 1st, 1877, Subdivisions 82, 92; Instructions June 10th, 1872, Subdivisions 80, 81, 82, 83.

by law, of the land applied for; and a like commission when the claim is finally established, and the certificate therefor issued as the basis of a patent.[1]

4th. The same commission on lands entered under any law to encourage the growth of timber on western prairies, as allowed when the like quantity of land is entered with money.[2]

5th. For locating military bounty-land warrants, issued since the 11th day of February, 1847, and for locating agricultural college-land scrip, the same commission, to be paid by the holder or assignee of each warrant or scrip as is allowed for sales of the public lands for cash, at the rate of one dollar and twenty-five cents per acre.

6th. In donation cases, for each final certificate for 160 acres of land, five dollars; 320 acres, ten dollars; 640 acres, fifteen dollars.

7th. In the location of lands by States and corporations, under grants from Congress for railroad and other purposes, (except for agricultural colleges) for each final location of 160 acres, to be paid by the State or corporation making such location, one dollar.

8th. For superintending public land sales at their respective offices, five dollars per diem, and to each receiver, mileage in going to and returning from depositing the public moneys received by him.[3]

9th. For filing and acting upon each application for patent or adverse claim filed for mineral lands, to be paid by the respective parties, five dollars.[4]

10th. Registers and receivers are allowed, jointly, at the rate of fifteen cents per hundred words for testimony reduced by them to writing for claimants, in establishing pre-emption and homestead rights.[5]

---

[1] Rev. Stats. 2238, Subdivisions 9-11, same as Sec. 12, Act of 1872, 17 U. S. Stats. 95. See Sec. 2334, Rev. Stats.
[2] See Registers' and Receivers' Fees for Affidavits, etc., Decision of the Commissioner, Dec. 10th, 1869, Copp's U. S. Mining Decisions, 26.
[3] Sec. 12, Act of May 10th, 1872, 17 U. S. Stats. 95.
[4] See Sec. 2334.
[5] For fees under Acts of 1866 and 1870, see Instructions of Commissioner, July 25th, 1870, Copp's U. S. Mining Decisions, 251; Act of 1870, fifteenth section.

11th. A like fee when such writing is done in the Land Office in establishing claims for mineral lands.

12th. Registers and receivers in California, Oregon, Washington, Nevada, Colorado, Idaho, New Mexico, Arizona, Utah, Wyoming, and Montana, fifty per cent. additional on fees and commissions named in Subdivisions 1, 3, and 10.

It was early decided under the Mining Act of 1866, that, as it made no specific provisions on the subject of fees, the latter must be the same as were specifically provided for like services under other acts of Congress.

A charge allowed registers and receivers per hundred words was held not limited to testimony in Courts, but was applicable to all written matter necessary to prepare the case for administrative action in the Land Office, whether in the form of affidavits, certificates, or other appropriate and necessary writing, if prepared by the register and receiver, or according to their direction and under their supervision.[1]

§ 261. **Payment pending contest.**—Agricultural claimants making payment to the local officers, pending a contest as to the mineral character of the lands, and before the Commissioner has acted, only acquire a vested right on condition that the Commissioner, or other superior authority, shall finally concur in the opinion of the local officers. To hold otherwise would be to deprive the Government of all protection against the hasty and ill-advised acts of its inferior officers.[2]

§ 262. **Decisions of the Land Department.—Their authoritative character.**—The action of the officers of the Land Office is not necessarily conclusive upon the parties. Courts of Equity may go behind them and inquire into proceedings by which titles are sought to be vested, and afford relief in proper cases. This may be considered the settled doctrine of the Supreme Court of the United States.[3]

The officers of the Government are the agents of the law. They cannot act beyond its provisions, nor make compromises

---

[1] Decision Commissioner, Dec. 10th, 1869, Copp's U. S. Mining Decisions, 26.

[2] Decision of Assistant Secretary Interior, April 19th, 1872, Copp's U. S. Mining Decisions, 88.

[3] Lindsay *v.* Howes, 2 Black. U. S. 557; Cunningham *v.* Ashley, 14 How. 377.

not sanctioned by it. The Courts will inquire into the facts of disputed entries, notwithstanding the decision of the register and receiver.[1]

In Garland v. Wynn,[2] Mr. Justice Catron, in delivering the opinion of the Court, said: "The general rule is, that where several parties set up conflicting claims to property, with which a special tribunal may deal, as between one party and the Government, regardless of the rights of others, the latter may come into the ordinary Courts of Justice, and litigate the conflicting claims. Such was the case of Comegys v. Vasse, 1 Peters U. S. 212. * * * Nor do the regulations of the Commissioner of the General Land Office, whereby a party may be heard to prove his better claim to enter, oust the jurisdiction of the Courts of Justice. We announce this to be the settled doctrine of this Court."

In Lyttle v. Arkansas,[3] the same member of the Court, delivering its opinion, says: "Another preliminary question is presented on this record, namely, whether the adjudication of the register and receiver. * * is subject to revision in Courts of justice, etc. * * We deem this question too well settled in the affirmative for discussion."[4]

In the case of Johnson v. Towsley,[5] the whole question was again reviewed. The following acts of Congress were examined: the Act of September 4th, 1841, (5 Stat. at Large, 455) and entitled, An Act to appropriate the proceeds of the public lands, and to grant pre-emption rights; and a subsequent Act of March 3d, 1843, (5 Stat. at Large, 620) entitled, An Act to authorize the investigation of alleged frauds, under the pre-emption laws, and for other purposes, Secs. 4, 5; the Act of June 12th, 1858 (11 Stat. at Large, 326, Sec. 10); the twenty-fifth section of the Judiciary Act of 1789, and the second section of the Act of February 5th, 1867, re-enacting that section.[6] The case of

---

[1] Barnard's Heirs v. Ashley's Heirs, 18 How. U. S. 43.
[2] 20 How. U. S. 8.
[3] 22 How. U. S. 192.
[4] See, also, Magwire v. Tyler, 1 Black U. S. 195; Cousin v. Blanc's Executor, 19 How. U. S. 202; Tate v. Carney, 24 How. U. S. 357; 3 Op. Attorneys-General, 93, 104, 664; 1 Op. Attorneys-General, 718; Wilcox v. Jackson, 13 Pet. U. S. 498; Doe v. Eslava, 9 How. 421.
[5] 13 Wall. U. S. 72.
[6] See the two acts in parallel columns in Trebilcock v. Wilson, 12 Wall. 687.

Barnard's Heirs v. Ashley's Heirs,[1] was also referred to. The Court said: "There has always existed, in the Courts of Equity, the power in certain classes of cases to inquire into and correct mistakes, injustice, and wrong, in both judicial and executive action, however solemn the form which the result of that action may assume, when it invades private rights; and by virtue of this power the final judgments of Courts of Law have been annulled or modified, and patents and other important instruments issuing from the crown, or other executive branch of the government, have been corrected, or declared void, or other relief granted. No reason is perceived why the action of the Land Office should constitute an exception to this principle. In dealing with the public domain under the system of laws enacted by Congress for their management and sale, that tribunal decides upon private rights of great value; and very often, from the nature of its functions, this is by a proceeding essentially ex parte, and peculiarly liable to the influence of frauds, false swearing, and mistakes; these are among the most ancient and well-established grounds of the special jurisdiction of Courts of Equity just referred to, and the necessity and value of that jurisdiction are nowhere better exemplified than in its application to cases arising in the Land Office. It is very well known that these officers do not confine themselves to determining, before a patent issues, who is entitled to receive it; but they frequently assume the right, long after a patent has issued and the legal title passed out of the United States, to recall or set aside the patent, and issue one to some other party, and if the holder of the first patent refuses to surrender it, they issue a second. In such a case as this, have the Courts no jurisdiction? If they have not, who shall decide the conflicting claims to the land? If the land officers can do this a few weeks or a few months after the first patent has issued, what limit is there to their power over rights? (Stark v. Starrs, 6 Wall. 402.) See, also, Lytle v. Arkansas, 22 How. 192; Garland v. Wynn, 20 How. 8; Lindsay v. Hawes, 2 Black, 559; Finley v. Williams, 9 Cranch, 164; McArthur v. Browder, 4 Wheat. 488; Hunt v. Wickliffe, 2 Peters, 201; Green v. Liter, 8 Cranch,

---

[1] 18 How. U. S. 45.

229; Minnesota v. Bachelder, 1 Wall. 109; Silver v. Ladd, 7 Wall. 219.

"This Court has at all times been careful to guard itself against an invasion of the functions confided by law to other departments of the Government; and, in reference to the proceedings before the officers intrusted with the charge of selling the public lands, it has frequently and firmly refused to interfere with them in the discharge of their duties, either by mandamus or injunction, so long as the title remained in the United States, and the matter was rightfully before those officers for decision. On the other hand, it has constantly asserted the rights of the proper Courts to inquire, after the title had passed from the Government, and the question become one of private right, whether, according to the established rules of equity, and the acts of Congress concerning the public lands, the party holding that title should hold absolutely as his own, or as trustee for another. And we are satisfied that the relations thus established between the Courts and the Land Department are not only founded on a just view of the duties and powers of each, but are essential to the ends of justice, and to a sound administration of the law."

The decisions of the Supreme Court of the United States were said to establish the following propositions:

1. That the judiciary will not interfere, by mandamus, injunction, or otherwise, with the officers of the Land Department in the exercise of their duties, while the matter remains in their hands for decision.

2. That their decision on the facts, which must be the foundation of their action, unaffected by fraud or mistake, is conclusive in the Courts.

3. But that after the title has passed from the Government to individuals, and the question has become one of private right, the jurisdiction of Courts of Equity may be invoked to ascertain if the patentee does not hold in trust for other parties. If it appear that the party claiming the equity has established his right to the land to the satisfaction of the Land Department, in the true construction of the acts of Congress, but that, by an

erroneous construction, the patent has been issued to another, the Court will correct the mistake.[1]

Decisions of Courts conclusive.—The Land Office can neither supervise nor disregard the decisions rendered by the Courts in cases of conflicting claims to the possession of mining property under local customs, and no patent can be issued in the face of such decisions.[2]

The circular instructions issued from the Land Office apply to all United States lands, whether surveyed or unsurveyed, containing mineral deposits in quantities sufficient to lead to their development as mines, and bring them under the operation of the local mining customs.[3]

§ 263. **Right of inspection.**—The Land Office does not attempt to interfere with the right of mine-owners to exercise the right of ownership and possession of mining premises claimed by them, so long as they comply with the requirements of law. Under such circumstances, the register is authorized in refusing to grant a motion for the privilege of a party to visit the interior of another's mine.[4]

§ 264. **Mining claims in river-beds.**—The mere fact that the banks of a stream are meandered is not conclusive of its navigability. The question is one of fact. Rivers are deemed navigable waters of the United States when they are used or are susceptible of being used in their ordinary condition as highways for commerce between the States. The shores of navigable rivers and the soil under them were not granted by the Constitution to the United States, but were reserved to the States respectively, and new States have the same rights, sovereignty and jurisdiction over this subject as the original ones.[5]

---

[1] Minnesota *v.* Batchelder, 1 Wall. 109; Silver *v.* Ladd, 7 Wall. 219. See also Secretary *v.* McGarrahan, 9 Wall. 298; Hestres *v.* Brennan, 50 Cal. 211; Vance *v.* Kohlberg, Id. 346; Weaver *v.* Fairchild, Id. 360; Hosmer *v.* Wallace, 47 Cal. 461; Parker *v.* Duff, Id. 554; Litchfield *v.* Register and Receiver, Woolw. C. C. 299; Shepley *v.* Cowan, 1 Otto. 330.

[2] In re Inimitable Company, Decision of Commissioner, Jan. 26th, 1869, Copp's U. S. Mining Decisions, 19.

[3] Decision of Commissioner, January 28th, 1869, Copp's U. S. Mining Decisions, 19.

[4] In re Brunswick Mine, Decision Oct. 19th, 1876, 3 Copp's Land-owner, 114.

[5] The Daniel Ball, 10 Wallace, U. S. 557; The Montello, 11 Ibid. 411; Pollard's

The Land Office will not in any way complicate the full jurisdictional rights in navigable rivers in territorial limits, but which in the future must fall within the boundaries of a new State, by an attempted sale of any portions of the beds of such streams. The ninth section of the Act approved May 18th, 1796, (1 Stats. at L. 468) furnishes a rule upon the subject of the proprietorship of the stream and the bed of non-navigable rivers: "In all cases where the opposite banks of any stream not navigable shall belong to different persons, the stream and the bed thereof shall become common to both."

§ 265. Timber on mineral land—Railroad companies.

—Mineral lands do not pass to the Central Pacific Railroad Company by virtue of its grant, but the timber being or growing upon mineral land, within ten miles of the center line of the road or branches, was granted to said railroad company, except so much as is necessary to support the improvements of mine-owners upon the given tracts. When patent issues for such mineral land, it is necessary to insert therein a clause excepting from the operation of the patent all timber being *or growing upon odd-numbered sections within the limits hereinbefore referred to, except* such "as is necessary to support his improvements as a miner." [1]

§ 266. Claim not within any mining district.

—In the event of a mining claim being situated outside of any regularly-constituted mining district, affidavit of the fact must be made, and secondary evidence of possessory title will be received, which may consist of the affidavit of the claimant, supported by those of any other parties cognizant of the facts relative to the location, occupation, and possession of such claim, and any deeds, certificates of location or purchase, or other evidence, which may be in the claimant's possession, and tend to establish his claim.[2]

Lessee v. Hagan, 3 How. U. S. 212; Goodtitle v. Kibbe, 9 Ibid. 471; Doe v. Beebe, 13 Ibid. 25; Railroad Company v. Schurmir, 7 Wall. U. S. 272; Decision of Commissioner, Nov. 5th, 1874, 1 Copp's Land-owner, 153.

[1] Act of July, 1862, 12 U. S. Stats. 489, Sec. 3; Act of July 2d, 1864, 13 U. S. Stats. 356; Decision of Commissioner, Nov. 12th, 1874; C. P. R. R. Co. v. Mammoth Blue Gravel Co., 1 Copp's Land-owner, 134; affirmed by Secretary S. C. 2 Id. 104.

[2] Decision of Commissioner, Nov. 12th, 1872, Copp's U. S. Mining Decisions, 147.

§ 267. Removal of machinery.—Where a party abandons a mining claim, he has the right to remove from the claim any machinery or buildings which he may have placed thereon, or any ore that he may have extracted from such mine. A party relocating an abandoned mine may, in prosecuting work thereon, either sink new shafts and run new tunnels, or continue the work upon such shafts or tunnels as may have been constructed by parties who had abandoned the same.[1]

§ 268. Criminal offenses concerning mineral lands, penalties, etc.—Every person who does himself, or causes or procures, or willingly aids and assists, to be falsely made, altered, forged, or counterfeited, any petition, certificate, order, report, decree, concession, denouncement, deed, patent, confirmation, diseño, map, expediente, or part of an expediente, or any title paper, or evidence of right, title, or claim to lands, mines, or minerals, in California, or any instrument of writing whatever in relation to lands or mines or minerals in the State of California, for the purpose of setting up or establishing against the United States any claim, right, or title to lands, mines, or minerals, within the State of California, or for the purpose of enabling any person to set up or establish any such claim; and every person who, for such purpose, utters or publishes as true and genuine any such false, forged, altered, or counterfeited paper, (enumerated as above) shall be punishable by imprisonment at hard labor not less than three years and not more than ten years, and by fine of not more than $10,000.[2]

By Rev. Stats. 2472, Sec. 2, Act of May 18th, 1858, 11 U. S. Stats. 291, similar punishments and penalties are applied to Mexican grants of lands and mines.

And by Rev. Stats. Sec. 2473, Sec. 3, Act May 18th, 1858, 11 U. S. Stats. 291, similar provisions are made applicable to suits prosecuted on such false and forged papers as those mentioned in the two preceding sections.[3]

Perjury.—If parties are guilty of perjury in the matter of making proof, and have falsely made oath in regard to the

---

[1] Decision of Commissioner, June 2d, 1876, 3 Copp's Land-owner, 50.
[2] Rev. Stats. 2471, Sec. 1, Act May 18th, 1858, 11 U. S. Stats. 290.
[3] Sec. 3, Act May 18th, 1858, 11 U. S. Stats. 291.

character of the land embraced by their entries, their cases should be brought to the attention of the grand jury. Any assistance which can be rendered by the Land Office is usually tendered, and copies of papers transmitted when needed.[1]

§ 269. **Various provisions.**—*Custody of Letters.*—All official letters sent to the Register and Receiver, as well as the official records of letters sent by them, are the property of the United States; as such they should be retained in their offices. Ex-Registers have no right to take away or retain any official records or documents.[2]

Removal of Papers.—Applicants for patents have the right to examine any and all papers that are filed with the Register and Receiver in the nature of protests and adverse claims to their applications for patents, but the local land officers are not permitted to allow papers which have been filed to be removed from the office.[3]

Warrants and scrip cannot be received in payment of mineral land, nor located thereon.[4]

---

[1] Decision of Commissioner, Dec. 11th, 1873, Copp's U. S. Mining Decis. 339.
[2] Ibid. April 14th, 1873, Ibid. 188.
[3] Ibid. April 14th, 1873, Ibid. 181.
[4] Decision of Commissioner, Jan. 30th, 1873, Dec. 1st, 1875, Feb. 23rd, 1872, Aug. 25th, 1876, Copp's U. S. Mining Decisions, 157, 2 Copp's Land-owner, 130, 178, 3 Ibid. 83.

THE REVISED STATUTES

OF THE

UNITED STATES

RELATING TO

MINERAL LANDS.

---

§ 2318. Mineral lands reserved.
§ 2319. Mineral lands open to purchase by citizens.
§ 2320. Length of mining claims upon veins or lodes.
§ 2321. Proof of citizenship.
§ 2322. Locators' rights of possession and enjoyment.
§ 2323. Owners of tunnels, rights of.
§ 2324. Regulations made by miners—Expenditures and improvements.
§ 2325. Patents for mineral lands, how obtained.
§ 2326. Adverse claim, proceedings on.
§ 2327. Description of vein claims on surveyed and unsurveyed lands.
§ 2328. Pending applications—Existing rights.
§ 2329. Conformity of placer claims to surveys, limit of.
§ 2330. Subdivision of ten-acre tracts, maximum of placer locations.
§ 2331. Conformity of placer claims to surveys, limitation of claims.
§ 2332. What evidence of possession, etc., to establish a right to a patent.
§ 2333. Proceedings for patent for placer claim, etc.
§ 2334. Surveyor-General to appoint surveyors of mining claims, etc.
§ 2335. Verification of affidavits, etc.
§ 2336. Where veins intersect, etc.
§ 2337. Patents for non-mineral lands, etc.
§ 2338. What conditions of sale may be made by local legislature.
§ 2339. Vested rights to use of water for mining, etc., right of way for canals.
§ 2340. Patents, pre-emptions, and homesteads subject to vested water rights.
§ 2341. Lands in which no valuable mines are discovered, open to homesteads.
§ 2342. Mineral lands, how set apart as agricultural lands.
§ 2343. Additional districts and officers, power of the President to provide.
§ 2344. Provisions of this chapter not to affect certain rights.

§ 2345. Mineral lands in certain States excepted.
§ 2346. Grants of land to States or corporations not to include mineral lands.
§ 2347. Entry of coal lands.
§ 2348. Pre-emption of coal lands.
§ 2349. Pre-emption claims of coal lands to be presented within sixty days.
§ 2350. Only one entry allowed.
§ 2351. Conflicting claims.
§ 2352. Rights reserved.

§ 2318. **Mineral lands reserved.**—In all cases lands valuable for minerals shall be reserved from sale, except as otherwise expressly directed by law.

§ 2319. **Mineral lands open to purchase by citizens.**—All valuable mineral deposits in lands belonging to the United States, both surveyed and unsurveyed, are hereby declared to be free and open to exploration and purchase, and the lands in which they are found to occupation and purchase by citizens of the United States, and those who have declared their intention to become such, under regulations prescribed by law, and according to the local customs or rules of miners in the several mining districts, so far as the same are applicable and not inconsistent with the laws of the United States.

Sec. 1 of the Act of 1872, 17 U. S. Stat. 91, was identical with the above.

Sec. 1 of the Statute of July 26th, 1866, read as follows: SEC. 1. That the mineral lands of the public domain, both surveyed and unsurveyed, are hereby declared to be free and open to exploration and occupation by all citizens of the United States, and those who have declared their intention to become citizens, subject to such regulations as may be prescribed by law, and subject also to the local customs or rules of miners in the several mining districts, so far as the same may not be in conflict with the laws of the United States. [14 U. S. Stat. 251.]

See Sec. 2329.

§ 2320. **Length of mining claims upon veins or lodes.**—Mining claims upon veins or lodes of quartz or other rock in place bearing gold, silver, cinnabar, lead, tin, copper, or other valuable deposits heretofore, located, shall be governed as to length along the vein or lode by the customs, regulations and laws in force at the date of their location. A mining claim located after the tenth day of May, eighteen hundred and seventy-two, whether located by one or more persons, may equal, but shall not exceed, one thousand five hundred feet in length along the vein or lode; but no location of a mining claim shall be made until the discovery of the vein or lode within the limits of the claim located. No claim shall extend more than three hundred feet on each side of the middle of the vein at the surface, nor shall any claim be limited by any mining regulation to less than twenty-five feet on each side of the middle of the vein at the surface, except where adverse rights existing on the tenth day of May, eighteen hundred and seventy-two, render such limitation necessary. The end-lines of each claim shall be parallel to each other.

Sec. 2 of the Act of 1872, 17 U. S. Stat. 91, was the same as the above.

Sec. 4 of the Statute of July 26th, 1866, read as follows: SEC. 4. That when such location and entry of a mine shall be upon unsurveyed lands, it shall and may be lawful, after the extension thereto of the public surveys, to adjust the surveys to the limits of the premises, according to the location and possession and plat aforesaid; and the surveyor-general may, in extending the surveys, vary the same from a rectangular form to suit the circumstances of the country, and the local rules, laws, and customs of miners: *Provided*, That no location hereafter made shall exceed two hundred feet in length along the vein for each locator, with an additional claim for discovery to the discoverer of the lode, with the right to follow such vein to any depth with all its dips, variations, and angles, together with a reasonable quantity of surface for the convenient working of the same, as fixed by local rules: *And provided further*, That no person may make more than one location on the same lode, and not more than three thousand feet shall be taken in any one claim by any association of persons. [14 U. S. Stat. 252.]

See Secs. 2323, 2357.

§ 2321. Proof of citizenship.

—Proof of citizenship, under this chapter, may consist, in the case of an individual, of his own affidavit thereof; in the case of an association of persons unincorporated, of the affidavit of their authorized agent, made on his own knowledge, or upon information and belief; and in the case of a corporation organized under the laws of the United States, or of any State or Territory thereof, by the filing of a certified copy of their charter or certificate of incorporation.

The last clause of Sec. 7 of the Act of 1872, 17 U. S. Stat. 94, was the same as the above, with the following addition: "and nothing herein contained shall be construed to prevent the alienation of the title conveyed by a patent for a mining claim to any person whatever," which language is now incorporated in the last clause of Sec. 2326.

See Sec. 2335.

§ 2322. Locators' rights of possession and enjoyment.

—The locators of all mining locations heretofore made, or which shall hereafter be made, on any mineral vein, lode, or ledge, situated on the public domain, their heirs and assigns, where no adverse claim exists on the tenth day of May, eighteen hundred and seventy-two, so long as they comply with the laws of the United States, and with State, territorial, and local regulations not in conflict with the laws of the United States governing their possessory title, shall have the exclusive right of possession and enjoyment of all the surface included within the lines of their locations, and of all veins, lodes, and ledges throughout their entire depth, the top or apex of which lies inside of such surface-lines extended downward vertically, although such veins, lodes, or ledges may so far depart from a perpendicular in their course downward as to extend outside the vertical side-lines of such surface locations. But their right of possession to such outside parts of such veins or ledges shall be confined to such portions thereof as lie between vertical planes drawn downward as above described, through the end-lines of their locations, so continued in their own direction that such planes will intersect such exterior parts of such veins or ledges. And nothing in this section shall authorize the locator or possessor of a vein or lode which extends in its downward course beyond the vertical lines of his claim to enter upon the surface of a claim owned or possessed by another.

Sec. 3 of the Act of 1872, 17 U. S. Stat. 91, was the same as the above.
See Secs. 2320, 2324.

§ 2323. Owners of tunnels, rights of.

Where a tunnel is run for the development of a vein or lode, or for the discovery of mines, the owners of such tunnel shall have the right of possession of all veins or lodes within three thousand feet from the face of such tunnel on the line thereof, not previously known to exist, discovered in such tunnel, to the same extent as if discovered from the surface; and locations on the line of such tunnel of veins or lodes not appearing on the surface, made by other parties after the commencement of the tunnel, and while the same is being prosecuted with reasonable diligence, shall be invalid; but failure to prosecute the work on the tunnel for six months shall be considered as an abandonment of the right to all undiscovered veins on the line of such tunnel.

Sec. 4 of the Act of 1872, 17 U. S. Stat. 92, was the same as the above.
See Sec. 2320.

§ 2324. Miners' regulations—Expenditures and improvements.—The miners of each mining district may make regulations not in conflict with the laws of the United States, or with the laws of the State or territory in which the district is situated, governing the location, manner of recording, amount of work necessary to hold possession of a mining claim, subject to the following requirements: The location must be distinctly marked on the ground so that its boundaries can be readily traced. All records of mining claims hereafter made shall contain the name or names of the locators, the date of the location, and such a description of the claim or claims located by reference to some natural object or permanent monument as will identify the claim. On each claim located after the tenth day of May, eighteen hundred and seventy-two, and until a patent has been issued therefor, not less than one hundred dollars' worth of labor shall be performed or improvements made during each year. On all claims located prior to the tenth day of May, eighteen hundred and seventy-two, ten dollars' worth of labor shall be performed or improvements made by the tenth day of June, eighteen hundred and seventy-four, and each year thereafter, for each one hundred feet in length along the vein, until a patent has been issued therefor; but where such claims are held in common, such expenditure may be made upon any one claim; and upon a failure to comply with these conditions, the claim or mine upon which such failure occurred shall be open to relocation in the same manner as if no location of the same had ever been made, provided that the original locators, their heirs, assigns, or legal representatives, have not resumed work upon the claim after failure and before such location. Upon the failure of any one of several co-owners to contribute his proportion of the expenditures required hereby, the co-owners who have performed the labor or made the improvements, may, at the expiration of the year, give such delinquent co-owner personal notice in writing, or notice by publication in the newspaper published nearest the claim, for at least once a week for ninety days, and if, at the expiration of ninety days after such notice in writing or by publication, such delinquent should fail or refuse to contribute his proportion of the expenditure required by this section, his interest in the claim shall become the property of his co-owners who have made the required expenditures.

"That section two thousand three hundred and twenty-four of the revised statutes be, and the same is hereby, amended, so that where a person or company has or may run a tunnel for the purposes of developing a lode or lodes, owned by said person or company, the money so expended in said tunnel shall be taken and considered as expended on said lode or lodes, whether located prior to or since the passage of said act; and such person or company shall not be required to perform work on the surface of said lode or lodes in order to hold the same as required by said act." [Amendment enacted Feb. 11th, 1875.]

Sec. 5 of the Act of May 10th, 1872, U. S. Stat. 92, substituted the words, "each year for each hundred feet," *instead of the words*, "by the tenth day of June, eighteen hundred and seventy-four, and each year thereafter," in the clause relating to expenditures; otherwise the section was the same.

An Act of Congress, approved March 1st, 1873, amended Sec. 5 of the Act of 1872, (17 U. S. Stat. 92) so as to read as follows: "That the time for the first annual expenditure on claims located prior to the passage of said act shall be extended to the tenth day of June, eighteen hundred and seventy-four."

An Act of Congress, approved June 6th, 1874, further extended said time for first annual expenditure to the 1st day of January, 1875.

See Secs. 2331, 2332.

§ 2325. Patents for mineral lands, how obtained.—A

patent for any land claimed and located for valuable deposits may be obtained in the following manner: Any person, association, or corporation authorized to locate a claim under this chapter, having claimed and located a piece of land for such purposes, who has, or have, complied with the terms of this chapter, may file in the proper land office an application for a patent, under oath, showing such compliance, together with a plat and field-notes of the claim or claims in common, made by or under the direction of the United States Surveyor-General, showing accurately the boundaries of the claim or claims, which shall be distinctly marked by monuments on the ground, and shall post a copy of such plat, together with a notice of such application for a patent, in a conspicuous place on the land embraced in such plat previous to the filing of the application for a patent, and shall file an affidavit of at least two persons that such notice has been duly posted, and shall file a copy of the notice in such land office, and shall thereupon be entitled to a patent for the land, in the manner following: The register of the land office, upon the filing of such application, plat, field-notes, notices, and affidavits, shall publish a notice that such application has been made, for the period of sixty days, in a newspaper to be by him designated as published nearest to such claim; and he shall also post such notice in his office for the same period. The claimant at the time of filing this application, or at any time thereafter, within the sixty days of publication, shall file with the register a certificate of the United States Surveyor-General that five hundred dollars' worth of labor has been expended or improvements made upon the claim by himself or grantors; that the plat is correct, with such further description by such reference to natural objects or permanent monuments as shall identify the claim, and furnish an accurate description, to be incorporated in the patent. At the expiration of the sixty days of publication the claimant shall file his affidavit, showing that the plat and notice have been posted in a conspicuous place on the claim during such period of publication. If no adverse claim shall have been filed with the register and the receiver of the proper land office at the expiration of the sixty days of publication, it shall be assumed that the applicant is entitled to a patent, upon the payment to the proper officer of five dollars per acre, and that no adverse claim exists; and thereafter no objection from third parties to the issuance of a patent shall be heard, except it be shown that the applicant has failed to comply with the terms of this chapter.

Sec. 6 of the Act of 1872, 17 U. S. Stat. 92, was the same as the above.

Sec. 2 of the Mining Statute of July 26th, 1866, read as follows: SEC. 2. That whenever any person, or association of persons, claim a vein or lode of quartz or other rock in place, bearing gold, silver, cinnabar, or copper, having previously occupied and improved the same according to the local custom or rules of miners in the district where the same is situated, and having expended in actual labor and improvements thereon an amount of not less than one thousand dollars, and in regard to whose possession there is no controversy or opposing claim, it shall and may be lawful for said claimant, or association of claimants, to file in the local land office a diagram of the same, so extended laterally or otherwise as to conform to the local laws, customs, and rules of miners, and to enter such tract and receive a patent therefor, granting such mine, together with the right to follow such vein or lode with its dips, angles, and variations to any depth, although it may enter the land adjoining, which land adjoining shall be sold subject to this condition. [14 U. S. Stat. 251.]

Sec. 3 of the Mining Statute of July 26th, 1866, read as follows: SEC. 3. That upon the filing of the diagram as provided in the second section of this act, and posting the same in a conspicuous place on the claim, together with a notice of intention to apply for a patent, the register of the land-office shall publish a notice of the same in a newspaper published nearest to the location of said claim, and shall also post such notice in his office for the period of ninety days; and after the expiration of said period, if no adverse claim shall have been filed, it shall be the duty of the surveyor-general, upon application of the party, to survey the premises and make a plat thereof, indorsed with his approval, designating the number and description of the location, the value of the labor and improvements, and the character of the vein exposed; and upon the payment to the proper officer of five dollars per acre, together with the cost of such survey, plat, and notice, and giving satisfactory evidence that said diagram and notice have been posted on the claim during said period of ninety days, the register of the land office shall transmit to the General Land Office said plat, survey, and description, and a patent shall issue for the same thereupon. But said plat, survey, or description shall in no case cover more than one vein or lode, and no patent shall issue for more than one vein or lode, which shall be expressed in the patent issued. [14 U. S. Stat. 252.]

See Secs. 2325, 2327, 2328, 2333.

§ 2326. Adverse claim, proceedings on.—Where an adverse claim is filed during the period of publication, it shall be upon oath of the person or persons making the same, and shall show the nature, boundaries, and extent of such adverse claim, and all proceedings, except the publication of notice and making and filing of the affidavit thereof, shall be stayed until the controversy shall have been settled or decided by a Court of competent jurisdiction, or the adverse claim waived. It shall be the duty of the adverse claimant, within thirty days after filing his claim, to commence proceedings in a Court of competent jurisdiction, to determine the question of the right of possession, and prosecute the same with reasonable diligence to final judgment; and a failure so to do shall be a waiver of his adverse claim. After such judgment shall have been rendered, the party entitled to the possession of the claim, or any portion thereof, may, without giving further notice, file a certified copy of the judgment-roll with the register of the land office, together with the certificate of the Surveyor-General that the requisite amount of labor has been expended or improvements made thereon, and the description required in other cases, and shall pay to the receiver five dollars per acre for his claim, together with the proper fees, whereupon the whole proceedings and the judgment-roll shall be certified by the register to the Commissioner of the General Land Office, and a patent shall issue thereon for the claim, or such portion thereof as the applicant shall appear, from the decision of the Court, to rightly possess.

If it appears, from the decision of the Court, that several parties are entitled to separate and different portions of the claim, each party may pay for his portion of the claim, with the proper fees, and file the certificate and description by the Surveyor-General, whereupon the register shall certify the proceedings and judgment-roll to the Commissioner of the General Land Office, as in the preceding case, and patents shall issue to the several parties according to their respective rights. Nothing herein contained shall be construed to prevent the alienation of the title conveyed by a patent for a mining claim to any person whatever.

Sec. 7 of the Act of 1872, 17 U. S. Stat. 93, was the same as the above, with the exception of the omission of the clause relating to proofs of citizenship, which was identical with Sec. 2321, Ante.

Sec. 6 of the Statute of July 26th, 1866, read as follows: SEC. 6. That whenever any adverse claimants to any mine, located and claimed as aforesaid, shall appear before the approval of the survey, as provided in the third section of this act, all proceedings shall be stayed until a final settlement and adjudication, in the Courts of competent jurisdiction, of the rights of possession to such claim, when a patent may issue as in other cases. [14 U. S. Stat. 252.]

See Sec. 2325.

§ 2327. Description of vein-claims on surveyed and unsurveyed lands.

—The description of vein or lode claims, upon surveyed lands, shall designate the location of the claim with reference to the lines of the public surveys, but need not conform therewith; but where a patent shall be issued for claims upon unsurveyed lands, the Surveyor-General, in extending the surveys, shall adjust the same to the boundaries of such patented claim, according to the plat or description thereof, but so as in no case to interfere with or change the location of any such patented claim.

Sec. 8 of the Act of 1872, 17 U. S. Stat. 94, was the same as the above.
See Sec. 2325.

§ 2328. Pending applications—Existing rights.

—Applications for patents for mining claims under former laws now pending may be prosecuted to a final decision in the General Land Office; but in such cases where adverse rights are not affected thereby, patents may issue in pursuance of the provisions of this chapter; and all patents for mining claims upon veins or lodes heretofore issued shall convey all the rights and privileges conferred by this chapter, where no adverse rights existed on the tenth day of May, eighteen hundred and seventy-two.

Sec. 9 of the Act of 1872, 17 U. S. Stat. 94, read: SEC. 9. That sections one, two, three, four and six of an Act entitled "An Act granting the right of way to ditch and canal-owners over the public lands, and for other purposes," approved July twenty-sixth, eighteen hundred and sixty-six, are hereby repealed, but such repeal shall not affect existing rights. Applications for patents for mining claims now pending may be prosecuted to a final decision in the General Land Office; but in such cases where adverse rights are not affected thereby, patents may issue in pursuance of the provisions of this act; and all patents for mining claims heretofore issued under the Act of July twenty-sixth, eighteen hundred and sixty-six, shall convey all the rights and privileges conferred by this act where no adverse rights exist at the time of the passage of this act. [For Secs. 1, 2, 3, 4, and 6 of the Act of 1866, repealed by Sec. 9 of the Act of 1872, see notes to Secs. 2319, 2320, 2325, and 2326, Ante.]

See Secs. 2325, 2326.

§ 2329. Conformity of placer claims to surveys—Limit of.

—Claims usually called "placers," including all forms of deposit, excepting veins of quartz, or other rock in place, shall be subject to entry and patent, under like circumstances and conditions, and upon similar proceedings, as are provided for vein or lode claims; but where the lands have been previously surveyed by the United States, the entry in its exterior limits shall conform to the legal subdivisions of the public lands.

The first clause of Sec. 12 of the Act of 1870, 16 U. S. Stat. 217, was substantially the same as the above. [See note to Sec. 2330.]

See Secs. 2319, 2331, 2334.

§ 2330. Subdivision of ten-acre tracts—Limit of placer locations.

—Legal subdivisions of forty acres may be subdivided into ten-acre tracts; and two or more persons, or associations of persons, having contiguous claims of any size, although such claims may be less than ten acres each, may make joint entry thereof; but no location of a placer claim, made after the ninth day of July, eighteen hundred and seventy, shall exceed one hundred and sixty acres for any one person or association of persons, which location shall conform to the United States surveys; and nothing in this section contained shall defeat or impair any bona fide pre-emption or homestead claim upon agricultural lands, or authorize the sale of the improvements of any bona fide settler to any purchaser.

Sec. 12 of the Act of 1870, 16 U. S. Stat. 217, read: SEC. 12. That claims usually called "placers," including all forms of deposit excepting veins of quartz, or other rock in place, shall be subject to entry and patent under this act, under like circumstances and conditions and upon similar proceedings as are provided for vein or lode claims: *Provided,* That where the lands have been previously surveyed by the United States, the entry in its exterior limits shall conform to the legal subdivisions of the public lands, no further survey or plat in such case being required, and the lands may be paid for at the rate of two dollars and fifty cents per acre: *Provided further,* That legal subdivisions of forty acres may be subdivided into ten-acre tracts; and that two or more persons or associations of persons, having contiguous claims of any size, although such claims may be less than ten acres each, may make a joint entry thereof: *And provided further,* That no location of a placer claim, hereafter made, shall exceed one hundred and sixty acres for any one person or association of persons, which location shall conform to the United States surveys; and nothing in this section contained shall defeat or impair any *bona fide* pre-emption or homestead claim upon agricultural lands, or authorize the sale of the improvements of any *bona fide* settler to any purchaser.

See Sec. 2334.

§ 2331. Survey of placer claims—Limitation of.

Where placer claims are upon surveyed lands, and conform to legal subdivisions, no further survey or plat shall be required, and all placer mining claims located after the tenth day of May, eighteen hundred and seventy-two, shall conform as near as practicable with the United States system of public land surveys, and the rectangular subdivisions of such surveys, and no such location shall include more than twenty acres for each individual claimant; but where placer claims cannot be conformed to legal subdivisions, survey and plat shall be made as on unsurveyed lands; and where by the segregation of mineral land in any legal subdivision a quantity of agricultural land less than forty acres remains, such fractional portion of agricultural land may be entered by any party qualified by law, for homestead or pre-emption purposes.

Sec. 10 of the Act of 1872, 17 U. S. Stat. 94, read: SEC. 10. That the act entitled "An act to amend an act granting the right of way to ditch and canal-owners over the public lands, and for other purposes," approved July ninth, eighteen hundred and seventy, shall be and remain in full force, except as to the proceedings to obtain a patent, which shall be similar to the proceedings prescribed by sections six and seven of this act, for obtaining patents to vein or lode claims; but where said placer claims shall be upon surveyed lands, and conform to legal subdivisions, no further survey or plat shall be required, and all placer mining claims hereafter located shall conform as near as practicable with the United States system of public land surveys, and the rectangular subdivisions of such surveys, and no such location shall include more than twenty acres for each individual claimant, but where placer claims cannot be conformed to legal subdivisions, survey and plat shall be made as on unsurveyed lands: *Provided,* That proceedings now pending may be prosecuted to their final determination under existing laws; but the provisions of this act, when not in conflict with existing laws, shall apply to such cases: *And provided also,* That where by the segregation of mineral land in any legal subdivision a quantity of agricultural land less than forty acres remains, said fractional portion of agricultural land may be entered by any party qualified by law for homestead or pre-emption purposes.

Sec. 16 of the Act of 1870, 16 U. S. Stat 214, read: SEC. 16. That so much of the Act of March third, eighteen hundred and fifty-three, entitled "An Act to provide for the survey of the public lands in California, the granting of pre-emption rights, and for other purposes," as provides that none other than township lines shall be surveyed where the lands are mineral, is hereby repealed. And the public surveys are hereby extended over all such lands: *Provided*, That all subdividing of surveyed lands into lots less than one hundred and sixty acres may be done by county and local surveyors at the expense of the claimants: *And provided further*, That nothing herein contained shall require the survey of waste or useless lands.

See Secs. 2320, 2334.

§ 2332. Evidence of possession to establish right to patent.

—Where such person or association, they and their grantors, have held and worked their claims for a period equal to the time prescribed by the statute of limitations for mining claims of the State or Territory where the same may be situated, evidence of such possession and working of the claims for such period shall be sufficient to establish a right to a patent thereto under this chapter, in the absence of any adverse claim; but nothing in this chapter shall be deemed to impair any lien which may have attached in any way whatever to any mining claim or property thereto attached prior to the issuance of a patent.

Sec. 13 of the Act of 1870, 16 U. S. Stats. 217, read: SEC. 13. That where said person or association, they and their grantors, shall have held and worked their said claims for a period equal to the time prescribed by the statute of limitations for mining claims of the State or Territory where the same may be situated, evidence of such possession and working of the claims for such period shall be sufficient to establish a right to a patent thereto under this act, in the absence of any adverse claim: *Provided, however*, That nothing in this act shall be deemed to impair any lien which may have attached in any way whatever to any mining claim or property thereto attached prior to the issuance of a patent.

See Sec. 2324.

§ 2333. Proceedings for patent for placer claim, etc.—

Where the same person, association, or corporation is in possession of a placer claim, and also a vein or lode included within the boundaries thereof, application shall be made for a patent for the placer claim, with the statement that it includes such vein or lode, and in such case[1] a patent shall issue for the placer

---

[1] Sec. 11 of the Act of 1872, 17 U. S. Stat. 94, was the same as the above, with the addition of the words following, in parenthesis, after the words "and in such case," fifth line: (subject to the provisions of this act and the act entitled "An act to amend an act granting the right of way to ditch and canal-owners over the public lands, and for other purposes," approved July ninth, eighteen

claim, subject to the provisions of this chapter, including such vein or lode, upon the payment of five dollars per acre for such vein or lode claim, and twenty-five feet of surface on each side thereof. The remainder of the placer claim, or any placer claim not embracing any vein or lode claim, shall be paid for at the rate of two dollars and fifty cents per acre, together with all costs of proceedings; and where a vein or lode, such as is described in section twenty-three hundred and twenty, is known to exist within the boundaries of a placer claim, an application for a patent for such placer claim which does not include an application for the vein or lode claim shall be construed as a conclusive declaration that the claimant of the placer claim has no right of possession of the vein or lode claim; but where the existence of a vein or lode in a placer claim is not known, a patent for the placer claim shall convey all valuable mineral and other deposits within the boundaries thereof.

§ 2334. Surveyor-general to appoint surveyors of mining claims.—The Surveyor-General of the United States may appoint in each land district containing mineral lands as many competent surveyors as shall apply for appointment to survey mining claims. The expenses of the survey of vein or lode claims, and the survey and subdivision of placer claims into smaller quantities than one hundred and sixty acres, together with the cost of publication of notices, shall be paid by the applicants, and they shall be at liberty to obtain the same at the most reasonable rates, and they shall also be at liberty to employ any United States deputy surveyor to make the survey. The Commissioner of the General Land Office shall also have power to establish the maximum charges for surveys and publication of notices under this chapter; and, in case of excessive charges for publication, he may designate any newspaper published in a land district where mines are situated, for the publication of mining notices in such district, and fix the rates to be charged by such paper; and, to the end that the Commissioner may be fully informed on the subject, each applicant shall file with the

hundred and seventy,) in lieu of the words, "subject to the provisions of this chapter."
See Sec. 2325.

register a sworn statement of all charges and fees paid by such applicant for publication and surveys, together with all fees and money paid the register and the receiver of the land office, which statement shall be transmitted, with the other papers in the case, to the Commissioner of the General Land Office.

Sec. 12 of the Act of 1872, 17 U. S. Stat. 95, was the same as the above, with the following addition: "The fees of the register and receiver shall be five dollars each for filing and acting upon each application for patent or adverse claim filed, and they shall be allowed the amount fixed by law for reducing testimony to writing, when done in the land office, such fees and allowances to be paid by the respective parties; and no other fees shall be charged by them in such cases. Nothing in this act shall be construed to enlarge or affect the rights of either party in regard to any property in controversy at the time of the passage of this act or of the act entitled 'An act granting the right of way to ditch and canal-owners over the public lands, and for other purposes,' approved July twenty-sixth, eighteen hundred and sixty-six, nor shall this act affect any right acquired under said act; and nothing in this act shall be construed to repeal, impair, or in any way affect the provisions of the act entitled 'An act granting to A. Sutro the right of way and other privileges to aid in the construction of a draining and exploring tunnel to the Comstock Lode, in the State of Nevada,' approved July twenty-fifth, eighteen hundred and sixty-six."

For fees of registers and receivers, see Sec. 2238.
See Secs. 2330, 2331, 2406.

§ 2335. Verification of affidavits, etc.

All affidavits required to be made under this chapter may be verified before any officer authorized to administer oaths within the land district where the claims may be situated, and all testimony and proofs may be taken before any such officer, and, when duly certified by the officer taking the same, shall have the same force and effect as if taken before the register and receiver of the Land Office. In cases of contest as to the mineral or agricultural character of land, the testimony and proofs may be taken as herein provided on personal notice of at least ten days to the opposing party; or if such party cannot be found, then by publication of at least once a week for thirty days in a newspaper, to be designated by the register of the Land Office as published nearest to the location of such land; and the register shall require proof that such notice has been given.

Sec. 13 of the Act of 1872, 17 U. S. Stat. 95, was the same as the above.
Sec. 14 of the Act of 1870, 16 U. S. Stat. 217, read: SEC. 14. That all ex parte affidavits required to be made under this act, or the act of which it is amendatory, may be verified before any officer authorized to administer oaths within the land district where the claims may be situated.
See Sec. 2321.

§ 2336. **Where veins intersect, etc.**—Where two or more veins intersect or cross each other, priority of title shall govern, and such prior location shall be entitled to all ore or mineral contained within the space of intersection; but the subsequent location shall have the right of way through the space of intersection for the purposes of the convenient working of the mine. And where two or more veins unite, the oldest or prior location shall take the vein below the point of union, including all the space of intersection.

Sec. 14 of the Act of 1872, 17 U. S. Stat. 96, was the same as the above.

§ 2337. **Patents for non-mineral lands, etc.**—Where non-mineral land not contiguous to the vein or lode is used or occupied by the proprietor of such vein or lode for mining or milling purposes, such non-adjacent surface ground may be embraced and included in an application for a patent for such vein or lode, and the same may be patented therewith, subject to the same preliminary requirements as to survey and notice as are applicable to veins or lodes; but no location hereafter made of such non-adjacent land shall exceed five acres, and payment for the same must be made at the same rate as fixed by this chapter for the superficies of the lode. The owner of a quartz mill or reduction works, not owning a mine in connection therewith, may also receive a patent for his mill site, as provided in this section.

Sec. 15 of the Act of 1872, 17 U. S. Stat. 96, was the same as the above.
See Secs. 2320, 2324.

§ 2338. **State or Territorial legislation concerning mineral lands.**—As a condition of sale, in the absence of necessary legislation by Congress, the local legislature of any State or Territory may provide rules for working mines, involving easements, drainage, and other necessary means to their complete development; and those conditions shall be fully expressed in the patent.

Sec. 5 of the Act of 1866, 14 U. S. Stat. 252, was the same as the above.

§ 2339. **Vested rights to use of water—Right of way for canals, etc.**—Whenever, by priority of possession, rights to the use of water for mining, agricultural, manufacturing, or other purposes, have vested and accrued, and the same are recognized and acknowledged by the local customs, laws, and the decisions of Courts, the possessors and owners of such vested rights shall be maintained and protected in the same; and the right of way for the construction of ditches and canals for the purposes herein specified is acknowledged and confirmed; but whenever any person, in the construction of any ditch or canal, injures or damages the possession of any settler on the public domain, the party committing such injury or damage shall be liable to the party injured for such injury or damage.

Sec. 9 of the Act of 1866, 14 U. S. Stat. 253, was the same as the above.
See Sec. 2324.

§ 2340. **Patents, etc., subject to vested water rights.**—All patents granted, or pre-emption or homesteads allowed, shall be subject to any vested and accrued water rights, or rights to ditches and reservoirs used in connection with such water rights, as may have been acquired under or recognized by the preceding section.

Sec. 17 of the Act of 1870, 16 U. S. Stat. 218, read: SEC. 17. That none of the rights conferred by sections five, eight, and nine of the act to which this act is amendatory shall be abrogated by this act, and the same are hereby extended to all public lands affected by this act; and all patents granted, or pre-emption or homesteads allowed, shall be subject to any vested and accrued water rights, or rights to ditches and reservoirs used in connection with such water rights as may have been acquired under or recognized by the ninth section of the act of which this act is amendatory. But nothing in this act shall be construed to repeal, impair, or in any way affect the provisions of the "Act granting to A. Sutro the right of way and other privileges to aid in the construction of a draining and exploring tunnel to the Comstock Lode, in the State of Nevada," approved July twenty-fifth, eighteen hundred and sixty-six.

See notes to Secs. 2338, 2339, 2344.

§ 2341. **Non-mineral lands open to homesteads.**—Wherever, upon the lands heretofore designated as mineral lands, which have been excluded from survey and sale, there have been homesteads made by citizens of the United States, or persons who have declared their intention to become citizens, which homesteads have been made, improved, and used for agricultural purposes, and upon which there have been no valuable mines of gold, silver, cinnabar, or copper discovered, and which are properly agricultural lands, the settlers or owners of such homesteads shall have a right of pre-emption thereto, and shall be entitled to purchase the same at the price of one dollar and twenty-five cents per acre, and in quantity not to exceed one hundred and sixty acres; or they may avail themselves of the provisions of chapter five of this title, relating to "Homesteads."

Sec. 10 of the Act of 1866, 14 U. S. Stat. 253, was substantially the same as the above, with the addition of the following words, after the words "one hundred and sixty acres," thirteenth line: "or said parties may avail themselves of the provisions of the Act of Congress, approved May twentieth, eighteen hundred and sixty-two, entitled 'An act to secure homesteads to actual settlers on the public domain,' and acts amendatory thereof."

See Sec. 2342.

§ 2342. Mineral lands, how set apart as agricultural.—Upon the survey of the lands described in the preceding section, the Secretary of the Interior may designate and set apart such portions of the same as are clearly agricultural lands, which lands shall thereafter be subject to pre-emption and sale as other public lands, and be subject to all the laws and regulations applicable to the same.

Sec. 11 of the Act of 1866, 14 U. S. Stat. 253, was the same as the above.
See Secs. 2341, 2258.

§ 2343. Power of the President to provide districts and officers.—The President is authorized to establish additional land districts, and to appoint the necessary officers under existing laws, wherever he may deem the same necessary for the public convenience in executing the provisions of this chapter.

Sec. 7 of the Act of 1866, 14 U. S. Stat. 252, was the same as the above.

§ 2344. Provisions of this chapter not to affect certain rights.—Nothing contained in this chapter shall be construed to impair, in any way, rights or interests in mining property acquired under existing laws; nor to affect the provisions of the Act entitled "An Act granting to A. Sutro the right of way and other privileges to aid in the construction of a draining and exploring tunnel to the Comstock Lode, in the State of Nevada," approved July twenty-fifth, eighteen hundred and sixty-six.

For Sec. 17 of the Act of 1870, 16 U. S. Stat. 218, see note to Sec. 2340, Ante.

Sec. 8 of the Act of 1866, 14 U. S. Stat. 253, read: SEC. 8. That the right of way for the construction of highways over public lands, not reserved for public uses, is hereby granted.

The last clause of Sec. 16 of the Act of 1872, 17 U. S. Stat. 96, read as follows: "*Provided*, That nothing contained in this act shall be construed to impair, in any way, rights or interests in mining property acquired under existing laws."

§ 2345. Mineral lands in certain States excepted.—The provisions of the preceding sections of this chapter shall not apply to the mineral lands situated in the States of Michigan, Wisconsin, and Minnesota, which are declared free and open to exploration and purchase, according to legal subdivisions, in like manner as before the tenth day of May, eighteen hundred and seventy-two. And any bona fide entry of such lands within the States named, since the tenth day of May, eighteen hundred and seventy-two, may be patented without reference to any of the foregoing provisions of this chapter. Such lands shall be offered for public sale in the same manner, at the same minimum price, and under the same rights of pre-emption as other public lands.

Act of Feb. 18th, 1873, 17 U. S. Stat. 465, is to the same effect.

§ 2346. **What grants not to include mineral lands.**—No act passed at the first session of the Thirty-eighth Congress, granting lands to States or corporations to aid in the construction of roads or for other purposes, or to extend the time of grants made prior to the thirtieth day of January, eighteen hundred and sixty-five, shall be so construed as to embrace mineral lands, which in all cases are reserved exclusively to the United States, unless otherwise specially provided in the act or acts making the grant.

Act of Jan. 30th, 1865, 13 U. S. Stat. 567, was the same as the above.

§ 2347. **Entry of coal lands.**—Every person above the age of twenty-one years, who is a citizen of the United States, or who has declared his intention to become such, or any association of persons severally qualified as above, shall, upon application to the register of the proper land office, have the right to enter, by legal subdivisions, any quantity of vacant coal lands of the United States not otherwise appropriated or reserved by competent authority, not exceeding one hundred and sixty acres to such individual person, or three hundred and twenty acres to such association, upon payment to the receiver of not less than ten dollars per acre for such lands, where the same shall be situated more than fifteen miles from any completed railroad, and not less than twenty dollars per acre for such lands as shall be within fifteen miles of such road.

Sec. 1, Act of 1873, 17 U. S. Stat. 607, is identical with the above.

§ 2348. Pre-emption of coal lands.—Any person or association of persons severally qualified, as above provided, who have opened and improved, or shall hereafter open and improve, any coal mine or mines upon the public lands, and shall be in actual possession of the same, shall be entitled to a preference-right of entry, under the preceding section, of the mines so opened and improved: *Provided*, That when any association of not less than four persons, severally qualified as above provided, shall have expended not less than five thousand dollars in working and improving any such mine or mines, such association may enter not exceeding six hundred and forty acres, including such mining improvements.

Sec. 2, Act of 1873, 17 U. S. Stat. 607. is identical with the above.

§ 2349. Pre-emption of coal lands—When claims to be presented.—All claims under the preceding section must be presented to the register of the proper land district within sixty days after the date of actual possession and the commencement of improvements on the land, by the filing of a declaratory statement therefor; but when the township plat is not on file at the date of such improvement, filing must be made within sixty days from the receipt of such plat at the district office; and where the improvements shall have been made prior to the expiration of three months from the third day of March, eighteen hundred and seventy-three, sixty days from the expiration of such three months shall be allowed for the filing of a declaratory statement, and no sale under the provisions of this section shall be allowed until the expiration of six months from the third day of March, eighteen hundred and seventy-three.

Sec. 3, Act of 1873, 17 U. S. Stat. 607, was the same as the above

§ 2350. Only one entry allowed.—The three preceding sections shall be held to authorize only one entry by the same person or association of persons; and no association of persons any member of which shall have taken the benefit of such sections, either as an individual or as a member of any other association, shall enter or hold any other lands under the provisions thereof; and no member of any association which shall have taken the benefit of such sections shall enter or hold any other lands under their provisions; and all persons claiming under section twenty-three hundred and forty-eight shall be required to prove their respective rights and pay for the lands filed upon within one year from the time prescribed for filing their respective claims; and upon failure to file the proper notice or to pay for the land within the required period, the same shall be subject to entry by any other qualified applicant.

Sec. 4, Act of 1873, 17 U. S. Stat. 607, was the same as the above.

§ 2351. Conflicting claims.—In case of conflicting claims upon coal lands where the improvements shall be commenced, after the third day of March, eighteen hundred and seventy-three, priority of possession and improvement, followed by proper filing and continued good faith, shall determine the preference-right to purchase. And also where improvements have already been made prior to the third day of March, eighteen hundred and seventy-three, division of the land claimed may be made by legal subdivisions, to include, as near as may be, the valuable improvements of the respective parties. The Commissioner of the General Land Office is authorized to issue all needful rules and regulations for carrying into effect the provisions of this and the four preceding sections.

Sec. 5, Act of 1873, 17 U. S. Stat. 607, was the same as the above.

§ 2352. Existing rights.—Nothing in the five preceding sections shall be construed to destroy or impair any rights which may have attached prior to the third day of March, eighteen hundred and seventy-three, or to authorize the sale of lands valuable for mines of gold, silver, or copper.

Sec. 6, Act of 1873, 17 U. S. Stat. 607, was the same as the above.

Miscellaneous Provisions.

§ 910. Possessory actions for recovery of mining titles.
§ 2238. Fees and commissions of registers and receivers.
§ 2258. Lands not subject to pre-emption.
§ 2386. Title to town-lots subject to mineral rights.
§ 2406. Public surveys extended over mineral lands.

§ 2471. Penalty for false making or altering instruments concerning mineral lands in California.
§ 2472. Penalty for false making or dating instruments concerning mineral lands on Mexican grants in California.
§ 2473. Penalty for presenting false or counterfeited papers, or prosecuting fraudulent suit for mineral lands in California.

§ 910. Possessory actions concerning mining titles.—
No possessory action between persons, in any Court of the United States, for the recovery of any mining title, or for damages to any such title, shall be affected by the fact that the paramount title to the land in which such mines lie is in the United States; but each case shall be adjudged by the law of possession.

Sec. 9, Act of Feb. 27th, 1865, 13 U. S. Stat. 441.

§ 2238. Registers' and receivers' fees and commissions.
—Registers and receivers, in addition to their salaries, shall be allowed each the following fees and commissions, namely:

1. A fee of one dollar for each declaratory statement filed and for services in acting on pre-emption claims.

2. A commission of one per centum on all moneys received at each receiver's office.

3. A commission to be paid by the homestead applicant, at the time of entry, of one per centum on the cash price, as fixed by law, of the land applied for; and a like commission when the claim is finally established, and the certificate therefor issued as the basis of a patent.

4. The same commission on lands entered under any law to encourage the growth of timber on western prairies, as allowed when the like quantity of land is entered with money.

5. For locating military bounty-land warrants, issued since the eleventh day of February, eighteen hundred and forty-seven, and for locating agricultural college land scrip, the same commission, to be paid by the holder or assignee of each warrant or scrip, as is allowed for sales of the public lands for cash, at the rate of one dollar and twenty-five cents per acre.

6. A fee, in donation cases, of five dollars for each final certificate for one hundred and sixty acres of land, ten dollars for three hundred and twenty acres, and fifteen dollars for six hundred and forty acres.

7. In the location of lands by States and corporations under grants from Congress for railroads and other purposes, (except for agricultural colleges) a fee of one dollar for each final location of one hundred and sixty acres; to be paid by the State or corporation making such location.

8. A fee of five dollars per diem for superintending public-land sales at their respective offices; and, to each receiver, mileage in going to and returning from depositing the public moneys received by him.

9. A fee of five dollars for filing and acting upon each application for patent or adverse claim filed for mineral lands, to be paid by the respective parties.

10. Registers and receivers are allowed, jointly, at the rate of fifteen cents per hundred words for testimony reduced by them to writing for claimants, in establishing pre-emption and homestead rights.

11. A like fee as provided in the preceding subdivision, when such writing is done in the land office, in establishing claims for mineral lands.

12. Registers and receivers in California, Oregon, Washington, Nevada, Colorado, Idaho, New Mexico, Arizona, Utah, Wyoming, and Montana, are each entitled to collect and receive fifty per centum on the fees and commissions provided for in the first, third, and tenth subdivisions of this section.

The Subdivisions 9 and 11, relating to mineral lands, are substantially the same as Sec. 12 of Act of May 10th, 1872, 17 U. S. Stat. 95.
See note to Sec. 2334.

§ 2258. Lands not subject to pre-emption.

The following classes of lands, unless otherwise specially provided for by law, shall not be subject to the rights of pre-emption, to wit:

1. Lands included in any reservation by any treaty, law, or proclamation of the President, for any purpose.

2. Lands included within the limits of any incorporated town, or selected as the site of a city or town.

3. Lands actually settled and occupied for purposes of trade and business, and not for agriculture.

4. Lands on which are situated any known salines or mines.

Sec. 10, Act of Sept. 4th, 1841, 5 U. S. Stat. 455.
See Sec. 3212.

§ 2386. Title to town lots subject to mineral rights.—
Where mineral veins are possessed, which possession is recognized by local authority, and to the extent so possessed and recognized, the title to town lots to be acquired shall be subject to such recognized possession and the necessary use thereof; but nothing contained in this section shall be so construed as to recognize any color of title in possessors for mining purposes, as against the United States.

Sec. 2, Act of March 3d, 1865, 13 U. S. Stat. 530.

§ 2406. Public surveys extended over mineral lands.—
There shall be no further geological survey by the Government, unless hereafter authorized by law. The public surveys shall extend over all mineral lands; and all subdividing of surveyed lands into lots less than one hundred and sixty acres may be done by county and local surveyors at the expense of claimants; but nothing in this section contained shall require the survey of waste or useless lands.

Sec. 9, Act of July 9th, 1870, 16 U. S. Stat. 218.
See Sec. 2334

§ 2471. Penalty for offenses concerning mineral lands in California.—
Every person who falsely makes, alters, forges, or counterfeits, or causes or procures to be falsely made, altered, forged, or counterfeited; or willingly aids and assists in the false making, altering, forging, or counterfeiting any petition, certificate, order, report, decree, concession, denouncement, deed, patent, confirmation, diseño, map, expediente, or part of an expediente, or any title paper, or evidence of right, title, or claim to lands, mines, or minerals in California, or any instrument of writing whatever in relation to lands or mines or minerals in the State of California, for the purpose of setting up or establishing against the United States any claim, right, or title to lands, mines, or minerals within the State of California, or for the purpose of enabling any person to set up or establish any such claim; and every person, who, for such purpose, utters or publishes as true and genuine any such false, forged, altered, or counterfeited petition, certificate, order, report, decree, concession, denouncement, deed, patent, confirmation, diseño, map, expediente or part of an expediente, title-paper, evidence of

W. C.—26.

right, title, or claim to lands or mines or minerals in the State of California, or any instrument of writing whatever in relation to lands or mines or minerals in the State of California, shall be punishable by imprisonment at hard labor not less than three years and not more than ten years, and by a fine of not more than ten thousand dollars.

Sec. 1, Act of May 18th, 1858, 11 U. S. Stat. 290.

§ 2472. Penalty for offenses concerning Mexican grants in California.—Every person who makes, or causes or procures to be made, or willingly aids and assists in making any falsely dated petition, certificate, order, report, decree, concession, denouncement, deed, patent, confirmation, diseño, map, expediente or part of an expediente, or any title-paper, or written evidence of right, title, or claim, under Mexican authority, to any lands, mines, or minerals in the State of California, or any instrument of writing in relation to lands or mines or minerals in the State of California, having a false date, or falsely purporting to be made by any Mexican officer or authority prior to the seventh day of July, eighteen hundred and forty-six, for the purpose of setting up or establishing any claim against the United States to lands or mines or minerals within the State of California, or of enabling any person to set up or establish any such claim; and every person who signs his name as governor, secretary, or other public officer acting under Mexican authority, to any instrument of writing falsely purporting to be a grant, concession, or denouncement under Mexican authority, and during its existence in California, of lands, mines, or minerals, or falsely purporting to be an informe, report, record, confirmation, or other proceeding on application for a grant, concession, or denouncement under Mexican authority, during its existence in California, of lands, mines, or minerals, shall be punishable as prescribed in the preceding section.

Sec. 2, Act of May 18th, 1858, 11 U. S. Stat. 291.

§ 2473. Penalty for prosecuting fraudulent suits, etc., in California.—Every person who, for the purpose of setting up or establishing any claim against the United States to lands, mines, or minerals within the State of California, presents, or causes or procures to be presented, before any Court, judge, commission, or commissioner, or other officer of the United States, any false, forged, altered, or counterfeited petition, certificate, order, report, decree, concession, denouncement, deed, patent, diseño, map, expediente or part of an expediente, title-paper, or written evidence of right, title, or claim to lands, minerals, or mines in the State of California, knowing the same to be false, forged, altered, or counterfeited, or any falsely dated petition, certificate, order, report, decree, concession, denouncement, deed, patent, confirmation, diseño, map, expediente or part of an expediente, title-paper, or written evidence of right, title, or claim to lands, mines, or minerals in California, knowing the same to be falsely dated; and every person who prosecutes in any Court of the United States, by appeal or otherwise, any claim against the United States for lands, mines, or minerals in California, which claim is founded upon, or evidenced by, any petition, certificate, order, report, decree, concession, denouncement, deed, patent, confirmation, diseño, map, expediente or part of an expediente, title-paper, or written evidence of right, title, or claim, which has been forged, altered, counterfeited, or falsely dated, knowing the same to be forged, altered, counterfeited, or falsely dated, shall be punishable as prescribed in section twenty-four hundred and seventy-one.

Sec. 3, Act of May 18th, 1858, 11 U. S. Stat. 291.

§ 5596. Repealing certain acts passed prior to December 1st, 1873.

All acts of Congress passed prior to said first day of December, one thousand eight hundred and seventy-three, any portion of which is embraced in any section of said revision, are hereby repealed, and the section applicable thereto shall be in force in lieu thereof; all parts of such acts not contained in such revision, having been repealed or superseded by subsequent acts, or not being general or permanent in their nature: *Provided*, That the incorporation into said revision of any general and permanent provision, taken from an act making appropriations, or from an act containing other provisions of a private, local, or temporary character, shall not repeal, or in any way affect any appropriation, or any provision of a private, local, or temporary character, contained in any of said acts; but the same shall remain in force; and all acts of Congress passed prior to said last named day, no part of which are embraced in said revision, shall not be affected or changed by its enactment.

Approved June 22d, 1874.

Instructions of the Land Department,

FEBRUARY 1st, 1877.

MINERAL LANDS OPEN TO EXPLORATION, OCCUPATION, AND PURCHASE.

1. It will be perceived that by the foregoing provisions of law the mineral lands in the public domain, surveyed or unsurveyed, are open to exploration, occupation, and purchase by all citizens of the United States, and all those who have declared their intention to become such.

STATUS OF LODE CLAIMS LOCATED PRIOR TO MAY 10TH, 1872.

2. By an examination of the several sections of the Revised Statutes it will be seen that the status of lode claims located *previous* to the 10th May, 1872, is not changed with regard to their *extent along the lode or width of surface*, such claims being restricted and governed both as to their *lateral and linear* extent by the State, Territorial, or local laws, customs, or regulations which were in force in the respective districts at the date of such locations.

3. Mining rights acquired under such previous locations are, however, enlarged by said Revised Statutes in the following respect, viz: The locators of all such previously taken veins or lodes, their heirs and assigns, so long as they comply with the laws of Congress and with State, Territorial, or local regulations not in conflict therewith, governing mining claims, are invested with the exclusive possessory right of all the surface included within the lines of their locations, and of all veins, lodes, or ledges throughout their entire depth, the top or apex of which lies inside of such surface lines extended downward vertically, although such veins, lodes, or ledges may so far depart from a perpendicular in their course downward as to extend outside the vertical side-lines of such locations at the surface, it being expressly provided, however, that the right of possession to such outside parts of said veins or ledges shall be confined to such portions thereof as lie between vertical planes drawn downward, as aforesaid, through the end-lines of their locations so continued in their own direction that such planes will intersect such exterior parts of such veins, lodes, or ledges; no right being granted, however, to the claimant of such outside portion of a vein or ledge to enter upon the surface location of another claimant.

4. It is to be distinctly understood, however, that the law limits the possessory right to veins, lodes, or ledges *other* than the one named in the original location, to such as were not *adversely claimed on May 10th, 1872*, and that where such other vein or ledge was so adversely claimed at that date, the right of the party so adversely claiming is in no way impaired by the provisions of the Revised Statutes.

5. In order to hold the possessory title to a mining claim located prior to May 10th, 1872, and for which a patent has not been issued, the law requires that *ten dollars* shall be expended annually in labor or improvements on each claim of *one hundred feet* on the course of the vein or lode until a patent shall have been

issued therefor; but where a number of such claims are held in common upon the same vein or lode, the aggregate expenditure that would be necessary to hold all the claims, at the rate of ten dollars per hundred feet, may be made upon any one claim; a failure to comply with this requirement in any one year subjecting the claim upon which such failure occurred to relocation by other parties, the same as if no previous location thereof had ever been made, unless the claimants under the original location shall have resumed work thereon after such failure and before such relocation. The first annual expenditure upon claims of this class should have been performed subsequent to May 10th, 1872, and prior to January 1st, 1875. From and after January 1st, 1875, the required amount must be expended *annually* until patent issues.

6. Upon the failure of any one of several co-owners of a vein, lode, or ledge, which has not been patented, to contribute his proportion of the expenditures necessary to hold the claim or claims so held in ownership in common, the co-owners who have performed the labor, or made the improvements, as required by said Revised Statutes, may, at the expiration of the year, give such delinquent co-owner personal notice in writing, or notice by publication in the newspaper published nearest the claim, for at least once a week for ninety days; and if upon the expiration of ninety days after such notice in writing, or upon the expiration of one hundred and eighty days after the first newspaper publication of notice, the delinquent co-owner shall have failed to contribute his proportion to meet such expenditure or improvements, his interest in the claim by law passes to his co-owners who have made the expenditures or improvements as aforesaid.

PATENTS FOR VEINS OR LODES HERETOFORE ISSUED.

7. Rights under patents for veins or lodes heretofore granted under previous legislation of Congress, are enlarged by the Revised Statutes so as to invest the patentee, his heirs or assigns, with title to all veins, lodes, or ledges throughout their entire depth, the top or apex of which lies within the end and side boundary-lines of his claim on the surface, as patented, extended downward vertically, although such veins, lodes, or ledges may so far depart from a perpendicular in their course downward as to extend outside the vertical side-lines of the claim at the surface. The right of possession to such outside parts of such veins or ledges to be confined to such portions thereof as lie between vertical planes drawn downward through the end-lines of the claim at the surface, so continued in their own direction that such planes will intersect such exterior parts of such veins or ledges, it being expressly provided, however, that all veins, lodes, or ledges, the top or apex of which lies inside such surface locations, *other* than the one named in the patent, which were *adversely claimed on the* 10*th May,* 1872, are excluded from such conveyance by patent.

8. Applications for patents for mining claims pending at the date of the Act of May 10th, 1872, may be prosecuted to final decision in the General Land Office, and where no adverse rights are affected thereby, patents will be issued, in pursuance of the provisions of the Revised Statutes.

MANNER OF LOCATING CLAIMS ON VEINS OR LODES AFTER MAY 10TH, 1872.

9. From and after the 10th May, 1872, any person who is a citizen of the United States, or who has declared his intention to become a citizen, may locate, record, and hold a mining claim of *fifteen hundred linear feet* along the course of any mineral vein or lode subject to location; or an association of persons, severally qualified as above, may make joint location of such claim of *fifteen hundred feet,* but in no event can a location of a vein or lode made subsequent to May

10th, 1872, exceed fifteen hundred feet along the course thereof, whatever may be the number of persons composing the association.

10. With regard to the extent of surface ground adjoining a vein or lode, and claimed for the convenient working thereof, the Revised Statutes provide that the lateral extent of locations of veins or lodes made after May 10th, 1872, shall in no case exceed *three hundred feet on each side of the middle of the vein at the surface*, and that no such surface rights shall be limited by any mining regulations to less than twenty-five feet on each side of the middle of the vein at the surface, except where adverse rights existing on the 10th May, 1872, may render such limitation necessary, the end-lines of such claims to be in all cases parallel to each other.

11. By the foregoing it will be perceived that no lode claim located after the 10th May, 1872, can exceed a parallelogram fifteen hundred feet in length by six hundred feet in width, but whether surface ground of that width can be taken, depends upon the local regulations, or State or Territorial laws, in force in the several mining districts; and that no such local regulations, or State or Territorial laws, shall limit a vein or lode claim to less than fifteen hundred feet along the course thereof, whether the location is made by one or more persons, nor can surface rights be limited to less than fifty feet in width, unless adverse claims existing on the 10th day of May, 1872, render such lateral limitation necessary.

12. It is provided by the Revised Statutes that the miners of each district may make rules and regulations not in conflict with the laws of the United States, or of the State or Territory in which such districts are respectively situated, governing the location, manner of recording, and amount of work necessary to hold possession of a claim. They likewise require that the location shall be so distinctly marked on the ground that its boundaries may be readily traced. This is a very important matter, and locators cannot exercise too much care in defining their locations at the outset, inasmuch as the law requires that all records of mining locations made subsequent to May 10th, 1872, shall contain the name or names of the locators, the date of the location, and such a *description of the claim or claims* located, by reference to some natural object or permanent monument, as will identify the claim.

13. The statutes provide that no lode claim shall be recorded until after the discovery of a vein or lode within the limits of the ground claimed; the object of which provision is evidently to prevent the incumbering of the district mining records with useless locations before sufficient work has been done thereon to determine whether a vein or lode has really been discovered or not.

14. The claimant should therefore, prior to recording his claim, unless the vein can be traced upon the surface, sink a shaft, or run a tunnel or drift, to a sufficient depth therein to discover and develop a mineral-bearing vein, lode, or crevice; should determine, if possible, the general course of such vein in either direction from the point of discovery, by which direction he will be governed in marking the boundaries of his claim on the surface, and should give the course and distance as nearly as practicable from the discovery-shaft on the claim, to some permanent, well-known points or objects, such, for instance, as stone monuments, blazed trees, the confluence of streams, point of intersection of well-known gulches, ravines, or roads, prominent buttes, hills, etc., which may be in the immediate vicinity, and which will serve to perpetuate and fix the *locus* of the claim, and render it susceptible of identification from the description thereof given in the record of locations in the district.

15. In addition to the foregoing data, the claimant should state the names of adjoining claims, or, if none adjoin, the relative positions of the nearest claims; should drive a post or erect a monument of stones at each corner of his surface-

ground, and at the point of discovery or discovery-shaft should fix a post, stake, or board, upon which should be designated the name of the lode, the name or names of the locators, the number of feet claimed, and in which direction from the point of discovery; it being essential that the location notice filed for record, in addition to the foregoing description, should state whether the entire claim of fifteen hundred feet is taken on one side of the point of discovery, or whether it is partly upon one and partly upon the other side thereof, and in the latter case, how many feet are claimed upon each side of such discovery-point.

16. Within a reasonable time, say twenty days after the location shall have been marked on the ground, notice thereof, accurately describing the claim in manner aforesaid, should be filed for record with the proper recorder of the district, who will thereupon issue the usual certificate of location.

17. In order to hold the possessory right to a location made since May 10th, 1872, not less than one hundred dollars' worth of labor must be performed, or improvements made thereon, within one year from the date of such location, and annually thereafter; in default of which the claim will be subject to relocation by any other party having the necessary qualifications, unless the original locator, his heirs, assigns, or legal representatives, have resumed work thereon after such failure and before such relocation.

18. The expenditures required upon mining claims may be made from the surface or in running a tunnel for the development of such claims, the Act of February 11th, 1875, providing that where a person or company has or may run a tunnel for the purpose of developing a lode or lodes owned by said person or company, the money so expended in said tunnel shall be taken and considered as expended on said lode or lodes, and such person or company shall not be required to perform work on the surface of said lode or lodes in order to hold the same.

19. The importance of attending to these details in the matter of location, labor, and expenditure, will be the more readily perceived, when it is understood that a failure to give the subject proper attention may invalidate the claim.

TUNNEL RIGHTS.

20. Sec. 2323 provides that where a tunnel is run for the development of a vein or lode, or for the discovery of mines, the owners of such tunnel shall have the right of possession of all veins or lodes within three thousand feet from the face of such tunnel on the line thereof, not previously known to exist, discovered in such tunnel, to the same extent as if discovered from the surface; and locations on the line of such tunnel of veins or lodes not appearing on the surface, made by other parties after the commencement of the tunnel, and while the same is being prosecuted with reasonable diligence, shall be invalid; but failure to prosecute the work on the tunnel for six months shall be considered as an abandonment of the right to all undiscovered veins or lodes on the line of said tunnel.

21. The effect of this is simply to give the proprietors of a mining tunnel run in good faith the possessory right to fifteen hundred feet of any blind lodes cut, discovered, or intersected by such tunnel, which were not previously known to exist, within three thousand feet from the face or point of commencement of such tunnel, and to prohibit other parties, after the commencement of the tunnel, from prospecting for and making locations of lodes on the *line thereof* and within said distance of three thousand feet, unless such lodes appear upon the surface or were previously known to exist.

22. The term "face," as used in said section, is construed and held to mean the first working face formed in the tunnel, and to signify the point at which

the tunnel actually enters cover, it being from this point that the three thousand feet are to be counted, upon which prospecting is prohibited as aforesaid.

23. To avail themselves of the benefits of this provision of law, the proprietors of a mining tunnel will be required, at the time they enter cover as aforesaid, to give proper notice of their tunnel location, by erecting a substantial post, board, or monument at the face or point of commencement thereof, upon which should be posted a good and sufficient notice, giving the names of the parties or company claiming the tunnel right; the actual or proposed course or direction of the tunnel; the height and width thereof, and the course and distance from such face or point of commencement to some permanent, well-known objects in the vicinity by which to fix and determine the *locus* in manner heretofore set forth applicable to locations of veins or lodes, and at the time of posting such notice they shall, in order that miners or prospectors may be enabled to determine whether or not they are within the lines of the tunnel, establish the boundary lines thereof by stakes or monuments placed along such lines at proper intervals, to the terminus of the three thousand feet from the face or point of commencement of the tunnel, and the lines so marked will define and govern as to the specific boundaries within which prospecting for lodes not previously known to exist is prohibited while work on the tunnel is being prosecuted with reasonable diligence.

24. At the time of posting notice and marking out the lines of the tunnel as aforesaid, a full and correct copy of such notice of location, defining the tunnel claim, must be filed for record with the mining recorder of the district, to which notice must be attached the sworn statement or declaration of the owners, claimants, or projectors of such tunnel, setting forth the facts in the case; stating the amount expended by themselves and their predecessors in interest in prosecuting work thereon; the extent of the work performed, and that it is bona fide their intention to prosecute work on the tunnel so located and described, with reasonable diligence, for the development of a vein or lode, or for the discovery of mines, or both, as the case may be. This notice of location must be duly recorded, and, with the said sworn statement attached, kept on the recorder's files for future reference.

25. By a compliance with the foregoing, much needless difficulty will be avoided, and the way for the adjustment of legal rights acquired in virtue of said Sec. 2323 will be made much more easy and certain.

26. This office will take particular care that no improper advantage is taken of this provision of law by parties making or professing to make tunnel locations, ostensibly for the purposes named in the statute; but really for the purpose of monopolizing the lands lying in front of their tunnels to the detriment of the mining interests and to the exclusion of bona fide prospectors or miners; but will hold such tunnel claimants to a strict compliance with the terms of the statutes and a *reasonable diligence* on their part in prosecuting the work is one of the essential conditions of their implied contract. Negligence or want of due diligence will be construed as working a forfeiture of their right to all undiscovered veins on the line of such tunnel.

MANNER OF PROCEEDING TO OBTAIN GOVERNMENT TITLE TO VEIN OR LODE CLAIMS.

27. By Sec. 2325 authority is given for granting titles for mines by patent from the Government to any person, association, or corporation having the necessary qualifications as to citizenship and holding the right of possession to a claim in compliance with law.

28. The claimant is required, in the first place, to have a correct survey of his claim made under authority of the surveyor-general of the State or Territory in

which the claim lies; such survey to show with accuracy the exterior surface boundaries of the claim, which boundaries are required to be distinctly marked by monuments on the ground. Four plats and one copy of the original field-notes, in each case, will be prepared by the Surveyor-General: one plat and the original field-notes to be retained in the office of the Surveyor-General; one copy of the plat to be given the claimant for posting upon the claim; one plat and a copy of the field-notes to be given the claimant for filing with the proper register, to be finally transmitted by that officer, with the other papers in the case, to this office; and one plat to be sent by the Surveyor-General to the register of the proper land district, to be retained on his files for future reference.

29. The claimant is then required to post a copy of the plat of such survey in a conspicuous place upon the claim, together with notice of his intention to apply for a patent therefor, which notice will give the date of posting, the name of the claimant, the name of the claim, mine, or lode; the mining district and county; whether the location is of record, and if so, where the record may be found; the number of feet claimed along the vein and the presumed direction thereof; the number of feet claimed on the lode in each direction from the point of discovery, or other well-defined place on the claim; the name or names of adjoining claimants on the same or other lodes; or if none adjoin, the names of the nearest claims, etc.

30. After posting the said plat and notice upon the premises, the claimant will file with the proper register and receiver a copy of such plat, and the field-notes of survey of the claim, accompanied by the affidavit of at least two credible witnesses that such plat and notice are posted conspicuously upon the claim, giving the date and place of such posting; a copy of the *notice* so posted to be attached to, and form a part of, said affidavit.

31. Attached to the field-notes so filed must be the sworn statement of the claimant that he has the possessory right to the premises therein described, in virtue of compliance by himself (and by his grantors, if he claims by purchase) with the mining rules, regulations, and customs of the mining district, State, or Territory in which the claim lies, and with the mining laws of Congress; such sworn statement to narrate briefly, but as clearly as possible, the facts constituting such compliance, the origin of his possession, and the basis of his claim to a patent.

32. This affidavit should be supported by appropriate evidence from the mining recorder's office as to his possessory right, as follows, viz: Where he claims to be a locator, a full, true, and correct copy of such location should be furnished, as the same appears upon the mining records; such copy to be attested by the seal of the recorder, or if he has no seal, then he should make oath to the same being correct, as shown by his records; where the applicant claims as a locator in company with others, who have since conveyed their interests in the lode to him, a copy of the original record of location should be filed, together with an abstract of title from the proper recorder, under seal or oath as aforesaid, tracing the co-locator's possessory rights in the claim to such applicant for patent; where the applicant claims only as a purchaser for valuable consideration, a copy of the location record must be filed, under seal or upon oath as aforesaid, with an abstract of title certified as above by the proper recorder, tracing the right of possession by a continuous chain of conveyances from the original locators to the applicant.

33. In the event of the mining records in any case having been destroyed by fire or otherwise lost, affidavit of the fact should be made, and secondary evidence of possessory title will be received, which may consist of the affidavit of the claimant, supported by those of any other parties cognizant of the facts relative to his location, occupancy, possession, improvements, etc.; and in such

case of lost records, any deeds, certificates of location or purchase, or other evidence which may be in the claimant's possession, and tend to establish his claim, should be filed.

34. Upon the receipt of these papers the register will, at the expense of the claimant, publish a notice of such application for the period of sixty days, in a newspaper published nearest to the claim, and will post a copy of such notice in his office for the same period. In all cases sixty days must intervene between the first and the last insertion of the notice in such newspaper.

35. The notices so published and posted must be as full and complete as possible, and embrace all the data given in the notice posted upon the claim.

36. Too much care cannot be exercised in the preparation of these notices, inasmuch as upon their accuracy and completeness will depend, in a great measure, the regularity and validity of the whole proceeding.

37. The claimant, either at the time of filing these papers with the register, or at any time during the sixty days' publication, is required to file a certificate of the Surveyor-General that not less than five hundred dollars' worth of labor has been expended or improvements made upon the claim by the applicant or his grantors; that the plat filed by the claimant is correct; that the field-notes of the survey, as filed, furnish such an accurate description of the claim, as will, if incorporated into a patent, serve to fully identify the premises, and that such reference is made therein to natural objects or permanent monuments as will perpetuate and fix the *locus* thereof.

38. It will be the more convenient way to have this certificate indorsed by the Surveyor-General, both upon the plat and field-notes of survey, filed by the claimant as aforesaid.

39. After the sixy days' period of newspaper publication has expired, the claimant will file his affidavit, showing that the plat and notice aforesaid remained conspicuously posted upon the claim sought to be patented during said sixty days' publication.

40. Upon the filing of this affidavit, the register will, if no adverse claim was filed in his office during the period of publication, permit the claimant to pay for the land according to the area given in the plat and field-notes of survey aforesaid, at the rate of five dollars for each acre and five dollars for each fractional part of an acre, the receiver issuing the usual duplicate receipt therefor; after which the whole matter will be forwarded to the Commissioner of the General Land Office, and a patent issued thereon, if found regular.

41. In sending up the papers in the case the register must not omit certifying to the fact that the notice was posted in his office for the full period of sixty days, such certificate to state distinctly when such posting was done and how long continued.

42. The consecutive series of numbers of mineral entries must be continued, whether the same are of lode or placer claims.

43. The Surveyor-General must continue to designate all surveyed mineral claims as heretofore by a progressive series of numbers, beginning with lot No. 37 in each township; the claim to be so designated at date of filing the plat, field-notes, etc., in addition to the local designation of the claim; it being required in all cases that the plat and field-notes of the survey of the claim must, in addition to the reference to permanent objects in the neighborhood, describe the *locus* of the claim with reference to the lines of public surveys by a line connecting a corner of the claim with the nearest public corner of the United States surveys, unless such claim be on unsurveyed lands at a remote distance from such public corner; in which latter case the reference by course and distance to permanent objects in the neighborhood will be a sufficient designation by which to fix the *locus* until the public surveys shall have been closed upon its boundaries.

ADVERSE CLAIMS.

44. Section 2326 provides for adverse claims; fixes the time within which they shall be filed to have legal effect, and prescribes the manner of their adjustment.

45. Said section requires that the adverse claim shall be filed during the period of publication of notice; that it must be on the oath of the adverse claimant; and that it must show the "*nature*," the "*boundaries*," and the "*extent*" of the adverse claim.

46. In order that this section of law may be properly carried into effect, the following is communicated for the information of all concerned:

47. An adverse mining claim must be filed with the register of the same land office with whom the application for patent was filed, or in his absence with the receiver, and within the sixty days' period of newspaper publication or notice.

48. The adverse notice must be duly sworn to by the person or persons making the same before an officer authorized to administer oaths within the land district, or before the register or receiver; it will fully set forth the nature and extent of the interference or conflict; whether the adverse party claims as a purchaser for valuable consideration, or as a locator; if the former, a certified copy of the original location, the original conveyance, a duly certified copy thereof, or an abstract of title from the office of the proper recorder should be furnished; or if the transaction was a mere verbal one, he will narrate the circumstances attending the purchase, the date thereof, and the amount paid, which facts should be supported by the affidavit of one or more witnesses, if any were present at the time; and if he claims as a locator, he must file a duly certified copy of the location from the office of the proper recorder.

49. In order that the "*boundaries*" and "*extent*" of the claim may be shown, it will be incumbent upon the adverse claimant to file a plat showing his claim, its relative situation or position with the one against which he claims, and the extent of the conflict. This plat must be made from an actual survey by a United States deputy surveyor, who will officially certify thereon to its correctness; and in addition, there must be attached to such plat of survey a certificate or sworn statement by the surveyor as to the approximate value of the labor performed or improvements made upon the claim by the adverse party or his predecessors in interest, and the plat must indicate the position of any shafts, tunnels, or other improvements, if any such exist, upon the claim of the party opposing the application, and by which party said improvements were made.

50. Upon the foregoing being filed within the sixty days as aforesaid, the register, or in his absence the receiver, will give notice in writing to *both parties* to the contest that such adverse claim has been filed, informing them that the party who filed the adverse claim will be required within thirty days from the date of such filing to commence proceedings in a Court of competent jurisdiction to determine the question of right of possession, and to prosecute the same with reasonable diligence to final judgment, and that should such adverse claimant fail to do so, his adverse claim will be considered waived, and the application for patent be allowed to proceed upon its merits.

51. When an adverse claim is filed as aforesaid, the register or receiver will indorse upon the same the precise date of filing, and preserve a record of the date of notifications issued thereon; and thereafter all proceedings on the application for patent will be suspended, with the exception of the completion of the publication and posting of notices and plat, and the filing of the necessary proof thereof, until the controversy shall have been adjudicated in Court, or the adverse claim waived or withdrawn.

52. The proceedings after rendition of judgment by the Court in such case are

so clearly defined by the act itself as to render it unnecessary to enlarge thereon in this place.

PLACER CLAIMS.

53. The proceedings to obtain patents for claims usually called placers, including all forms of deposit, are similar to the proceedings prescribed for obtaining patents for vein or lode claims; but where said placer-claim shall be upon surveyed lands, and conform to legal subdivisions, no further survey or plat will be required, and all placer-mining claims located after May 10th, 1872, shall conform as nearly as practicable with the United States system of public-land surveys and the rectangular subdivisions of such surveys, and no such location shall include more than twenty acres for each individual claimant; but where placer-claims cannot be conformed to legal subdivisions, survey and plat shall be made as on unsurveyed lands. But where such claims are located previous to the public surveys, and do not conform to legal subdivisions, survey, plat, and entry thereof may be made according to the boundaries fixed by the local laws.

54. The proceedings for obtaining patents for veins or lodes having already been fully given, it will not be necessary to repeat them here; it being thought that careful attention thereto by applicants and the local officers will enable them to act understandingly in the matter, and make such slight modifications in the notice, or otherwise, as may be necessary in view of the different nature of the two classes of claims, placer-claims being fixed, however, at two dollars and fifty cents per acre, or fractional part of an acre.

55. By Section 2330, authority is given for the subdivision of forty-acre legal subdivisions into *ten-acre* lots, which is intended for the greater convenience of miners in segregating their claims both from one another and from intervening agricultural lands.

56. It is held, therefore, that under a proper construction of the law these ten-acre lots in mining districts should be considered and dealt with, to all intents and purposes, as legal subdivisions, and that an applicant having a legal claim which conforms to one or more of these ten-acre lots, either adjoining or cornering, may make entry thereof, after the usual proceedings, without further survey or plat.

57. In cases of this kind, however, the notice given of the application must be very specific and accurate in description, and as the forty-acre tracts may be subdivided into ten-acre lots, either in the form of squares of ten-by-ten chains or of parallelograms five-by-twenty chains, so long as the lines are parallel and at right angles with the lines of the public surveys, it will be necessary that the notice and application state specifically what ten-acre lots are sought to be patented, in addition to the other data required in the notice.

58. Where the ten-acre subdivision is in the form of a square, it may be described, for instance, as the "S. E. ¼ of the S. W. ¼ of N. W. ¼," or, if in the form of a parallelogram as aforesaid, it may be described as the "W. ½ of the W. ½ of the S. W. ¼ of the N. W. ¼, (or the N. ½ of the S. ½ of the N. E. ¼ of the S. E. ¼) of section ———, township ———, range ———," as the case may be; but, in addition to this description of the land, the notice must give all the other data that are required in a mineral application, by which parties may be put on inquiry as to the premises sought to be patented. The proof submitted with applications for claims of this kind must show clearly the character and the extent of the improvements upon the premises.

59. The proceedings necessary for the adjustment of rights where a known vein or lode is embraced by a placer claim are so clearly defined by Section 2333 as to render any particular instructions upon that point at this time unnecessary.

60. When an adverse claim is filed to a placer application, the proceedings are the same as in the case of vein or lode claims already described.

QUANTITY OF PLACER GROUND SUBJECT TO LOCATION.

61. By Section 2330 it is declared that no location of a placer claim, made after July 9th, 1870, shall exceed one hundred and sixty acres for any one person or association of persons, which location shall conform to the United States surveys.

62. Section 2331 provides that all placer mining claims, located after May 10th, 1872, shall conform as nearly as practicable with the United States system of public surveys, and the subdivisions of such surveys, and no such location shall include more than twenty acres for each individual claimant.

63. The foregoing provisions of law are construed to mean that after the 9th day of July, 1870, no location of a placer claim can be made to exceed one hundred and sixty acres, whatever may be the number of locators associated together, or whatever the local regulations of the district may allow; and that from and after May 10th, 1872, no location made by an individual can exceed twenty acres, and no location made by an association of individuals can exceed one hundred and sixty acres, which location of one hundred and sixty acres cannot be made by a less number than eight bona fide locators, but that whether as *much* as twenty acres can be located by an individual, or one hundred and sixty acres by an association, depends entirely upon the mining regulations in force in the respective districts at the date of the location; it being held that such mining regulations are in no way enlarged by the statutes, but remain intact and in full force with regard to the *size* of locations, in so far as they do not permit locations in excess of the limits fixed by Congress, but that where such regulations permit locations in excess of the maximums fixed by Congress as aforesaid, they are restricted accordingly.

64. The regulations hereinbefore given as to the manner of marking locations on the ground, and placing the same on record, must be observed in the case of placer locations, so far as the same are applicable; the law requiring, however, that where placer claims are upon *surveyed* public lands, the locations must hereafter be made to conform to the legal subdivisions thereof (as near as practicable).

65. With regard to the proofs necessary to establish the possessory right to a placer claim, Section 2332 provides that "where such person or association, they and their grantors, have held and worked their claims for a period equal to the time prescribed by the statute of limitations for mining claims of the State or Territory where the same may be situated, evidence of such possession and working of the claims for such period shall be sufficient to establish a right to a patent thereto under this chapter, in the absence of any adverse claim."

66. This provision of law will greatly lessen the burden of proof, more especially in the case of old claims located many years since, the records of which, in many cases, have been destroyed by fire, or lost in other ways during the lapse of time, but concerning the possessory right to which all controversy or litigation has long been settled.

67. When an applicant desires to make his proof of possessory right in accordance with this provision of law, you will not require him to produce evidence of location, copies of conveyances, or abstracts of title, as in other cases, but will require him to furnish a duly certified copy of the statute of limitations of mining claims for the State or Territory, together with his sworn statement giving a clear and succinct narration of the facts as to the origin of his title, and likewise as to the continuation of his possession of the mining ground covered by his application; the area thereof, the nature and extent of the min-

ing that has been done thereon; whether there has been any opposition to his possession or litigation with regard to his claim; and if so, when the same ceased; whether such cessation was caused by compromise or by judicial decree, and any additional facts within the claimant's knowledge having direct bearing upon his possession and bona fides which he may desire to submit in support of his claim.

68. There should likewise be filed a certificate, under seal of the Court having jurisdiction of mining cases within the judicial district embracing the claim, that no suit or action of any character whatever involving the right of possession to any portion of the claim applied for is pending, and that there has been no litigation before said Court affecting the title to said claim or any part thereof for a period equal to the time fixed by the Statute of Limitations for mining claims in the State or Territory as aforesaid, other than that which has been finally decided in favor of the claimant.

69. The claimant should support his narrative of facts relative to his possession, occupancy, and improvements by corroborative testimony of any disinterested person or persons of credibility who may be cognizant of the facts in the case, and are capable of testifying understandingly in the premises.

70. It will be to the advantage of claimants to make their proofs as full and complete as practicable.

MILL SITES.

71. Section 2337 provides that, "where non-mineral land not contiguous to the vein or lode is used or occupied by the proprietor of such vein or lode for mining or milling purposes, such non-adjacent surface ground may be embraced and included in an application for a patent for such vein or lode, and the same may be patented therewith, subject to the same preliminary requirements as to survey and notice as are applicable to veins or lodes: but no location hereafter made of such non-adjacent land shall exceed five acres, and payment for the same must be made at the same rate as fixed by this chapter for the superficies of the lode. The owner of a quartz mill or reduction works, not owning a mine in connection therewith, may also receive a patent for a mill site as provided in this section."

72. To avail themselves of this provision of law, parties holding the possessory right to a vein or lode, and to a piece of non-mineral land not contiguous thereto, for mining or milling purposes, not exceeding the quantity allowed for such purpose by the local rules, regulations, or customs, the proprietors of such vein or lode may file in the proper land office their application for a patent, under oath, in manner already set forth herein, which application, together with the plat and field-notes, may include, embrace, and describe, in addition to the vein or lode, such non-contiguous mill site, and after due proceedings as to notice, etc., a patent will be issued conveying the same as one claim.

73. In making the survey in a case of this kind, the lode claim should be described in the plat and field-notes as "Lot No. 37, A," and the mill site as "Lot No. 37, B," or whatever may be its appropriate numerical designation; the course and distance from a corner of the mill site to a corner of the lode claim to be invariably given in such plat and field-notes, and a copy of the plat and notice of application for patent must be conspicuously posted upon the mill site as well as upon the vein or lode, for the statutory period of sixty days. In making the entry no separate receipt or certificate need be issued for the mill site, but the whole area of both lode and mill site will be embraced in one entry, the price being five dollars for each acre and fractional part of an acre embraced by such lode and mill site claim.

74. In case the owner of a quartz-mill or reduction-works is not the owner or

claimant of a vein or lode, the law permits him to make application therefor in the same manner prescribed herein for mining claims, and after due notice and proceedings, in the absence of a valid adverse filing, to enter and receive a patent for his mill site at said price per acre.

75. In every case there must be satisfactory proof that the land claimed as a mill site is not mineral in character, which proof may, where the matter is unquestioned, consist of the sworn statement of the claimant, supported by that of one or more disinterested persons capable from acquaintance with the land to testify understandingly.

76. The law expressly limits mill-site locations made from and after its passage to *five acres*, but whether so *much* as that can be located depends upon the local customs, rules, or regulations.

77. The registers and receivers will preserve an unbroken consecutive series of numbers for all mineral entries.

PROOF OF CITIZENSHIP OF MINING CLAIMANTS.

78. The proof necessary to establish the citizenship of applicants for mining patents must be made in the following manner: In case of an incorporated company, a certified copy of their charter or certificate of incorporation must be filed. In case of an association of persons unincorporated, the affidavit of their duly authorized agent, made upon his own knowledge, or upon information and belief, setting forth the residence of each person forming such association, must be submitted. This affidavit must be accompanied by a power of attorney from the parties forming such association, authorizing the person who makes the affidavit of citizenship to act for them in the matter of their application for patent.

79. In case of an individual or an association of individuals who do not appear by their duly authorized agent, you will require the affidavit of each applicant, showing whether he is a native or naturalized citizen, when and where born, and his residence.

80. In case an applicant has declared his intention to become a citizen, or has been naturalized, his affidavit must show the date, place, and the Court before which he declared his intention, or from which his certificate of citizenship issued, and present residence.

81. The affidavit of citizenship may be taken before the register and receiver, or any other officer authorized to administer oaths within the district.

APPOINTMENT OF DEPUTY SURVEYORS OF MINING CLAIMS—CHARGES FOR SURVEYS AND PUBLICATIONS—FEES OF REGISTERS AND RECEIVERS, ETC.

82. Section 2334 provides for the appointment of surveyors of mineral claims, authorizes the Commissioner of the General Land Office to establish the rates to be charged for surveys and for newspaper publications, prescribes the fees allowed to the local officers for receiving and acting upon applications for mining patents and for adverse claims thereto, etc.

83. The Surveyors-General of the several districts will, in pursuance of said law, appoint in each land district as many *competent* deputies for the survey of mining claims as may seek such appointment; it being distinctly understood that all expenses of these notices and surveys are to be borne by the mining claimants and not by the United States; the system of making *deposits* for mineral surveys, as required by previous instructions, being hereby revoked as regards *field-work;* the claimant having the option of employing *any* deputy surveyor within such district to do his work in the field.

84. With regard to the *platting* of the claim and other *office work* in the Surveyor-General's office, that officer will make an estimate of the cost thereof,

which amount the claimant will deposit with any assistant United States treasurer, or designated depository, in favor of the United States Treasurer, to be passed to the credit of the fund created by "individual depositors for surveys of the public lands," and file with the Surveyor-General duplicate certificates of such deposit in the usual manner.

85. The Surveyors-General will endeavor to appoint mineral deputy surveyors so that one or more may be located in each mining district, for the greater convenience of miners.

86. The usual oaths will be required of these deputies and their assistants as to the correctness of each survey executed by them.

87. The law requires that each applicant shall file with the register and receiver a sworn statement of all charges and fees paid by him for publication of notice and for survey; together with all fees and money paid the register and receiver, which sworn statement is required to be transmitted to this Office, for the information of the Commissioner.

88. Should it appear that excessive or exorbitant charges have been made by any surveyor or any publisher, prompt action will be taken with the view of correcting the abuse.

89. The fees payable to the register and receiver for filing and acting upon applications for mineral land patents are five dollars to each officer, to be paid by the applicant for patent at the time of filing, and the like sum of five dollars is payable to each officer by an adverse claimant at the time of filing his adverse claim.

90. All fees or charges under this law may be paid in United States currency.

91. The register and receiver will, at the close of each month, forward to this office an abstract of mining applications filed, and a register of receipts, accompanied with an abstract of mineral lands sold, and an abstract of adverse claims filed.

92. The fees and purchase-money received by registers and receivers must be placed to the credit of the United States in the receiver's monthly and quarterly account, charging up in the disbursing account the sums to which the register and receiver may be respectively entitled as fees and commissions, with limitations in regard to the legal maximum.

HEARINGS TO ESTABLISH THE CHARACTER OF LANDS.

93. Sec. 2335 provides that all affidavits required under this chapter may be verified before *any* officer authorized to administer oaths within the land district where the claims may be situated, and all testimony and proofs may be taken before any such officer, and when duly certified by the officer taking the same shall have the same force and effect as if taken before the register and receiver of the Land Office.

94. In cases of contest as to the mineral or agricultural character of land, the testimony and proofs may be taken, as hereinbefore provided, on personal notice of at least ten days to the opposing party, or if such party cannot be found, then by publication of notice for at least once a week for thirty days, in a newspaper to be designated by the register of the land office as published nearest to the location of such land, and the register shall require proof that such notice has been given.

95. Testimony for the purpose of disproving the mineral character of the lands may be taken before any officer authorized to administer oaths within the land district, and where the residence of the parties who claim the land to be mineral is known, such evidence may be taken without publication, ten days after the mineral claimants or affiants shall have been personally notified of the time and place of such hearing; but in cases where such affiants or claimants

cannot be served with personal notice, or where the land applied for is returned as mineral upon the township plat, or where the same is now or may hereafter be suspended for non-mineral proof, by order of this Office, then the party who claims the right to enter the land as agricultural will be required, at his own expense, to publish a notice once each week for five consecutive weeks in the newspaper of largest circulation published in the county within which said land is situated, or if no newspaper is published within such county, then in a newspaper published in an adjoining county; the newspaper in either case to be designated by the register; which notice must be clear and specific, giving the name and address of the claimant, the designation of the subdivision embraced by his filing, the names of any miners or mining companies whose claims or improvements are upon the land or in the immediate vicinity thereof, the names of the parties who filed the affidavits that the land is mineral, and finally the notice should name a day, which shall not be less than thirty days from the date of the first insertion of said notice in such newspaper, upon which testimony will be taken to determine the facts as to the mineral or non-mineral character of the land. The notice must also state before what officer such hearing will be held, and the place of such hearing. A copy of this notice must be posted in a conspicuous place, upon each forty-acre subdivision claimed, during the publication of the notice, proof of which must be made under oath by at least two persons, who will state when the notice was posted and where posted.

96. At the hearing there must be filed the affidavit of the publisher of the paper that the said notice was published for the required time, stating when and for how long such publication was made, a printed copy thereof to be attached and made a part of the affidavit. In every case where practicable, in addition to the foregoing, *personal* notice must be served upon the mineral affiants, and upon any parties who may be mining upon or claiming the land.

97. At the hearing the claimants and witnesses will be thoroughly examined with regard to the character of the land; whether the same has been thoroughly prospected; whether or not there exists within the tract or tracts claimed any lode or vein of quartz or other rock in place, bearing gold, silver, cinnabar, lead, tin, or copper, or other valuable deposit, which has ever been claimed, located, recorded, or worked; whether such work is entirely abandoned, or whether occasionally resumed; if such lode does exist, by whom claimed, under what designation, and in which subdivision of the land it lies; whether any placer mine or mines exist upon the land; if so, what is the character thereof—whether of the shallow surface description, or of the deep cement, blue lead, or gravel deposits; to what extent mining is carried on when water can be obtained, and what the facilities are for obtaining water for mining purposes; upon what particular ten-acre subdivisions mining has been done, and at what time the land was abandoned for mining purposes, if abandoned at all.

98. The testimony should also show the agricultural capacities of the land, what kind of crops are raised thereon, and the value thereof; the number of acres actually cultivated for crops of cereals or vegetables, and within which particular ten-acre subdivisions such crops are raised; also, which of these subdivisions embrace his improvements, giving in detail the extent and value of his improvements, such as house, barn, vineyard, orchard, fencing, etc.

99. It is thought that bona fide settlers upon lands really agricultural will be able to show, by a clear, logical, and succinct chain of evidence, that their claims are founded upon law and justice; while parties who have made little or no permanent agricultural improvements, and who only seek title for speculative purposes, on account of the mineral deposits known to themselves to be contained in the land, will be defeated in their intentions.

100. The testimony should be as full and complete as possible; and in addition

to the leading points indicated above, everything of importance bearing upon the question of the character of the land should be elicited at the hearing.

101. Where the testimony is taken before an officer who does not use a seal, other than the register and receiver, the official character of such officer must be attested by a clerk of a Court of Record, and the testimony transmitted to the register and receiver, who will thereupon examine and forward the same to this Office, with their joint opinion as to the character of the land as shown by the testimony.

102. When the case comes before this Office such an award of the land will be made as the law and the facts may justify; and in cases where a survey is necessary to set apart the mineral from the agricultural land in any forty-acre tract, the necessary instructions will be issued to enable the agricultural claimant, *at his own expense*, to have the work done, at his option, either by United States deputy, county, or other local surveyor; the survey in such case may be executed in such manner as will segregate the portion of land actually containing the mine, and used as surface ground for the convenient working thereof, from the remainder of the tract, which remainder will be patented to the agriculturist to whom the same may have been awarded, subject, however, to the condition that the land may be entered upon by the proprietor of any vein or lode for which a patent has been issued by the United States for the purpose of extracting and removing the ore from the same, where found to penetrate or intersect the land so patented as agricultural, as stipulated by the mining act.

103. Such survey when executed must be properly sworn to by the surveyor, either before a notary public, officer of a Court of Record, or before the register or receiver, the deponent's character and credibility to be properly certified to by the officer administering the oath.

104. Upon the filing of the plat and field-notes of such survey, duly sworn to as aforesaid, you will transmit the same to the Surveyor-General for his verification and approval; who, if he finds the work correctly performed, will properly mark out the same upon the original township plat in his office, and furnish authenticated copies of such plat and description both to the proper local land office and to this Office, to be affixed to the duplicate and triplicate township plats respectively.

105. In cases where a portion of a forty-acre tract is awarded to an agricultural claimant, and he causes the segregation thereof from the mineral portion, as aforesaid, such agricultural portion will not be given a numerical designation as in the case of surveyed mineral claims, but will simply be described as the "Fractional ——— quarter of the ——— quarter of section ———, in township ———, of range ———, meridian, containing ——— acres, the same being exclusive of the land adjudged to be mineral in said forty-acre tract."

106. The surveyor must correctly compute the area of such agricultural portion, which computation will be verified by the Surveyor-General.

107. After the authenticated plat and field-notes of the survey have been received from the Surveyor-General, this Office will issue the necessary order for the entry of the land, and in issuing the receiver's receipt and register's patent certificate you will invariably be governed by the description of the land given in the order from this Office.

108. The fees for taking testimony and reducing the same to writing, in these cases, will have to be defrayed by the parties in interest. Where such testimony is taken before any other officer than the register and receiver, the register and receiver will be entitled to no fees.

109. If, upon a review of the testimony at this Office, a ten-acre tract should be found to be properly mineral in character, that fact will be no bar to the execution of the settler's legal right to the remaining *non-mineral* portion of his claim, if contiguous.

110. No fear need be entertained that miners will be permitted to make entries of tracts ostensibly as mining claims, which are not mineral, simply for the purpose of obtaining possession and defrauding settlers out of their valuable agricultural improvements; it being almost an impossibility for such a fraud to be consummated under the laws and regulations applicable to obtaining patents for mining claims.

111. The fact that a certain tract of land is decided upon testimony to be mineral in character, is by no means equivalent to an award of the land to a miner. A miner is compelled by law to give sixty days' publication of notice, and posting of diagrams and notices, as a preliminary step; and then, before he can enter the land, he must show that the land yields mineral; that he is entitled to the possessory right thereto in virtue of compliance with local customs or rules of miners, or by virtue of the Statute of Limitations; that he or his grantors have expended, in actual labor and improvements, an amount of not less than five hundred dollars thereon, and that the claim is one in regard to which there is no controversy or opposing claim. After all these proofs are met, he is entitled to have a survey made at his own cost, where a survey is required, after which he can enter and pay for the land embraced by his claim.

J. A. WILLIAMSON,
Commissioner.

[NOTE.—The following Opinion was not received in time to incorporate in the body of the work. See reference to it, and decision of the Supreme Court of Nevada in the same case, which is affirmed, on pages 67-74, Sec. 35.]

SUPREME COURT OF THE UNITED STATES, } No. 878.
OCTOBER TERM, 1876.

Solomon Heydenfeldt, Plaintiff in Error,
vs.
The Daney Gold and Silver Mining Company.
} In error to the Supreme Court of the State of Nevada.

Mr. Justice Davis delivered the opinion of the Court.

This is an action of ejectment to recover a specific portion of the west half of the southwest quarter of section sixteen, township sixteen, range twenty-one east, in Lyon County, Nevada. The land in controversy is rich in minerals, and was not surveyed by the United States until the year 1867. Prior to the date of the survey, or the approval of it, the defendant's grantors and predecessors in interest had for mining purposes entered upon the land, and claimed and occupied it according to the mining laws and the custom of miners in the locality. This possession and claim of ownership have been continuous and uninterrupted, and the defendant has expended over eighty thousand dollars in the construction of improvements for carrying on the business of mining on the land.

The plaintiff claims title from the State by patent. It is dated the 14th day of July, 1868, and was issued on the assumption that sections sixteen and thirty-six, whether surveyed or unsurveyed, and whether containing minerals or not, were granted to the State for the support of common schools, by the seventh section of the Nevada Enabling Act, approved March 21st, 1864, 13 Stat. 32.

This interpretation of that act is denied by the General Government, and the defendant has a patent of the 2d of March, 1874, from the United States for the land in controversy, issued in conformity with the laws of Congress on the subject of mining. Which is the better title is the point for decision. It has been the settled policy of the Government to promote the development of the mining resources of the country, and as mining is the chief industry in Nevada, the question presented for decision is of great interest to the people of that State.

Opinion of the Supreme Court.

[NOTE.—The following Opinion was not received in time to incorporate in the body of the work. See reference to it, and decision of the Supreme Court of Nevada in the same case, on pages 67–74, Sec. 35.]

SUPREME COURT OF THE UNITED STATES, } No. 878.
OCTOBER TERM, 1876.

Solomon Heydenfeldt, Plaintiff in Error,
      vs.
The Daney Gold and Silver Mining Company.

} In error to the Supreme Court of the State of Nevada.

Mr. Justice Davis delivered the opinion of the Court.

This is an action of ejectment to recover a specific portion of the west half of the southwest quarter of section sixteen, township sixteen, range twenty-one east, in Lyon County, Nevada. The land in controversy is rich in minerals, and was not surveyed by the United States until the year 1867. Prior to the date of the survey, or the approval of it, the defendant's grantors and predecessors in interest had for mining purposes entered upon the land, and claimed and occupied it according to the mining laws and the custom of miners in the locality. This possession and claim of ownership have been continuous and uninterrupted, and the defendant has expended over eighty thousand dollars in the construction of improvements for carrying on the business of mining on the land.

The plaintiff claims title from the State by patent. It is dated the 14th day of July, 1868, and was issued on the assumption that sections sixteen and thirty-six, whether surveyed or unsurveyed, and whether containing minerals or not, were granted to the State for the support of common schools, by the seventh section of the Nevada Enabling Act, approved March 21st, 1864, 13 Stat. 32.

This interpretation of that act is denied by the General Government, and the defendant has a patent of the 2d of March, 1874, from the United States for the land in controversy, issued in conformity with the laws of Congress on the subject of mining. Which is the better title is the point for decision. It has been the settled policy of the Government to promote the development of the mining resources of the country, and as mining is the chief industry in Nevada, the question presented for decision is of great interest to the people of that State.

The seventh section of that act is as follows: "That sections numbered sixteen and thirty-six in every township, and where such sections have been sold or otherwise disposed of by any act of Congress, other lands equivalent thereto, in legal subdivisions of not less than one quarter-section, and as contiguous as may be, shall be and are hereby granted to said State for the support of common schools."

It is true that there are words of present grant in this law, but in construing it we are not to look at any single phrase in it, but to its whole scope, in order to arrive at the intention of the makers of it. "It is better always," says Sharswood, Judge, "to adhere to a plain, common-sense interpretation of the words of

[420]

a statute, than to apply to them refined and technical rules of grammatical construction." (Gyges' Estate, 65 Pa. St. 312.)

If a literal interpretation of any part of it would operate unjustly or lead to absurd results, and be contrary to the evident meaning of the act taken as a whole, it will be rejected. And there is no better way of discovering the true meaning of a law, when there are expressions in it which are rendered ambiguous by their connection with other clauses, than by considering the necessity for it and the causes which induced the legislature to pass it. With these rules as our guide it is not difficult, we think, to give a true construction to the law in controversy.

Congress, at the time, was desirous that the people of the Territory of Nevada should form a State government and come into the Union. The terms on which this admission could be obtained were proposed, and, as was customary in the enabling acts for new States, the particular sections of the public lands to be donated to the State for the use of common schools were specified. These sections had not been surveyed, nor had Congress then made, or authorized to be made, any disposition of the public lands within the Territory of Nevada.

But this condition of things did not stand in the way of Congress making proper provision on the subject. Some provision was necessary in order to place Nevada in this respect on an equal footing with States recently admitted. But the people were not interested in getting the identical sixteenth and thirty-sixth sections in every township. Indeed, it could not be known until after survey where these sections would fall, and a grant of quantity put Nevada in as good a condition as other States, which had received the benefit of this bounty. A grant operating at once and attaching prior to the surveys by the United States, would deprive Congress of the power of disposing of any part of the public domain until there was a segregation by survey of the land granted. In the meantime further improvements would be arrested, and the persons who before the surveys were made had occupied and improved the country would lose their possessions and labor, in case it turned out that they had settled upon the granted lands. Congress was fully advised of the condition of a new community like Nevada; of the evil effects of such legislation upon its prosperity, and of all antecedent legislation upon the subject of the public lands within the bounds of the proposed new State. In the light of this information, and surrounded by these circumstances, Congress made the grant in question. That it is ambiguous is very clear, for the different parts of it cannot be reconciled, if the words used are to receive their usual meaning. Schulenberg v. Harriman, 21 Wallace, 44, establishes the rule that "unless there are other clauses in a statute restraining the operation of words of present grant, these must be taken in their natural sense." This is a correct rule, and we do not seek to depart from it, but there are words of qualification in this grant.

And these words restrict the operation of the words of present grant. If their literal meaning be taken, they refer to past transactions; but evidently they were not used in this sense, for there had been no lands in Nevada sold or disposed of by any act of Congress, and why indemnify the State against a loss that could not occur? There could be no loss, and there was no occasion of making provision for substituted lands if the grant took effect absolutely on the admission of the State into the Union, and the title to the lands then vested in the State. Congress cannot be supposed to have intended a vain thing, and yet it is quite certain that the language of the qualification was intended to protect the State against a loss that might happen through the action of Congress in selling or disposing of the public domain. It could not, as we have seen, apply to past sales or dispositions, and to have any effect at all must be held to apply to the future.

This interpretation, although seemingly contrary to the letter of the statute, is within its reason and spirit. It accords with a wise public policy, gives to Nevada all she had any right to ask for, and acquits Congress of passing a law which in its effects would be unjust to the people of the Territory. Besides, no other construction is consistent with the statute as a whole, and this alone answers the evident intention which the makers of it had in view, and this was to grant to the State *in præsenti* a quantity of lands equal in amount to the sixteenth and thirty-sixth sections, the grant to take effect when the status of the lands was fixed by survey and they were capable of identification. Congress, however, reserved until this was done the power of disposition, and if in the exercise of this power the whole or any part of a sixteenth or thirty-sixth section had been disposed of, the State was to be compensated by other lands equal in quantity and as near as may be in quality. By this means the State was indemnified against loss, and the people ran no risk of losing the labor of years. While the State suffered no injury, Congress was left free to dispose of the public domain in any way it saw fit, to promote the interests of the people.

It is argued that, conceding the construction given this grant to be correct, this defense cannot be sustained, because the land in controversy was not actually sold by direction of Congress until after the survey. This position ignores a familiar rule in the construction of statutes, that they must be so construed as to admit all parts of them to stand if possible. (1 Bouvier's Institutes, p. 42, Sec. 7.)

The language used is, "sold or otherwise disposed of by an act of Congress," and the point made by the plaintiff would reject a part of these words from the statute.

To limit the qualification to the grant in this way would defeat one of the main purposes Congress had in view. Congress knew, as did the whole country, that Nevada was possessed of great mineral wealth, and that mineral lands should be disposed of differently from those which were fit only for agriculture. No method for doing this had then been provided, but Congress said to the people of the Territory, "You shall, if you decide to come into the Union, have for the use of schools a quantity of land equal to two sections in every township, and the identical sections themselves, if on survey no one else has any claim to them, but until this decision is made and the lands surveyed we reserve the right either to sell them or dispose of them in any other way that commends itself to our judgment." This right of disposition is subject to no limitations, and the wisdom of not surrendering it is apparent. The whole country is interested in the development of its mineral wealth, and to accomplish this object adequate protection was required for those engaged in this business. This protection was furnished by the Act of Congress of July 26th, 1866, (14 U. S. Stats. p. 251) which was passed before the land in controversy was surveyed. This act disposes of the mineral lands of the United States to actual occupants and claimants, and provides a method for the acquisition of title from the United States. And these defendants occupied the land prior to the survey and were entitled to purchase, and the patent subsequently obtained from the Government relates back to the time of the original location and entry, and perfects their title.

These views dispose of this case, but there is another ground equally conclusive. Congress, on the 4th of July, 1866, (14 Stat. p. 85) passed an act concerning lands granted to the State of Nevada, and, among other things, reserved from sale all mineral lands in the State, and authorized the lines of surveys to be changed from rectangular, so as to exclude them. This was, doubtless, intended by Congress as a construction of the grant in this case; but whether that construction be correct or not, and whatever may be the effect of the grant in its original shape, it was clearly competent for the grantee to accept it in its modified form, and agree to any construction put upon it by the grantor. The State, through its legislature, (see Act of February 13th, 1867) ratified the construction given to it by Congress, and accepted it with the conditions annexed.

We agree with the Supreme Court of Nevada that this acceptance "was a recognition by the legislature of the State of the validity of the claim made by the Government of the United States to the mineral lands."

It is objected that the constitution of Nevada inhibits such legislation, but the Supreme Court of the State, in the case we are reviewing, held that it did not, (10 Nevada Reports, p. 314) and we think their reasoning on this subject is conclusive.

We see no error in the record, and the judgment is affirmed.

Table of Cases and Statutes Cited.

NOTE.—In this Table of Cases the following abreviations are used :

D. C........................Decision of Commissioner of General Land Office.
D. A. C..Decision of Acting Commissioner.
D. S..Decision of Secretary of Interior.
D. A. S..Decision of Acting Secretary.
D. A. G............Decision or Opinion of Attorney-General of United States.
D. A. A. G.................Decision or Opinion of Assistant Attorney-General.

The reference is to the page of the volume.

A.

Ah Yew v. Choate, 24 Cal. 562, pp. 43, 331.
Ajax, or Big Indian Lode, In re, D. C. p. 226.
Alford v. Barnum, 45 Cal. 482, pp. 76, 77.
Alger Lode, In re, D. C. pp. 226, 227.
American Company v. Bradford, 27 Cal. 360, p. 265.
Antelope Lode, In re, D. S. p. 223.
Atchison v. Peterson, 20 Wall. 510, pp. 279, 280.
Attorney-General's Opinions—
 Feb. 11th, 1862, p. 2.
 Sept. 30th, 1870, p. 203.
 July 21st, 1871, pp. 9, 31, 211.
 Aug. 7th, 1871, p. 98.
 July 21st, 1871, p. 93.
 Nov. 24th, 1871, p. 170.
Ayers v. Foley, D. S. p. 223.

B.

Bagnell v. Broderick, 13 Pet. 436, pp. 109, 179, 271.
Ballancer v. Forsyth, 13 How. 18, p. 179.
Bank of Commerce Lode, D. A. C. p. 219.
Bank of the U. S. v. Deveaux, 5 Cranch, 84, p. 92.
Barnard's Heirs v. Ashley's Heirs, 18 How. 43, pp. 371, 372.
Barry v. Gamble, 8 Mo. 88, p. 139.
Basey v. Gallagher, 20 Wall. 685, pp. 283, 284.
Bates v. Chambers, D. C. p. 192.
Bealey v. Shaw, 6 East, 208, p. 272.
Bear River & A. W. & M. Co. v. N. Y. M. Co. 8 Cal. 327, p. 265.
Beard v. Federy, 3 Wall. 479, p. 179.
Becker v. Central City, Colorado, D. C. pp. 307, 308.
Beckner v. Coates, D. A. C. p. 91.
Bigelow v. Willson, 1 Pick. 485, p. 164.
Bissell v. Bissell, 11 Barb. 96, p. 164.
Blanchard v. Sprague, 3 Sum. 535, p. 269.
Boston Quicksilver Mine, In re, D. A. S. p. 149.
Brashear v. Mason, 6 How. 92, p. 138.
Broder v. Natoma W. & M. Co. 50 Cal. 621, p. 261.
Brown v. Lewis, D. C. p. 176.
 v. Quartz M. Co. 15 Cal. 155, p. 87.
Brunswick Mine, D. C. pp. 114, 374.
Brush v. Ware, 15 Pet. 93, p. 179.
Burr v. Lewis, 6 Tex. 76, p. 64.
Butte Canal & D. Co. v. Vaughn, 11 Cal. 143, p. 265.
Butte Table Mt. Co. v. Morgan, 19 Cal. 609, p. 265.

C.

Cal. & Oregon R. R. In re, D. C. 228, p. 338.
Cann. v. Warren, 1 Houst. 188, p. 164.
Carleton v. Byington, 16 Iowa, 588, p. 164.
Carron v. Curtis, D. C. p. 302.
Carothers v. Wheeler, 1 Oregon, 194, p. 164.
Cascade Lode, In re, p. 162.
Central Pacific Railroad Co. v. Mammouth Blue Gravel Co. D. C. & D. S. p. 375.

[425]

Cerro Bonito Quicksilver Mine, D. S. p. 358.
Chambers v. Pitt, D. S. p. 207.
Chicago & C. C. G. & S. M. Co. In re, D. A. C. p. 140.
Chouteau v. Moloney, 16 How. 203, pp 45, 46.
City Rock, and Utah Claimants v. Pitts, D. C. p. 98.
Clark v. Calkins, D. C. p. 221.
  v. Ellis, D. S. p. 332.
Clear Creek Q. Mine, D. C. p. 9.
Cole v. Cole, 33 Me. 542, p. 195.
Colman v. Clements, 23 Cal. 245, p. 266.
Columbia M. Co. v. Holter, 1 Mont. 296, p. 289.
Comegys v. Vasse, 1 Peters, 212, p. 371.
Consolidated Channel Co. v. Central P. R. R. 51 Cal. 269, p. 262.
Coml. & R. R. Bank of Vicksburg v. Slocum, 14 Pet. 60, p. 92.
Commissioner of the General Land Office, Decisions of—
  June 6th, 1868, p. 8.
  Aug. 15th, 1868, p. 358.
  Aug. 27th, 1868, p. 3.
  Sept. 1st, 1868, p. 93.
  Jan. 21st, 1869, p. 178.
  Jan. 28th, 1869, pp. 7, 185, 374.
  July, 1869, p. 33.
  Aug. 15th, 1869, p. 358.
  Aug. 25th, 1869, p. 80.
  Nov. 6th, 1869, p. 102.
  Nov. 20th, 1869, p. 261.
  Dec. 10th, 1869, p. 370.
  Jan. 14th, 1870, p. 223.
  March, 8th, 1870, p. 291.
  April 15th, 1870, p. 112.
  April 17th, 1870, p. 76.
  April 18th, 1870, p. 132.
  May 24th, 1870, p. 67.
  Aug. 17th, 1870, p. 121.
  Aug. 27th, 1870, p. 241.
  Sept. 11th, 1870, p. 159.
  Sept. 14th, 1870, p. 178.
  March 1st, 1871, p. 338.
  March 14th, 1871, p. 77.
  April 16th, 1871, p. 258.
  June 7th, 1871, p. 98.
  Aug. 4th, 1871, p. 106.
  Aug. 25th, 1871, p. 59.
  Aug. 26th, 1871, p. 241.
  Oct. 21st, 1871, pp. 301, 337.
  Nov. 24th, 1871, p. 319.
  Dec. 2d, 1871, p. 319.
  Dec. 7th, 1871, p. 319.
  Dec. 29th, 1871, p. 227.
  Jan. 22d, 1872, p. 319.
  Jan. 24th, 1872, p. 331.
  Feb. 12th, 1872, p. 80.
  Feb. 23d, 1872, p. 377.
  Feb. 27th, 1872, p. 178.
  March 11th, 1872, pp. 319, 321, 325.
  March 26th, 1872, pp. 319, 326.
  March, 27th, 1872, pp. 32, 33.
  April 26th, 1872, p. 319.
  Aug. 6th, 1872, p. 246.
  Aug. 19th, 1872, p. 304.
  Aug. 21st, 1872, p. 367.

Commissioner, Decisions of—Cont'd.
  Aug. 27th, 1872, p. 115.
  Sept. 9th, 1872, p. 118.
  Sept. 14th, 1872, p. 120.
  Sept. 20th, 1872, p. 113.
  Nov. 12th, 1872, p. 375.
  Dec. 10th, 1872, p. 200.
  Dec. 26th, 1872, p. 140.
  Jan. 1st, 1873, p. 115.
  Jan. 22d, 1873, p. 175.
  Jan. 30th, 1873, p. 377.
  Feb. 3d, 1873, p. 96.
  April 14th, 1873, p. 377.
  April 16th, 1873, p. 255.
  April 18th, 1873, p. 81.
  May 19th, 1873, p. 240.
  May 20th, 1873, pp. 103, 255.
  June 17th, 1873, p. 106.
  June 26th, 1873, p. 38
  July 10th, 1873, p. 79.
  July 15th, 1873, pp. 79, 100.
  July 23th, 1873, p. 148.
  July 30th, 1873, p. 350.
  Sept. 11th, 1873, p. 96.
  Sept. 25th, 1873, p. 109.
  Oct. 8th, 1873, p. 25.
  Oct. 23d, 1873, p. 147.
  Oct. 31st, 1873, p. 226.
  Nov. 11th, 1873, p. 301.
  Nov. 18th, 1873, p. 102.
  Dec. 11th, 1873, p. 377.
  Jan. 6th, 1874, pp. 140, 159.
  April 20th, 1874, pp. 119, 120.
  April 27th, 1874, p. 80.
  July 21st, 1874, pp. 163, 222.
  Oct. 23d, 1874, pp. 79, 80.
  Nov. 3d, 1874, p. 349.
  Dec. 2d, 1874, pp. 119, 120.
  Dec. 14th, 1874, p. 98.
  Dec. 17th, 1874, p. 176.
  Jan. 30th, 1875, pp. 79, 80, 121.
  Feb. 11th, 1875, p. 100.
  March 11th, 1875, p. 115.
  June 28th, 1875, p. 79.
  July 29th, 1875, p. 47.
  Aug. 4th, 1875, pp. 309, 339.
  Aug. 14th, 1875, p. 329.
  Aug. 17th, 1875, p. 144.
  Oct. 21st, 1875, p. 254.
  Oct. 28th, 1875, p. 144.
  Nov. 5th, 1875, p. 57.
  Dec. 1st, 1875, p. 377.
  Dec. 3d, 1875, p. 79.
  Jan. 3d, 1876, p. 335.
  Jan. 27th, 1876, p. 154.
  March 7th, 1876, p. 162.
  March 25th, 1876, p. 247.
  June 2d, 1876, p. 376.
  June 10th, 1876, p. 174.
  June 13th, 1876, p. 125.
  June 14th, 1876, p. 349.
  June 21st, 1876, p. 301.
  July 18th, 1876, p. 96.
  July 21st, 1876, p. 39.
  Aug. 25th, 1876, p. 377.
  Aug. 28th, 1876, p. 174.
  Nov. 23d, 1876, p. 305.
  Jan. 4th, 1877, p. 167.
Cooper v. Roberts, 18 How. 173, pp. 44 65, 66.

TABLE OF CASES.

Cornell v. Moulton, 3 Denio, 12, p. 164.
Corning v. Troy Iron & Nail Factory, 40 N. Y. 191, p. 272.
Corning Tunnel Mining & Reduction Co. v. Bell, D. S. 110.
Corning Tunnel Mining & Reduction Co. v. Pell, D. S. 185.
Cotton v. U. S. 11 How. 229, pp. 44, 271.
Cousin v. Blanc's Executors, 19 How. 202, p. 371.
Covington Drawbridge Co. v. Shepherd, 20 How. 233, p. 92.
Craig v. Leslie, 3 Wheat. 563, pp. 97. 173.
  v. Radford, 3 Wheat. 594, p. 143.
Crismon v. U. P. R. R. Co. D. C. p. 351.
Crooker v. Bragg, 10 Wend. 260, p. 272.
Cross v. DeValle, 1 Wall. 1, pp. 97, 173.
Crown Point Lode, D. S. pp. 173–177.
C. T. M. Co. v. Bell, D. S. p. 206.
Cunningham v. Ashley, 14 How. 377, pp. 139, 370.

D.

Daney G. & S. M. Co. v. Sapphire M. Co. D. S. p. 178.
Daniel Ball, The, 10 Wall. 557, p. 374.
Daniel Peters Lode, D. C. p. 178.
Dardanelles M. Co. v. Cal. M. Co. D. A. S. p. 185.
  v. Bosphorus Lode, D. C. p. 186.
Dartmouth College v. Woodward, 4 Wheat. 636, p. 91.
Davenport v. Lamb, 13 Wall. 418, p. 179.
Davis v. Fuller, 12 Vt. 190, p. 272.
Decatur v. Paulding, 14 Pet. 497, p. 138.
Delaney v. Thomas, D. C. p. 66.
Delogny v. Rentoul, 2 Mart. 175, p. 195.
Doe v. Beebe, 13 How. 25, p. 375.
Doe v. Eslava, 9 How. 421, p. 371.
Doll v. Meador, 16 Cal. 296, p. 64.
Dredge v. Forsyth, 2 Black. 563, p. 179.
Dunkirk Lode, D. C. 31.
Dutch Flat Placer Cañon Claim, D. C. p. 147.

E.

Earl Mine v. Mt. Pleasant Mine, D. A. S. pp. 149, 219.
Easton v. Salisbury, 21 How. 426, p. 63.
Eddy v. Simpson, 3 Cal. 249, p. 264.
Elliot v. Fitchburg R. R. Co. 10 Cush. 193, p. 272.
Embrey v. Owen, 6 Ex. 353, p. 272.
Empire Mining Co. 1 D. C. pp. 146, 147.
Equator M. & Smelting Co. v. Marshall S. M. Co., D. A. C., D. S., and D. A. G. p. 187.
Equator Lode, D. C. pp. 171, 252.
Eureka Lode, In re, 186.
Eureka M. Co. v. Jenny Lind Co. D. A. A. G. p. 195.
Evans v. Randall, D. S. p. 197.
Ewing v. Hartman, D. C. p. 332.
Excelsior Lode, D. S. p. 186.

F.

Fairfax v. Hunter, 7 Cranch, 603, pp. 96, 143.
Fairmount Lode and Mill Site, D. C. p. 221.
Farwell v. Rogers, 4 Cush. 460, p. 64.
Fenn v. Holme, 21 How. 481, p. 339.
Feniau Star Lode, D. C. p. 221.
Field v. Seabury, 19 How. 323, p. 179.
Finley v. Williams, 9 Cranch, 164, p. 372.
Finney v. Berger, 50 Cal. 248, p. 64.
Flagstaff Case, pp. 22, 24, 162, 170, 167–226.
Fletcher v. Peck, 6 Cranch, 87, pp. 42, 691.
Foley v. Harrison, 15 How. 447, p. 64.
Foscalina v. Doyle, 47 Cal. 437, p. 134.
Four Twenty M. Co. v. Bullion M. Co. D. S. 3 Sawy. 634, pp. 154, 188, 197, 208, 225.
French v. Fyan, 3 Otto, 119, p. 134.
Fuller v. Hampton, 5 Conn. 416, p. 195.

G.

Gaines v. Nicholson, 9 How. 365, p. 66.
  v. Thompson, 7 Wall. 352, p. 138.
Galloway v. Finley, 12 Pet. 264, p. 179.
Gardner v. Newburgh, 2 Johns. Ch. 166, p. 272.
Garland v. Wynn, 20 How. 6, pp. 371, 372.
Gibson v. Chouteau, 13 Wall. 92, p. 139.
Golconda Mine, D. C. p. 256.
Gold Hill Q. M. Co. v. Ish, 5 Oregon, 104, pp. 5, 40, 328, 329.
Gorst v. Lowndes, 11 Sim. 434, p. 164.
Goodtitle v. Kibbe, 9 How. 471, p. 375.
Gorham v. Wing, 10 Mich. 486, p. 164.
Gould v. Conde Lode, D. C. p. 13.
Gouverneur's Heirs v. Robertson, 11 Wheat. 332, pp. 97, 143.
Green v. Liter, 8 Cranch, 229, p. 372.
Gregg v. Tesson, 1 Black, 150, p. 179.
Griffith v. Bogert, 18 How. 162, p. 164.
Grogan v. Knight, 27 Cal. 517, p. 66.
Gus Belmont Lode, D. C. pp. 13, 103.

H.

Hall v. Litchfield, D. A. C. p. 50.
Harris v. Shoutz, p. 259.
Harris Lode, In re, D. C. p. 225.
Hawley Consolidated M. Co. v. Memnon, D. S. p. 190.

TABLE OF CASES.

Helmic Mine, In re, D. C., D. A. S. p. 31.
Henrietta Lode, In re, p. 186.
Henshaw v. Bissell, 18 Wall. 255, p. 149.
Hercules Lode and Seven Thirty, In re, D. C. pp. 20, 140, 146, 174, 180.
Hestres v. Brennan, 50 Cal. 211, p. 374.
Heydenfelt v. Dancy G. & S. M. Co. 10 Nov. 290, p. 74.
Hidden Treasure Lode, D. C. p. 48.
Higgins v. Houghton, 25 Cal. 252, pp. 44, 62, 64, 66, 69.
Hobart v. Ford, 6 Nev. 77, p. 287.
Hill v. King, 8 Cal. 336, p. 265.
   v. Smith, 27 Cal. 483, pp. 265, 280.
Hoffman v. Stone, 7 Cal. 49, pp. 264, 287.
Holland v. Gulielmi, D. C. p. 358.
Hoofnagle v. Anderson, 7 Wheat. 212, p. 179.
Hooper v. Scheimer, 23 How. 235, p. 179.
Hosmer v. Wallace, 47 Cal. 461, p. 374.
How v. Missouri, 12 How. 126, p. 66.
Huff v. Doyle, 3 Otto, 558, p. 59.
Hunt v. Wickliffe, 2 Pet. 201, p. 372.

I.

Idaho Lode, In re, D. C. p. 140.
Inimitable Co. In re, D. C. p. 374.
Instructions of Land Department—
  Jan. 14th, 1867, pp. 5, 11, 12, 15, 21, 23, 27, 257, 313.
  June 25th, 1867, pp. 19, 36, 37.
  May 16th, 1868, pp. 310, 311.
  July, 1869, p. 36.
  July 25th, 1870, p. 369.
  Aug. 8th, 1870, pp. 6, 15, 16, 17, 95, 236, 239.
  June 8th, 1870, p. 183.
  May 6th, 1871, pp. 76, 235, 315, 327.
  Aug. 3d, 1871, p. 98.
  Sept. 7th, 1871, p. 96.
  March 20th, 1872, p. 326.
  March 26th, 1872, p. 96.
  June 10th, 1872, pp. 1, 9, 83, 99, 104, 105, 108, 111, 112, 118, 122, 128, 129, 130, 131, 132, 164, 228.
  April 15th, 1873, pp. 342, 343, 344, 345, 346, 347, 348, 349.
  Aug. 11th and 14th, 1873, p. 352.
  Nov. 20th, 1873, pp. 124, 127, 158.
  Nov. 29th, 1875, p. 354.
  Dec. 1st, 1875, p. 354.
  June 9th, 1874, 354.
  Feb. 19th, 1875, p. 246.
  March 11th, 1875, p. 120.
  June 17th, 1875, 303.
  Feb. 1st, 1877, pp. 63, 97, 104, 105, 108, 109, 161, 303, 326, 367, 368.
Irvine v. Marshall, 20 How. 558, pp. 139, 271.
Irwin v. Phillips, 5 Cal. 140, p. 264.

J.

Jackson v. Beach, 1 Johns. Cas. 401, pp. 96, 143.
Jefferson M. Co. v. Penn. M. Co. D. C. pp. 153, 163, 222.
Jenny Lind M. Co. v. Eureka M. Co. D. S. pp. 163, 164, 186, 192.
Johnson v. Towsley, 13 Wall. 72, pp. 109, 179, 371.
   v. Jordan, 2 Met. 239, p. 272.
Jones & Matteson Lode, D. C. p. 220.
Josephs v. U. S. 1 Nott. & H. 197, p. 38.
Jourdan v. Barrett, 4 How. 185, p. 271.
Judd v. Fulton, 10 Barb. 117, p. 164.
Julia Gold and Silver M. Co. D. C.; D. S. p. 250.

K.

Kansas Lode, D. C. p. 147.
Kelly v. Taylor, 23 Cal. 14, p. 158.
Kelsey Lode, D. C. p. 8.
Kempton Mine, D. C. & D. S. pp. 90, 141, 142, 152, 156, 161, 179, 227.
Kendall v. U. S. p. 108.
Kernan v. Griffith, 27 Cal. 87, p. 137.
Keystone Case, D. C. D. S., pp. 64, 65, 349.
Kidd v. Laird, 15 Cal. 161, pp. 264, 265.
Kimball v. Gearhart, 12 Cal. 27, p. 266.
Kimm v. Osgood's Adm. 19 Mo. 60, p. 164.
King David Lode, In re, D. S. p. 186.
King of the West Lode, D. C. pp. 186, 196, 201, 223.
King of the West v. City Rock Lode, D. S., D. A. S. pp. 219, 220.
Kissell v. St. Louis Public School, 18 How. 19, p. 66.

L.

Lady Allen Lode, D. C. p. 143.
Lafayette's Heirs v. Kenton, 18 How. 197, p. 179.
Lake Quicksilver M. Co. D. C. p. 152.
Lang v. Phillips, 27 Ala. 311, p. 164.
Lessieur v. Price, 12 How. 59, p. 66.
Lindsey v. Haws, 2 Black, 554, pp. 139, 370, 372.
   v. Miller, 6 Pet. 672, p. 139.
Litchfield v. The Register and Receiver, Woolw. 299, pp. 138, 374.
Little Fred Mine, D. C. p. 121.
Live Oak Quartz Mine, D. A. C. p. 280.
Lobdell v. Simpson, 2 Nev. 274, p. 280.
Louisville R. R. Co. v. Letson, 2 How. 550, p. 92.
Lykens Valley Coal Co. v. Dock, 62 Penn. 231, St. p. 89.
Lyons v. Hunt, 11 Ala. 295, p. 164.
Lytle v. Arkansas, 22 How. 193, pp. 371, 372.

TABLE OF CASES. 429

M.

Magnolia M. Co. v. Magn. E. & W. Co. D. C. p. 227.
Maney v. Carter, 4 Conn. 635, p. 195.
Mann v. Wilson, 23 How. 458, p. 179.
Marshall v. B. & O. R. R. Co. 16 How. 327, p. 92.
Marvin v. Richmond, 3 Denio, 58, p. 195.
Mason v. Hill, 5 B. & Ad. 22, p. 272.
Magwire v. Tyler, 1 Black, 195, p. 371.
McArthur v. Browder, 4 Wheat. 488, pp. 179, 372.
McDonald v. Askew, 29 Cal. 200, p. 264.
    v. Bear R. Co. 13 Cal. 220, p. 264.
McGarrahan v. New Idria M. Co. 49 Cal. 333, p. 134.
McGarrity v. Byington, 12 Cal. 426, p. 266.
McGillivray v. Evans, 27 Cal. 92, p. 265.
McKenna v. Dillon, D. A. S. p. 335.
McKibben Lode, D. C. p. 13.
McKinney v. Smith, 21 Cal. 374, p. 265.
McLaughlin v. Powell, 50 Cal. 64, p. 75.
McMurdy v. Streeter, D. S. pp. 162, 186.
Melton v. Lambard, 51 Cal. 258, p. 219.
Middleton v. Low, 30 Cal. 596, p. 66.
Miller v. Kerr, 7 Wheat. 1, p. 179.
Mills v. Rolls & Ross, D. A. S. p. 332.
Minnesota v. Batchelder, 1 Wall. 109, p. 373.
Minter v. Crommelin, 18 How. 88, p. 57.
Mono Mine Co. v. Gisborn, D. C. and D. A. G. p. 192.
Montana Fluming & M. Co. D. C. p. 220.
Montana Lode, In re, D. C. p. 186.
Montello, The, 11 Wall. 411, p. 374.
Morrow v. Kingsbury, Cal. N. R. p. 64.
Morse v. Streeter, D. C. and D. S. p. 219.
Morton v. Greene, 21 Wall. 660, p. 50.
    v. Nebraska, 21 Wall. 660, p. 49.
Mount v. Bogart, Anth. 259, A. S. p. 149.
Mountain City Lode, In re, D. C. p. 226.
Mountain Tiger Lode, D. A. C. pp. 188, 367.
Mountjoy Lode, D. C. p. 28.
Mt. Pleasant Mine and Earl Mine, D. A. S. p. 149.

N.

Nagler's Application, D. C. p. 304.
Nelson's Lessee v. Moon, 3 McLean, 319, p. 179.
Nevada C. P. R. R. Co. D. C. p. 328.
Nevada Water Co. v. Powell, 34 Cal. 109, p. 264.
Newark Mill and M. Co. v. Meinke, D. S. p. 303.
New Idria Mine, McGarrahan's Case, D. A. S., D. A. A. G., D. A. G., pp. 22, 33, 152.

New Orleans v. DeArmas, 9 Pet. 223, p. 179.
    v. United States, 10 Pet. 662, p. 179.
Northern Light and Fair View Mine, D. C. pp. 150, 162.
Noteware v. Sterns, 1 Mont. 311, p. 289.

O.

Oliver v. Piatt, 3 How. 333, p. 179.
Omaha Gold Quartz Mine, D. A. C. p. 223.
Ophir S. M. Co. v. Carpenter, 4 Nev. 534, p. 285.
Orr v. Hodgson, 4 Wheat. 453, pp. 96, 143.
Ortman v. Dixon, 13 Cal. 34, p. 264.
Osterman v. Baldwin, 6 Wall. 116, pp. 96, 143.
Overman Silver M. Co. v. Dardanelles S. M. Co. D. C. and D. S. pp. 202, 368.
Owens v. Jackson, 9 Cal. 322, p. 64.

P.

Packer v. Heaton, 9 Cal. 568, p. 266.
Page v. Weymouth, 47 Me. 238, p. 164.
    v. Wiliams, 2 Dev. & B. 55, pp. 52, 72.
Parker v. Duff, 47 Cal. 554, pp. 134, 374.
Patterson v. Lynch, Sawy. N. R. p. 76.
    v. Tatum, 3 Sawy. 164, p. 179.
Pelican Lode, D. C. and D. S. p. 188.
Penn. Quartz Mines, D. C. pp. 153, 222.
People v. Shearer, 30 Cal. 645, p. 4.
    v. Stratton, 25 Cal. 242, p. 76.
    v. Williams, 35. Cal. 673, p. 88.
Philadelphia Lode v. Pride, D. C. pp. 156, 157.
Phœnix Water Co. v. Fletcher, 23 Cal. 481, p. 265.
Polk v. Wendal, 9 Cranch, 99, p. 57.
Pollard's Lessee v. Hagan, 3 How. 212, p. 374.
Pope v. Headen, 5 Ala. 433, p. 164.
Porcupine Mine, D. C. p. 191.
Prince of Wales Lode, D. S. p. 162.
Pugh v. Wheeler, 2 Dev. & B. 50, p. 272.
Pulliam v. Hunter, D. S. p. 334.

R.

Railroad v. Fremont, 9 Wall. 90, p. 66.
    v. Schurmier, 7 Wall. 272, pp. 272, 375.
    v. Smith, 9 Wall. 99, pp. 16, 66, 108.
Read v. Caruthers, 47 Cal. 181, p. 76.
Reeside v. Walker, 11 How. 272, p. 138.
Red Pine Mines, D. C. p. 201.
Red Warrior Lode, D. A. C. p. 25.
Reichart v. Felps, 6 Wall. 160, pp. 57, 63.

430 TABLE OF CASES.

Richardson v. Kier, 34 Cal. 63, p. 265.
Robinson v. Forrest, 29 Cal. 317, pp. 64-76.
Rockwell Lode, D. A. C. pp. 188, 367.
Rogers v. Cooney, 7 Nev. 213, p. 89.
Root v. Shields, 1 Woolw. 340, p. 179.
Rupley v. Welch, 23 Cal. 452, p. 264.
Russell v. Beebe, Hemp. 704, p. 38.
Rutherford v. Green's Heirs, 2 Wheat. 196, p. 66.

S.

Saco Lode, D. C. p. 218.
Sacramento M. Co. v. Last Chance M. Co. D. C. pp. 223, 252.
Samson v. Smiley, 13 Wall. 91, p. 179.
San Augustine M. Co. D. C. p. 102.
Sanborn v. Neilson, 4 N. H. 501, p. 195.
Santa Rita del Cobre Mine, D. C. p. 150.
San Xavier Mine, D. C. p. 133.
Schedda v. Sawyer, 4 McL. 181, p. 179.
Schulenberg v. Harriman, 21 Wall. 62, p. 69.
Searle Lode, D. C. pp. 227, 248.
Secretary v. McGarrahan, 9 Wall. 298, pp. 138, 374.
Secretary of Interior, Decisions of—
 May 20th, 1870, p. 74.
 Oct. 28th, 1870, p. 186.
 April, 1871, p. 203.
 Aug. 4th, 1871, p. 9.
 Dec. 25th, 1871, p. 167, Feb. 12th, 1872, p. 331.
 April 19th, 1872, p. 370.
 July 10th, 1872, p. 334.
 Dec. 11th, 1872, p. 198.
 Feb. 27th, 1873, p. 250.
 April 28th, 1873, pp. 64, 69.
 July 19th, 1873, p. 367.
 Nov. 6th, 1873, pp. 64, 150.
 Nov. 12th, 1873, pp. 162, 165.
 Jan. 2d, 1875, p. 96.
 March 22d, 1875, pp. 159, 172, 205.
 April 1st, 1875, pp. 149, 154, 168, 223.
 July 28th, 1875, p. 227.
 March 24th, 1876, p. 336.
 May 20th, 1876, p. 67.
 July 16th, 1876, p. 125.
 July 29th, 1876, pp. 97, 98.
 Feb. 17th, 1877, pp. 219, 220.
Seven Thirty and Hercules Lodes, D. S. p. 249.
Seymour v. Woods, D. C. p. 187.
Seven Thirty Lode, In re, D. C. p. 150.
Sheets v. Selden, 2 Wall. 177, p. 164.
Shepley v. Cowan, 1 Otto. 330, p. 374.
Sheridan Lode, In re, D. S. p. 190.
Sherman v. Buick, 45 Cal. 656; 3 Otto, 209, pp. 58, 66, 319.
Silver v. Ladd, 7 Wall. 219, pp. 373, 374.
Silver Ore Lode, In re, D. C. p. 101.
Slide Lode, In re, D. C. pp. 113, 186.
Smith v. Stewart, D. A. S. pp. 302, 303.
Soulard v. United States, 4 Peters, 511, p. 55.

South Comstock G. & S. M. Co. In re, D. C. pp. 102, 306.
St. John v. Kidd, 26 Cal. 263, p. 266.
Stark v. Starrs, 6 Wall. 402, pp. 139, 149, 179, 372.
State v. Berryman, 8 Nev. 270, p. 88.
 v. Gasconade County Ct. 33 Mo. 102, p. 164.
 v. Schwerle, 5 Pick. 279, p. 164.
State of Nev. & C. P. R. R. Co. of Cal. D. C. p. 328.

STATUTES CITED, COMMENTED ON, AND CONSTRUED.

Act of May 18th, 1796, 1 U. S. Stat. 466, pp. 48, 50, 259.
 May 10th, 1800, 2 U. S. Stat. 73, pp. 48, 50.
 April 30th, 1802, 2 U. S. Stat. 173, p. 48.
 March 26th, 1804, 2 U. S. Stat. 277, pp. 48, 51.
 March 2d, 1805, 2 U. S. Stat. 324, p. 51.
 April 21st, 1806, 2 U. S. Stat. 391, p. 51.
 March 3d, 1807, 2 U. S. Stat. 445, pp. 1, 40, 43.
 March 3d, 1807, 2 U. S. Stat. 548, p. 51.
 April 18th, 1818, 3 U. S. Stat. 429, p. 48.
 March 2d, 1819, 3 U. S. Stat. 489, p. 51.
 March 6th, 1820, 3 U. S. Stat. 545, pp. 48, 55.
 April 20th, 1822, 3 U. Stat. 665, p. 51.
 June 23d, 1836, 5 U. S. Stat. 58, pp. 49, 55.
 June 23d, 1836, 5 U. S. Stat. 59, pp. 49, 65.
 Sept. 4th, 1841, 5 U. S. Stat. 455, pp. 38, 41, 56.
 March 3d, 1845, 5 U. S. Stat. 789, p. 49.
 July 11th, 1846, 9 U. S. Stat. 37, pp. 37, 40, 44.
 Aug. 6th, 1846, 9 U. S. Stat. 58, p. 49.
 March 1st, 1847, 9 U. S. Stat. 146, p. 44.
 March 3d, 1847, 9 U. S. Stat. 181, pp. 40, 44.
 March 3d, 1849, 9 U. S. Stat. 396, p. 41.
 Sept. 9th, 1850, 9 U. S. Stat. 452, p. 42.
 Sept. 26th, 1850, 9 U. S. Stat. 472, p. 44.
 March 3d, 1853, 10 U. S. Stat. 248, pp. 41, 43, 58, 59, 64, 65.
 July 22d, 1854, 10 U. S. Stat. 308, pp. 49, 52.
 March 3d, 1857, 11 U. S. Stat. 186, p. 54.
 May 4th, 1858, 11 U. S. Stat. 269, p. 49.

TABLE OF CASES. 431

Act of Feb. 14th, 1859, 11 U. S. Stat. 383, p. 49.
    May 30th, 1862, 12 U. S. Stat. 410, pp. 43, 64.
    July 1st, 1862, 12 U. S. Stat. 489, p. 42.
    July 2d, 1862, 12 U. S. Stat. 503, p. 43.
    July 17th, 1862, 12 U. S. Stat. 597, p. 98.
    March 21st, 1864, 13 U. S. Stat. 32, p. 68.
    April 19th, 1864, 13 U. S. Stat. 47, pp. 49, 54.
    July 1st, 1864, 13 U. S. Stat. 343, pp. 40, 45.
    Jan. 30th, 1865, 13 U. S. Stat. 567, p. 39.
    March 3d, 1865, 13 U. S. Stat. 529, pp. 40, 45.
    Jan. 30th, 1865, 13 U. S. Stat. 567, p. 39.
    Feb. 27th, 1865, 13 U. S. Stat. 441, p. 41.
    May 5th, 1866, 14 U. S. Stat. 43, p. 47.
    July 4th, 1866, 14 U. S. Stat. 85, p. 70.
    July 13th, 1866, 14 U. S. Stat. 94, p. 42.
    July 23d, 1866, 14 U. S. Stat. 218, p. 41.
    July 25th, 1866, 14 U. S. Stat. 242, pp. 41, 47, 289.
Act of July 26th, 1866, 14 U. S. Stat. 251.
    Sec. 1, pp. 2, 78.
    Sec. 2, pp. 5, 6, 93, 127.
    Sec. 3, pp. 21,127.
    Sec. 4, pp. 26, 90, 115.
    Sec. 5, p. 257.
    Sec. 6, pp. 33, 181.
    Sec. 7, p. 353.
    Sec. 8, p. 289.
    Sec. 9, p. 259.
    Sec. 10, p. 299.
    Sec. 11, p. 310.
Act of 1870, 16 U. S. Stats. 217.
    Sec. 9, p. 243.
    Sec. 12, pp. 228, 229, 236.
    Sec. 13, pp. 230, 231.
    Sec. 14, p. 153.
    Sec. 16, pp. 230, 236.
    Sec. 17, pp. 259, 289.
Act of May 10th, 1872, 17 U. S Stat. 92.
    Sec. 1, p. 78.
    Sec. 2, p. 99.
    Sec. 3, pp. 107, 108.
    Sec. 4, p. 110.
    Sec. 5, pp. 116, 117, 122, 123.
    Sec. 6, pp. 127, 128, 150.
    Sec. 7, pp. 94, 181, 206, 208.
    Sec. 8, p. 243.
    Sec. 9, p. 354.
    Sec. 10, pp. 38, 230.
    Sec. 11, pp. 144, 151, 231, 233.
    Sec. 12, p. 243.
    Sec. 13, pp. 153, 334.
    Sec. 14, p. 247.
    Sec. 15, p. 253.
    Sec. 16, p. 289.

Feb. 18th, 1873, 17 U. S. Stat. 465, p. 39.
March 1st, 1873, 17 U. S. Stat. 92, p. 115.
March 3d, 1873, 17 U. S. Stat. 607, pp. 340-352.
June 6th, 1874, 17 U. S. Stat. 92, p. 115.
Feb. 11th, 1875, 18 U. S. Stat. 315, pp. 110, 115, 117, 121.
March 3d, 1875, 18 U. S. Stat. 474, p. 49.
March 3d, 1875, 18 U. S. Stat. 470, p. 209.
May 5th, 1876, 19 U. S. Stat. 52.
Jan. 12th, 1877, 19 U. S. Stat.
Revised Statutes of the United States:
    Sec. 910, p. 354.
    Sec. 2258, pp. 38, 57, 47, 48, 302.
    Sec. 2289, pp. 48, 57, 302.
    Sec. 2318, p. 38.
    Sec. 2319, pp. 78, 79.
    Sec. 2320, pp. 99, 231.
    Sec. 2321, pp. 93, 94, 97.
    Sec. 2322, pp. 107, 108.
    Sec. 2323, pp. 110, 115.
    Sec. 2324, pp. 110, 115, 116, 119, 121, 122, 123.
    Sec. 2325, pp. 79, 93, 127, 159, 161, 182, 231, 243.
    Sec. 2326, pp. 180, 181, 206, 207, 208, 227.
    Sec. 2327, p. 243.
    Sec. 2328, p. 353.
    Sec. 2329, pp. 78, 228.
    Sec. 2330, pp. 228, 232, 237.
    Sec. 2331, pp. 119, 229, 230, 237, 232.
    Sec. 2332, pp. 230, 239.
    Sec. 2333, pp. 47, 144, 151, 231, 233.
    Sec. 2334, pp. 230, 243.
    Sec. 2335, pp. 94, 153, 362.
    Sec. 2336, pp. 247, 250.
    Sec. 2337, p. 253.
    Sec. 2338, p. 257.
    Sec. 2339, pp. 258, 259, 267.
    Sec. 2340, pp. 259, 289.
    Sec. 2341, p. 299.
    Sec. 2342, pp. 310, 318.
    Sec. 2343, p. 353.
    Sec. 2344, p. 289.
    Sec. 2345, p. 39.
    Sec. 2346, p. 39.
    Sec. 2347, p. 340.
    Sec. 2348, p. 340.
    Sec. 2349, p. 341.
    Sec. 2350, p. 341.
    Sec. 2351, p. 342.
    Sec. 2352, p. 342.
    Sec. 2386, p. 304.
    Sec. 2392, pp. 304, 306.
    Sec. 2406, p. 243.
    Sec. 910, p. 354.
Stephenson v. Smith, 7 Mo. 610, p. 139.
Stockton & V. R. R. Co. v. City of Stockton, 41 Cal. 147, p. 264.
Stoddard v. Chambers, 2 How. 317, pp 63, 179.
Strawbridge v. Curtiss, 3 Cranch, 267, p. 92.
Summers v. Dickinson, 9 Cal. 554, p. 64.
Sutro Tunnel Co. In re, D. A. C. p. 291.

TABLE OF CASES.

T.

Tartar v. Spring Creek W. & M. Co. 5 Cal. 397, p. 282.
Taylor v. Smith, In re, D. C. pp. 226, 258.
Terry v. Megerle, 24 Cal. 624, p. 66.
Teschemacher v. Thompson, 18 Cal. 11, p. 134.
Thomas v. Richards, D. S. p. 225.
Thompson v. Lee, 8 Cal. 275, pp. 265, 266.
Thorne v. Moshor, 20 N. J. Eq. 257, p. 164.
Tiernan v. Salt Lake M. Co. D. S. p. 200.
Titcomb v. Kirk, 51 Cal. 288, p. 262.
Tong v. Hall, D. S. p. 335.
Township of Butte, In re, D. C. p. 305.
Trafton v. Nougues, U. S. C. C. p. 210.
Treadway v. Wilder, 8 Nev. 92, p. 309.
Tremaine v. Brydon, D. A. C. p. 337.
Turner v. Am. B. Union, 5 McLean, 344, p. 38.
Tyler v. Wilkinson, 4 Mason, 397, p. 277.

U.

Unicorn Lode, In re, D. C. p. 187.
Union M. & M. Co. v. Dangberg, 2 Sawy. 450, p. 270.
    v. Ferris, 2 Sawy. 176, p. 270.
Union Water Co. v. Crary, 25 Cal. 504, p. 265.
United States v. Ames, 1 Wood. & M. 76, p. 272.
    v. Arredondo, 6 Peters, 736, p. 179.
    v. Castillero, 2 Black. 17, p. 44.
    v. Comr, 5 Wall. 563, p. 138.
    v. Gear, 3 How. 120, pp. 43, 271.
    v. Gratiot, 14 Peters, 526, pp. 5, 41, 179.
    v. Guthrie, 17 How. 284, p. 138.
    v. Hughes, 11 How. 552; 4 Wall. 232, pp. 179, 271.
    v. Parrott, 1 McAllister, 271, pp. 42, 43, 44.
    v. R. R. Bridge Co. 6 McLean, 517, p. 38.
    v. Seaman, 17 How. 230, p. 138.
    v. Stone, 2 Wall. 526, p. 179.

V.

Vance v. Kohlberg, 50 Cal. 346, p. 374.
Vansickle v. Haines, 7 Nev. 249, pp. 269, 271, 272, 274, 285.
Van Valkenburg v. McCloud, 21 Cal. 330, pp. 64, 66.
Veeder v. Guppy, 3 Wis. 520, p. 66.
Vespasian Lode, In re, D. C. p. 140.

W.

Wadsworth v. Tillotson, 15 Conn. 372, p. 272.
Walsh v. Boyle, 30 Md. 262, p. 164.
Wandering Boy Lode, In re, D. C. pp. 146, 165, 191, 192; Id. D. S. p. 173, 178.
Wandering Boy M. v. Highland Chief M. D. S. p. 149.
War Eagle Mine, In re, D. C. pp. 103, 191, 192.
Washington Lode, In re, D. A. C. p. 176.
Weaver v. Conger, 10 Cal. 233, p. 265.
    v. Eureka Lake Co. 15 Cal. 271, pp. 265, 266, 268.
    v. Fairchild, 50 Cal. 360, p. 374.
Webster Lode, In re, D. S. p. 184.
Weeks v. Hull, 19 Conn. 376, p. 164.
Wellington Mine, In re, D. S, p. 149.
Wesko v. Leet, D. A. S. p. 150.
West v. Cochran, 17 How. 413, p. 66.
White v. Cannon, 6 Wall. 443, p. 179.
Whitney v. Whitney, 14 Mass. 92, p. 72.
Wiggin v. Peters, 1 Met. 127, p. 164.
Wilcox v. Jackson, 13 Pet. 498, pp. 38, 139, 179.
Wilcoxon v. McGhee, 12 Ill. 381, pp. 272, 274.
Wilkinson v. Gaston, 9 Q. B. 141, p. 164.
Wiseman v. McNulty, 25 Cal. 230, p. 266.
Wood v. Hyde, D. C. p. 220.
Woolman v. Garringer, 1 Mont. 535, pp. 283, 284.
Wyoming Mine, In re, D. A. A. G. p. 148.

Y.

Yosemite Mine N. E. Ex. In re, D. A. C. p. 142.

Z.

Zella Lode, In re, D. A. C. pp. 151, 188, 367.

INDEX.

[NOTE.—The reference in this Index is to the page of the volume.]

A.

Abandonment—of adverse claims, 221.
    of surface ground, 222, 252.
Act of 1866—the repealed sections, 1.
    section 1—license without title, 2.
    duties of registers and receivers, 5.
    title and patent—the second section, 5.
    limitation of the right to obtain patents, 6.
    applicants for patent, 6.
    the evidence, 7.
    citizenship required, 10.
    entry and diagram, 11.
    defects in the instructions, 13.
    the application, 14.
    publication of the notice, 16.
    the duties of claimants, registers, and receivers, 17.
    Surveyor-General's duty, 19.
    what a patent conveyed, 19.
    diagram, notice, survey, and patent, 20.
    notice, 21.
    survey, 22.
    posting the notice of application to make the entry, 23.
    effects of irregularities—notice of application—requisites, 24.
    fees of surveyors, 25.
    size of locations—adjustment of surveys, 26.
    duties of deputy surveyors, 26.
    following the vein to any depth, 27.
    mode of survey—quantity and restriction to one claim, 28.
    deviation from rectangular form of survey, 29.
    number of feet located, 30.
    adverse claims and contests, 33.
    proceedings on adverse claims, 35.
    miscellaneous, 36.
Acts of Congress—relative to mines previous to Act of 1866, 46.
Adjustment—of surveys—Act of 1866, 26.
Adverse occupation—as against a patent, 138.

Adverse claims—and contests—Act of 1866, 33, 35, 182.
    proceedings in Court, 180, 227.
    adverse claims, 180.
    adverse claims under Act of 1866, 182.
    adverse claims under statutes now in force—details of procedure, 183.
    who may file, 184.
    verification of adverse claim, 185.
    verification of adverse claims by agents of companies, 186.
    time of filing, 187.
    commencing second suit—dismissal of former suit, 187.
    what constitutes an adverse claim, 188.
    necessary allegations, 189.
    what adverse claimant must show, 189.
    form of adverse claim, 195.
    prima facie adverse claim, 195–196.
    sufficient filing, 197.
    adverse claim must be accompanied by certified survey, 198.
    the object of giving notice by publication, 198.
    jurisdiction of the Land Office over adverse claims, 200.
    notice of suit, 201.
    authority of register to dismiss, 201.
    proceedings in Court—proper party to commence suit, 202.
    possession as equivalent to adverse claim—parties to institute suit, 203.
    what are Courts of competent jurisdiction, 205.
    contests in Court—jurisdiction, 206.
    jurisdiction of State Courts, 207.
    transfer of causes to United States Courts—jurisdiction of mining causes, 209.
    cancelation of entry pending suit, 218.
    stay of proceedings, 218.
    filing consent to judgment, 219.
    laches in bringing suit, 220.
    prosecution of suits—reasonable diligence, 221.
    abandonment of portion of adverse claim, 221.
    abandonment of surface ground, 222.
    cross-applications—delay, 223.
    fees on filing adverse claim, 223.
    amendment of adverse claim, 224.
    evidence of, 225.
    withdrawal of protest by cotenant, 225.
    questions for adjudication by the Courts, 225.
    papers to be filed, 225.
    negligence, 226.
    caveat against issuing patents, 226.
    public highways—adverse claims, 227.
    suit decided, 227.
    rights of foreign corporations, 227.
Affidavits—of citizenship, 97.
    proper party to make, 152.
    verification of, 153.
    mineral, 311.
        " on timber land, 312.
Agent—verification of adverse claims by, 186.
Agricultural entry—withdrawal from, 315.

INDEX. 435

Agricultural patents—excepting clauses, 47.
    minerals discovered after, 148.
    covering mines already worked.
Agricultural and mineral lands—310-339.
Agricultural land—proof, 332.
    mines on, 336.
Alien—application by, 90.
    soldiers, 98.
    as grantee, 142.
Allegations—of adverse claim, 189.
Amendment—of adverse claim, 224.
Annual expenditure—placer claims, 118.
    lode claims, 120.
    tunnel claims, 115-121.
Appeals—354, 359, 360, 367.
Applicant—identity of, 140.
Application—under Act of 1866, 14.
    by aliens, 90.
    who may make, 140.
    united, 143.
    one cannot embrace several claims, 144.
    for several lodes and a mill site, 151.
    errors in, 167-176.
Appointment—of surveyors, 242.
    of deputies, 243.
Approval—of survey, 177.
Assignment—of patents, 178.
Association—unincorporated, 143.
Authority—of register to dismiss adverse **claims, 201.**
Authority—of Land Office decisions, 370.

B.

Borax deposits—80.
Boundaries—and survey, 122.
    plat must show accurately, 158.
    exterior to be shown by survey, 159.
    of placer claims—228.
Burden of proof—character of land, 331.

C.

Canals and Ditches—257-290.
    See WATER RIGHTS.
Cancelation—of entry, pending suit, 218.
Caveat—against issuing patents, 226.
Certified survey—must accompany adverse **claim, 198.**
Certificates—of naturalization, 97.
    as to improvements, 124, 241.
Cinnabar—and copper deposits, 241.
Citizenship—required, 10.
    and proof thereof, 78, 93, 98.

Citizenship—*Continued.*
    and who is a citizen, 91.
    affidavits of, 97.
    certificates of naturalization, 97.
    proof of only required of applicants, 177.

Claims—dimensions of, 99.
    definition of, 118.
    through executor, 142.
    partly in one district and partly in another, 151.
    adverse, 180-227.
        See ADVERSE CLAIMS.
    not in any mining district, 375.

Classification—of mineral veins, 82.

Coal lands—right of entry and of pre-emption—presentation of claims—limitation of entry—conflicting claims—existing rights, 340.
    entry of coal lands, 340.
    pre-emption of coal lands, 340.
    when claims are to be presented, 341.
    only one entry allowed, 341.
    conflicting claims, 342.
    existing rights, 342.
    departmental regulations and instructions, 342.
    restrictions as to purchase, 349.
    school sections containing coal, 349.
    coal lands and town sites, 350.
    actual possession of coal mines upon railroad sections, 350.
    coal lands in Minnesota, Wisconsin, and Michigan, 352.
    coal lands, sale and pre-emption before mining acts, 45.

Co-claimants—contribution by, 121.

Committee, Congressional—delaying action at request of, 152.

Companies—agents—verification by—adverse claims, 186.

Compromises—between miners and settlers, 357.

Conditions—in patent, 258, 261, 290.

Conflicting claims—coal lands, 342.

Conflicting patents—145.

Conflicts—as to surface ground, 248.
    between mineral and town-site claimants, 306.
    between mill-site and homestead claimants, 302.
    lode and placer claims, 240.

Contact deposits—81.

Contests—under Act of 1866, 33.
    in Court, jurisdiction and hearings, 206, 354.
    payment pending, 370.

Contribution—by co-claimants, 121.

Consent—to judgment, 219.

Copper and cinnabar deposits—241.

Corporation—entries by, 148.
    foreign, application by, 98.
    and adverse claims, 227.

Cotenant—protest by—withdrawal, 225.

Counting time—of publication, 164.

Courses and distances—125.

Court—proceedings in, 180, 202, 205, 206, 227.
    See ADVERSE CLAIMS.

Court—*Continued.*
    U. S., transfers to, 209.
    of competent jurisdiction, 205.
Criminal offenses—376.
Cross-applications—223.
Custody of letters—377.
Customs and regulations—116.

D.

Definitions—78-89.
    of claim, 118.
Defects—in published notice, 167.
    in instructions, Act of 1866, 13.
Delay—in adverse claim, 223.
Delaying action—at request of Congressional Committee, 152.
Deposits—valuable, 79.
    of borax, 80.
Deraigning title—140.
Description—in patent, errors in, 146.
    in notice, placer claims, 231.
Deviation—from rectangular form of survey, 29.
Diagram—entry, etc.—Act of 1866, 11, 20.
Diligence—work on tunnel, 113.
    suits on adverse claims, 220-221.
    adverse claims, 223.
Dimensions—of placer claims, 228-241.
    of claims and locations upon veins or lodes, 99-106.
    length and width of lode claims, 99.
    veins or lodes of quartz or other rock in place, 100.
    locations previous to the mining acts of Congress—limitations and size, 101.
    width of lode claims—rights granted by the patent, 102.
    survey must conform to the patent, 103.
    manner of locating claims on veins or lodes subsequently to May 10th, 1872, 103.
    several locations may be made, 106.
    local regulations, 106.
Dismissal—of suit on adverse claims, 187.
    of adverse claims by register, 201.
Ditches—and eminent domain, 262.
    in railroad grants, 261.
    and canals, 258.
    on public lands, 287.
Ditch-owners—and miners, 261.
Diversion—of water on patented lands, 270.
Discrepancies—between final survey, patent, and original application and published notice, 167.
    between published notice and diagram filed, 169.
        " published notice, diagram, and posted notice, 170.
        " final survey and patent and application, 171.
        " survey and diagram filed, 173.
        " survey and notice, matter of description, 173.
Drainage—257-290.

Duties—of registers and receivers, 5.
    of Surveyor-General—Act of 1866, 19, 242.
    of proceedings for patent, 132.
        See PATENT.

E.

Easements—257-290.
Eminent domain—for ditch companies' use, 262.
Entries—of mineral lands by settlers and corporations,
    pending suit, 218.
    of coal lands, 340.
    of placer claims, 228-241.
Entry—and diagram—Act of 1866, 11.
    of coal lands, 340-352.
Errors—in description in patent, 146.
    in survey, etc., 167-175.
    and defects in patent and application, 167-176.
Eruptive masses—81.
Evidence—359, 360, 367.
    of ownership, 140.
    parol, to aid location notice, 158
    of adverse claim, 225.
    of possession of placer claims, 230.
    under Act of 1866, 7.
Exceptions—359, 360, 367.
Excepting clauses—in placer and agricultural patents, 47.
Exceptions—and reservations of minerals in grants by the Government, 38.
    See RESERVATIONS.
Executor—claim through an, 142.
Exemplified copies—of patents, 178.
Existing rights—353.
Expenditures—upon tunnels, 113, 115, 121.
    and improvements, 116.
    annual placer claims, 118.
        " lode " 120.
    relocated mines, 121.
Exploration—and purchase of valuable mineral deposits, and the occupation
    and purchase of mineral lands—citizenship and proof thereof, 78.
    right to purchase, 78.
    valuable deposits, 79.
    the general rule stated, 80.
    borax deposits, 80.
    mineral deposits, 81.
    what is a mineral vein? 82.
    mineral veins, classifications, 82.
    eruptive masses, 81.
    contact deposits, 81.
    impregnations, 81.
    fahlbands, 81.
    stockwerke, 81.
    gash veins, 82.
    segregated veins, 82.

INDEX. 439

Exploration—*Continued.*
    fissure veins, 83.
    rock in place, 86.
    lode, 87.
    vein, 87.
    quartz ledge, 87.
    spur, 87.
    feeder, 87.
    float ore, 87.
    silver-bearing ore, 88.
    tailings, 88.
    who may acquire patents, 89.
    application by aliens, 90.
    citizenship, 91.
    proof of citizenship, 93.
    affidavit of citizenship, 97.
    foreign corporation, 98.
    restriction as to proof, 98.
Exploration and purchase—of valuable deposits, 78.
Extension—of time, 119.
Exterior boundaries—to be shown by survey, 159.

F.

Fahlbands—81.
Feeder—defined, 87.
Fees—adverse claim, 223.
    registers and receivers, 368.
Feet located—number of, Act of 1866, 30.
Filing—adverse claims, 184.
    adverse sufficient, 197.
Fissure veins—83.
Five-acre lots—241.
Fixed monuments—125.
Float ore—defined, 87.
Flumes—over public lands, 286.
Following the vein—27.
Foreign corporation—application by, 98.
    and adverse claims, 227.
Form—of adverse claim, 195.
Form of survey—deviation from rectangular, 29.
Fraud—in pre-emption entry, 336.

G.

Gash veins—82.
Government title—to mineral lands, procuring, 127-178.
    See PATENT.
Grants—from Indians, 45.

H

Hearings—and contests, 329, 354.
    and publication of notice, 329.

Highways—public and adverse claims, 226.
Homestead and town sites—homestead rights on non-mineral lands—town-site entries, 299.
    non-mineral lands—open to homesteads, 299.
    pre-emption of homesteads on agricultural lands formerly designated as mineral, 300.
    homestead entries including mineral deposits, 301.
    rights of pre-emptioners and homestead claimants, 302.
    conflicts between homestead and mill-site claimants, 303.
    title to town lots subject to mineral rights, 303.
    conflicts between mineral and town-site claimants, 306.
    town sites and coal lands, 350.

I.

Identity—of applicant, 140.
    of lodes, 249.
Illegal location—invalidates subsequent proceedings, 150.
Illinois lead case—43.
Impeachment—of patent, 134.
Implied license—44.
Impregnations—81.
Improvements—and expenditures upon lode, 116-241.
    certificates as to, 124.
    of mill sites, 255.
Indians—grants from, 45.
Inspection—of mine, 374.
Interference—of claims, 250.
Intersection of veins—247.
    conflicts as to surface ground, 248.
    identity of lodes, 249.
    interference of claims, 250.
    abandonment of surface ground, 252.
Irregularities—effect of, 24.

J.

Judgment—consent, 219.
Jurisdiction—of Land Office over adverse claims, 200.
    Courts of competent—adverse claims, 205.
    of State Courts, 207, 209.
    of United States Courts, 209.
    transfer to United States Courts, 209.

L.

Laches—in bringing suit, 220-223.
Land Office—Jurisdiction over adverse claims, 200.
    practice, 351.
    decisions, authority of, 370.

Lead case—Illinois, 43.
    mines, sale of, 44.
Length and width—of lode laims, 99.
Letters—custody of, 377.
License—without title, 2.
    implied, 44.
Liens—241.
Local legislatures—authority to make laws to govern the mines, 257.
Local regulations—106.
Local water rights—260.
Locations—tunnel, 110, 112
    patenting, 112.
    and survey boundaries, 122.
    by a minor, 150.
    size of, 26, 30, 99, 100, 101,
    previous to mining acts, 99, 100.
    mode of making, after May 10th, 1872.
    several, 103.
    priority of, 109.
    illegal, invalidates subsequent proceedings, 150.
    notice of its efficiency, 154.
    notice, parol evidence admissible to aid, 158.
    of mill sites, 253, 254.
Locator's right—of possession and enjoyment of surface ground and lode, 107, 109.
    grantee of several may obtain patent for whole tract, 144.
    right of possession and enjoyment, 107.
    status of lode claims located prior to May 10th, 1872, 108.
    patents for veins or lodes previously issued, 108.
    priority of location, 109.
Lode claims—annual expenditure, 120.
    procuring patent, 127-178.
        See PATENT.
    length and width of, 99, 102.
    locations upon, 99, 100.
        See LOCATION.
    possession and enjoyment of, 101.
    status of, before May 10th, 1872, 108.
    patents for, 108.
    identity of, 249.
    and placer claims, conflicts, 240.

M.

Machinery—removal of, 317.
Masses—eruptive, 81.
Mill-site—and several lodes, application for, 151.
    patents for non-mineral lands, 253.
    location of mill-sites, 253.
    procuring patent, 254.
    mill-site must be non-mineral in character, 255.
    improvements, 255.
    mill-sites and railroad grants, 255.

Mineral lands—the first mining act, 1-37.
    See ACT OF 1866.
  reservations and exceptions of mineral lands in grants by the Government, 38-77.
    See RESERVATIONS AND EXCEPTIONS.
  right of exploration and purchase of valuable mineral deposits and the occupation and purchase of mineral lands—citizenship and proof thereof, 78-98.
    See EXPLORATION AND PURCHASE, CITIZENSHIP.
  what is mineral land, 329.
  dimensions of claims and locations upon veins or lodes, 99-106.
    See DIMENSIONS OF CLAIMS, LOCATIONS.
  locator's right of possession and enjoyment of the surface ground and of the lode, 107-109.
    See LOCATOR'S RIGHTS OF POSSESSION, ETC.
  tunnel rights, 110-115.
    See TUNNELS.
  regulations and customs—expenditures and improvements—surveys and boundaries, 116-125.
    See REGULATIONS, EXPENDITURES, SURVEYS, AND BOUNDARIES.
  patents to mineral lands—mode of procuring Government title, 126-179.
    See PATENTS.
  adverse claims, proceedings in Court, 180-227.
    See ADVERSE CLAIMS.
  placer claims—survey, entry, and patent—dimensions of claims—subdivisions of ten-acre tracts—evidence of possession—mode of obtaining patent, 228-241.
    See PLACER CLAIMS.
  public surveys over mineral lands—duties of Surveyor-General—appointment of deputies, 242-246.
    See SURVEYS.
  intersection of veins, 247-252.
    See INTERSECTION OF VEINS.
  mill-sites—patents for non-mineral lands, 253-256.
    See MILL SITES.
  water and other vested rights—right of way for canals and ditches—easements—drainage—State and Territorial legislation—Sutro Tunnel Act, 257-298.
    See WATER RIGHTS, VESTED RIGHTS, EASEMENTS, DRAINAGE, CANALS AND DITCHES, SUTRO TUNNEL ACT.
  homesteads and town sites, 299-309.
    See HOMESTEADS, TOWN SITES.
  segregation of mineral and agricultural lands—withdrawal from agricultural entry, 310-339.
    See SEGREGATION, WITHDRAWAL FROM AGRICULTURAL ENTRY.
  coal lands, 340-352.
    See COAL LANDS.
  miscellaneous provisions, 353-377.
    See POWER OF PRESIDENT, PENDING APPLICATIONS, POSSESSORY ACTIONS, LAND OFFICE PRACTICE, APPEALS, ETC., FEES OF REGISTERS AND RECEIVERS, LAND OFFICE DECISIONS CRIMINAL OFFENSES.

Mineral vein—what is, 81, 87.
  classifications, 82.
  gash veins, 82.

Mineral vein—*Continued.*
    segregated, 82.
    fissure, 83.
Miners and settlers—compromises, 337.
Mining ditch—in railroad grant, 261.
Miscellaneous provisions—power of the President as to appointments, 353.
    pending applications—existing rights, 353.
    possessory actions relative to mines, 354.
    practice before the Land Department—hearings, contests, and appeals—witnesses and testimony, 354.
    appeals, exceptions, evidence, 359, 360, 367.
    fees of registers and receivers, 368.
    payment pending contest, 370.
    decisions of the Land Department—their authority, 370.
    right of inspection of mine, 374.
    mining claims in river beds, 374.
    timber on mineral lands—railroad companies, 375.
    claims not within any mining district, 375.
    removal of machinery, 376.
    criminal offenses, 376.
    hearings and contests, 354.
    perjury, 376.
    custody of letters, 377.
    removal of papers, 377.
    warrants and scrip, 377.
Mode of survey—Act of 1866, 28.
Monuments—fixed, 125.

N.

Nature—of patent, 138.
Naturalization—certificates of, 97.
Neglect—of co-claimants to contribute, 121.
Negligence—adverse claims, 225, 226.
Nevada—school lands containing minerals in, 67.
New trial—as ground for staying proceedings, 219.
Newspaper—in which to make publication of notice, 165.
Notice—Act of 1866, 20, 21, 24.
    publication of—Act of 1866, 16.
    revised statutes, 161.
    posting, 23.
    of location—its sufficiency, 154.
        "   parol evidence admissible to aid, 158.
    of publication—newspaper in which to make, 165.
    published—defects in, 167.
    of publication—object of, 198.
    of suit on adverse claim, 201.
    description in—placer claims, 231.
Number—of feet located—Act of 1866, 30.
    of patents, 149.

O.

Occupation and purchase—of mineral lands, 78.
Ownership—evidence of, 140.

P.

Papers—to be filed—adverse claims, 225.
    removal of, 377.
Parol evidence—to aid location notice, 158.
Party—to commence suit on adverse claim, 202, 203.
Patents—for veins or lodes previously issued, 108.
    for tunnels, 110–112.
    for non-mineral lands, 253.
    for mill sites, 254.
    under Act of 1866, 5.
    of placer claims, 231, 228–241.
    limitation of right to obtain, 6.
    applicants for, 6.
    conditions in, 258, 261, 290.
    subject to vested rights, 259.
    what it conveyed under Act of 1866, 19, 20.
    excepting clauses, 47.
    who may acquire, 89.
    application by aliens, 90.
    rights granted by, 102.
    and survey, 103.
    for placer claims—excepting clauses, 47
    annual expenditure, 118.
Patents to mineral lands—mode of procuring Government title, 127–178.
    patents for vein or lode claims, how obtained, 127.
    details of procedure, 128.
    duties of registers and receivers, 132.
    nature of the patent, 133.
    impeachment of patent, 134.
    adverse possession, as against a patent, 138.
    what is granted, 139.
    who may apply, 140.
    evidence of ownership, deraigning title, identity of applicant, transfers, 140.
    claim through an executor—where an alien is grantee of a claim, 142.
    united applications—unincorporated associations, 143.
    several claims cannot be embraced in one application, 144.
    grantee of several locators may obtain patent for the whole tract, 144.
    conflicting patents, 145.
    errors in description in patent—relinquishment—calls for the relinquishment of land inadvertently patented, 146, 147.
    second patent—entries of mineral lands by settlers and corporations, 148.
    minerals discovered after agricultural patent, 148.
    setting aside patent, 148.
    number of patents, 149.
    protests against issuance of patents—status of protestants, 149.
    an illegal location invalidates subsequent proceedings, 150.
    location by a minor, 150.
    application for several lodes and a mill site—claim partly in one district and partly in another, 151.
    delaying action at request of Congressional Committees, 152.
    the affidavit—proper party to make it, 152.

Patents to mineral lands—*Continued.*
    verification of affidavits, 153.
    the location notice, 154.
    parol evidence to aid the notice, 158.
    plat must show the boundaries of the claim, 158.
    surveys to show exterior boundaries, 159.
    specific surface ground, 159.
    posting on claim and proof thereof, 160.
    publication of the notice, 161.
    time of publication, 163.
    counting the sixty days, 164.
    proof of publication, 165.
    the newspaper in which the notice is to be published, 165.
    defects in the published notice, 167.
    discrepancies between final survey and patent and the application and published notice, 167.
    discrepancies between the published notice and the notice and diagram filed, 169.
    discrepancies between the published notice and the diagram and posted notice, 170.
    discrepancies between the final survey and patent and the application, 171.
    new survey, pending another application, 173.
    discrepancies between survey and diagram, 173.
    discrepancies between survey and notice, matter of description, 173.
    errors in survey, 174, 175.
    when application will be rejected, 176.
    sworn statement, 176.
    approval of survey—jurisdiction of Surveyor-General, 177.
    proof of citizenship, 177.
    bona fide application for patent, 178.
    portion of claim, 178.
    exemplified copies, 178.
    assignment of patents, 178.
    refunding purchase-money, 178.
    caveat against issuing patents, 226.
Payment—pending contest, 370.
Pending applications—353.
Perjury—376.
Placer claims—survey, entry, and patent—dimensions of claims—subdivisions of ten-acre tracts—evidence of possession—mode of obtaining patent, 228.
    conformity of placer claims to surveys—limits and boundaries, 228.
    subdivision of ten-acre tracts—extent of placer locations, 228.
    survey of placer claims—limitations, 229.
    evidence of possession—sufficient to establish right to patent, 230.
    proceedings for patent for placer claims, 231.
    details of procedure, 232.
    description in the notice, 232.
    entry and survey of placer claims under the Act of 1866, 233.
    survey of placer claims under the Acts of 1866, 1870, 235, 236.
    quantity of placer ground subject to location, 237.
    proofs necessary to establish possessory rights, 239.
    placer ground located after May 10th, 1872, 240.
    conflicting claims—placer and lode claims, 240.

Placer claims—*Continued.*
    cinnabar and copper deposits, 241.
    publication, 241.
    liens, 241.
    surveyed lands, 241.
    five-acre lots, 241.
    certificates of improvement, 241.
Placer patents—excepting clauses, 47.
    claims—annual expenditure, 118.
Plat and field-notes—of survey to show amount of expenditure, 121.
    and boundaries of claim, 158.
Policy—of the Government in reserving mineral lands, 39.
    See RESERVATIONS.
Possession—of placer claims, 230.
Possession and enjoyment—of surface ground and lode, 107.
Possession—as equivalent to an adverse claim, 203.
Possessory—actions, 354.
    water rights, 259.
    rights—placer claims, 239.
Posting on claim—proof of, 160.
    notice, Act of 1866, 23.
Practice—before Land Department, 354.
Pre-emption—of coal lands, 340-352.
President's right—of appointment, 353.
Prima facie—adverse claim, 195.
Prior appropriation—275-281.
Priority—of location, 109.
Proceedings—stay of pending suit, 218.
    in Court, 180-227.
        See ADVERSE CLAIM.
Procedure—on obtaining patent, 128.
    See PATENT.
Proof of citizenship—78-98.
    only required of applicants, 177.
    restriction as to, 98.
Proof—of posting on claim, 160.
    of publication, 165.
    by adverse claimant, 189.
    burden of—character of land, 331.
    character of land, 331-334.
    of possessory rights—placer claims, 239.
Protests—against issuance of patents, 149.
    and adverse claims, 203.
    by cotenant, withdrawal, 225.
Protestants—status of, 149.
Public highway—and adverse claims, 226.
Publication of notice—Act of 1866, 16.
    Act of 1872, 161.
    time of, 163.
    counting time, 164.
    proof of, 165.
    newspaper in which to make, 165.
    of notice, object of, 198.
    placer claims, 241.

Published notice—defects in, 167.
Purchase—of valuable deposits, 78.
Purchase-money—refunding, 178.

Q.

Quantity—and restrictions to one claim—Act of 1866, 28.
    of placer ground, 237.
Quartz ledge—what is, 87.
Questions—presented in suits on adverse claims, 225.

R.

Railroad grants—reservations in, 41.
    mineral lands in, 75, 337, 375.
    mineral lands and coal mines, 350.
    mill-sites, 255.
Rectangular form of survey—deviation from, 29.
Refunding purchase-money—178.
Refusal—of co-claimants to contribute, 121.
Registers and receivers—duties—Act of 1866, 17.
    proceedings for patent, 132.
        See PATENT.
    as agents for applicants, 162.
    authority to dismiss adverse claims, 201.
    fees, 368.
Regulations—local, 106.
    coal lands, 342.
Regulations and customs—expenditures and improvements, 116.
    definition of "claim," 118.
    annual expenditure not required on placer claims, 118.
    extension of time—re-location, 119.
    annual expenditure on lode claims, etc., 120
    work done on a tunnel, 121.
    neglect or refusal of claimants to contribute, 121.
    relocated mines, expenditures, 121.
    amount of expenditures shown upon plat and field-notes of survey, 121.
    location and survey, boundaries, 122.
    improvements, certificates as to, 124.
    fixed monuments, courses, and distances, 125.
Relinquishment—146.
    calls for, 147.
Relocations—119.
    expenditures, 121.
Removal—of papers, 377.
Reservations and exceptions—of mineral lands in Government grants, 38.
    mineral lands reserved, 38.
    mineral lands in certain States not excepted, 39.
    exceptions from the operation of the act, 39.
    certain grants not to include mineral lands, 39.
    the policy of the Government in reserving or excepting mineral lands, 39.
    reservation in railroad grants, 41.

Reservations and exceptions—*Continued.*
    the Government never parted with the right to the mines, 42.
    reservation in grants to the States, 43.
    the Illinois lead case, 43.
    implied license, 44.
    the sale of lead mines, 44.
    sale and pre-emption of coal lands, 45.
    grants from Indian tribes in America, 45.
    further acts of Congress, 46.
    excepting clauses in placer and agricultural patents, 47.
    saline lands, 48.
    school lands containing minerals, 58.
    school lands containing minerals in Nevada, 67.
    mineral lands in railroad grants, 75.
Restrictions—to one claim, Act of 1866, 28.
    as to proof of citizenship, 98.
Rights—granted by patent, 102.
    of possession and enjoyment of surface ground and lode, 107.
    of tunnel-owners, 110, 113.
River beds—claims in, 374.

S.

Saline lands, 48.
School sections—containing coal, 349.
School lands—containing minerals, 58, 67.
Scrip and warrants—377.
Second patent—148.
Segregated veins—82.
Segregation of mineral and agricultural lands—withdrawal from agricultural entry, 310.
    manner of setting apart mineral lands as agricultural, 310.
    segregation of agricultural from mineral lands, 310.
    mineral affidavits, 311.
    mineral affidavits on timber land, 312.
    segregation under Acts of 1866 and 1870, 313, 315.
    withdrawal of certain lands from agricultural entry, 315.
    surveyors' returns, 326.
    their prima facie accuracy, 328.
    hearings to determine the character of land—publication, 329.
    what is mineral land, 329.
    burden of proof, 331.
    evidence as to agricultural character of land, 332.
    the testimony, 333.
    proof as to mineral character of land, 334.
    discovery of mines on agricultural lands, 336.
    agricultural patent covering mines already worked, 336.
    fraud in pre-emption entry, 336.
    compromises between miners and settlers, 337.
    attempt by railroad to disprove mineral character of lands, 337.
    non-mineral proof by settlers on lands within railroad limits, 338, 339.
Setting aside patent—148.
Settlers—entries of mineral lands by, 148.

Settlers—*Continued*.
>and miners, 337.
>non-mineral proof, 333-339.

Several claims—in one application, 144.
Silver-bearing ore—defined, 88.
Size—of locations—Act of 1866, 26, 99, 100, 101.
Soldiers—alien, applications, 98.
Spur—defined, 87.
State Courts—jurisdiction, 207.
>and Territorial legislation, 257-290.

Statement—sworn, 176.
States—reservations in grants to, 43.
Stay—of proceedings, 218, 219.
Stockwerke—81.
Sufficiency—of location notice, 154.
Suits—on adverse claims, 187.
>questions presented, 225.
>laches in bringing, 221.
>" notice of, 201-202.
>party to commence, 202.
>proper Courts, 205.
>entry pending, 218.
>stay of proceedings, 218.

Surface ground—possession and enjoyment of, 107.
>specific, 159.
>conflicts as to, 248.
>abandonment of, 222-252.

Surveys over mineral lands—surveys of mining claims—duties of Surveyor-General—appointment of deputies, 242.
>appointment of surveyors of mining claims by Surveyor-General, 242.
>public surveys extended over mineral lands, 243.
>description of vein claims on surveyed and unsurveyed lands, 243.
>appointment of deputies, 243.
>charges for surveys and publications, 244.
>special instructions to deputies, 244.
>authority of deputies outside the district, 246.
>plat and field-notes to show amount of expenditure, 121.
>of placer claims, 228-241.
>and boundaries, 122.
>under Act of 1866, 20, 22.
>adjustment of, 26, 28.
>rectangular form, deviation from, 29.
>conforming to patent, 103.
>to show exterior boundaries, 159.
>approval of by Surveyor-General, 177.
>certified, must accompany adverse claim, 198.

Surveyors' fees—Act of 1866, 25.
>returns, 326.
>" their prima facie accuracy, 328.

Surveyor-General's duties—19, 242.
>jurisdiction of, 177.
>approval of survey by, 177.

Suspension—of proceedings—new trial as ground of, 219.
Sutro Tunnel Act—289.
Sworn statement—176.

W. C.—29.

T.

Tailings—defined, 88.
Testimony—character of land, 331, 332, 333, 334, 354.
Timber—on mineral lands, 375.
Time—extension of, 119.
    of publication, 163.
    " counting, 164.
    of filing adverse claims, 187.
    of commencing suit on adverse claims, 202.
Title—deraigning, 140.
Title and patent—under Act of 1866, 5.
Town sites—299-306.
    subject to mineral rights, 303, 306.
Transfers—140.
    of causes to United States Courts, 209.
Tunnel rights—110, 115.
    locations, patenting, 112, 113.
    expenditures upon, 115, 121.
    owners of—rights of, 110.
    expenditures upon, regarded as expenditure upon a lode, 113.
    rights—diligence—expenditure, 115.

U.

Unincorporated associations—143.
United applications—143.

V.

Valuable deposits—exploration and purchase of, 78, 79.
Vein—following the, 27.
    mineral, what is, 81.
    gash, 82.
    segregated, 82.
    fissure, 83.
    intersection of, 217.
Verification—of adverse claims, 185, 186.
Vested rights—257, 290.
    See WATER RIGHTS.

W.

Warrants and scrip—377.
Water and other vested rights—right of way for canals and ditches—easements—drainage—State and Territorial legislation—patents subject to vested rights—Sutro Tunnel Act, 257.
    State and Territorial legislation—easements, drainage, etc., 257.
    conditions inserted in the patent, 258.
    vested rights to use of water—right of way for canals, 258.

Water and other vested rights—*Continued.*
    patents subject to vested water rights, 259.
    possessory water rights confirmed, 259.
    local water rights protected, 260.
    conditions as to vested water rights inserted in patent, 261.
    mining ditch in railroad grant, 261.
    conflicting rights of ditch-owners and miners, 261.
    exercise of eminent domain for a private ditch company's use, 262.
    water rights in California under the Codes, 264.
    existing water rights obtained by patent, how affected, 266.
    effect of the acts upon previous diversion of water upon patented lands, 270.
    recognition of the doctrine of prior appropriation, 275.
    effect of the statute upon prior appropriation without Government title, 281.
    construction of flumes over public lands, 286.
    rights of ditch-owners on public lands, 287.
    Sutro-Tunnel Act, 289.
    conditions inserted in patents for mines on Comstock Lode, Nevada, 290.
Width—of lode claims, 99, 102.
Withdrawal—of protest by cotenant, 225.
    from agricultural entry, 310-315.
Witnesses—354.

www.ingramcontent.com/pod-product-compliance
Lightning Source LLC
Chambersburg PA
CBHW030323020526
44117CB00030B/636